Ape Language

Animal Intelligence
A Series of Columbia University Press
Herbert S. Terrace, Series Editor

APE LANGUAGE

From Conditioned Response to Symbol

E. Sue Savage-Rumbaugh

Foreword by Herbert S. Terrace

Columbia University Press
New York 1986

Library of Congress Cataloging-in-Publication Data

Savage-Rumbaugh, E. Sue, 1946–
Ape language.

Bibliography: p.
Includes index.
1. Chimpanzees—Psychology. 2. Human–animal communication—Data processing. 3. Animal communication—Data processing. 4. Mammals—Psychology. I. Title.
QL737.P96S25 1986 156′.36 85-29161
ISBN 0-231-06198-6

Columbia University Press
New York

Printed in the United States of America

Book design by Ken Venezio

This Book is Dedicated to Sherman and Austin

I thank them for all the times they worked so hard to teach me how to be a better chimpanzee. My own life has been immeasurably enriched from my neverending attempts to see things through a chimpanzee's eyes. I hope that what I shared with them about being human proves to be of equal benefit to them. This book is about what I taught them. If they were able to write a book about what they taught me, it would surely be a far better book than this one.

Contents

The Animal Intelligence Series

Both the scientist and the layman approach evidence of animal intelligence with deep-seated and contradictory attitudes. One point of view, usually attributed to Descartes, regards animals as mechanical beasts. Behavior is elicited automatically by stimuli emanating from the animal's internal or external environments. However complex or elaborate the animal's behavior, Descartes and his followers have argued that it can always be reduced to some configuration of reflexes in which thought plays no role.

Darwin's theory of evolution acknowledges the possibility that animals have the capacity to think. In comparing man and animals, Darwin observed that it is just as logical to argue that the human mind evolved from animal minds as it is to argue that human anatomical and physiological structures evolved from their animal counterparts. Animals may not think and solve problems as rapidly as humans do. It is also unlikely that, given any amount of time, an animal could solve a wide variety of problems that are trivially easy for humans. Darwin nevertheless believed that animals engaged in various forms of thinking that were truly homologous to human thought.

Until recently, there has been little concrete basis for choosing between the contradictory positions of Darwin and Descartes. Research that sought to demonstrate thinking in animals encountered many conceptual and technical difficulties. The basic problem was to devise experiments in which an animal's behavior could not be explained by reference to reflex theory. During the past decade that problem has been overcome in numerous studies that have demonstrated the existence of cognitive processes in birds, rats, dolphins, sea lions, monkeys, apes, and a large variety of other species.

As but one of many instructive examples, consider a monkey or a pigeon who is shown a long series of photographs. After seeing such a list, the subject is presented with a "probe" photograph. On half of the trials, the probe is identical to one of the photographs shown previously; on the remaining trials, the photograph is novel. The subject can earn a small food reward by pressing a "yes" button when the probe photograph matches one of the previously presented photographs, and a "no" button when it does not. This task is performed with a high degree of accuracy even when the probe is delayed for as much as a minute.

The ability of a pigeon or a monkey to perform such a task poses two kinds of question. What was the stimulus for the correct choice and how did the subject think of the photographs it was shown before the probe was presented? In order to answer the first question, we must recognize that the probe by itself cannot suffice as a cue for a correct choice. In addition, the subject must somehow recall the photographs he has seen. A correct choice can occur only if the subject is able to compare its representations of the previously shown photographs with the probe on hand. Such a mental comparison does occur: the greater the number of photographs on the list, the longer it takes for the subject to decide if it had seen the probe. It should not be necessary to belabor the point that conditioned reflex theory cannot explain an animal's ability to reflect about recently seen photographs.

How the subject thinks of the photographs it has just seen is a more difficult problem. Were the subject of such an experiment human, it would seem reasonable to hypothesize that the photographs were represented by verbal labels—"a car," "a tree," "a building," "a river," and so on. It has yet to be shown, however, that even the most intelligent of animals could learn enough about language to name their thoughts and to use such names as mnemonics. We will consider shortly recent attempts to teach apes and dolphins to "talk." For the moment, however, we must come to grips with the fact that most if not all of animal thought is nonverbal. Although this makes it difficult to determine the nature of animal thought it has the virtue of placing thinking in an evolutionary perspective—one that suggests the highly verbal thinking with which we are so familiar evolved from more fundamental nonverbal forms of thinking. In short, the study of animal thinking provides a way of simplifying the complex phenomenon of thinking in a way that factors out its verbal component and that also allows us to take a giant step toward realizing Darwin's hope of establishing a comparative psychology of intellectual processes.

What, then, of the many recent studies that have reported evidence of linguistic competence in animals with highly developed brains? At the very least, these studies have shown that, after intensive training programs, an ape can learn to use an arbitrary symbol as a means of spontaneously requesting specific rewards. As the first volume of this series, *Ape Language* by E. Sue Savage-Rumbaugh, will reveal, this is no small achievement. It is also likely that future studies will reveal other aspects of an animal's competence to use symbols. During the foreseeable future, however, there is no basis for anticipating that an ape, or any other nonhuman creature, will be able to talk to us directly about its thoughts. The best evidence we have that animals think—apes and dolphins included—remains indirect evidence which must be gleaned from their behavior.

Such evidence is being collected, with increasing frequency, by a variety of ingenious techniques in psychological laboratories throughout the world. Consider some examples that will be the topics of future volumes of this series. Pigeons are able to learn various kinds of "natural" concepts (e.g., trees, fish, people, water, and so on) simply by contrasting photographs containing an exemplar of the concept with photographs that do not. Rats can learn to infer whether the magnitude of a reward it encounters at the end of a runway is a member of a series of increasing or decreasing magnitudes of rewards. Having made that inference it adjusts its running speed accordingly. Monkeys can look at a particular sequence of colors and, moments later, produce them in the appropriate sequence. Dolphins can respond to complex symbols instructing them to perform a particular action with respect to a particular object and to then move that object to a particular location. They can do so even when given novel instructions.

Having discovered that animals do actually think, psychologists are ready to take the next step and inquire as to the nature of animal thought. Readers of this and of future volumes will discover why the pursuit of knowledge regarding the nature of animal intelligence should produce many unique dividends. I have already noted the significance of providing an evolutionary perspective for human intellectual processes. Indeed, the coming decades may prove to be the beginning of an era of psychology that will parallel the era of the blossoming of biology—the period in which it began to assimilate the theory of evolution into its conceptual framework. Scientific and intellectual growth aside, there is another special insight to be gained by studying animal thinking. For thousands of years humans have believed that they are the only creatures capable of thought. Until we encounter some (intelligent) extraterrestrial creature, what better way of breaking out of the limited circle of human thought than to discover how other species with whom we share this planet think about their worlds?

New York City
October 11, 1985

H. S. Terrace

Foreword

Until recently, conversations with animals occurred only in works of fiction. Countless children (and adults) have been charmed by tales in which one animal talked to another or to some human companion—witness Aesop's fables, Alice in Wonderland, Dr. Doolittle, and Tarzan, to name but a few examples.

During the first half of the twentieth century, psychologists, both in the United States and Russia, sought to create the conditions in which such conversations would actually occur. In each instance, an infant chimpanzee was separated from its natural mother and reared like a human infant by a surrogate human family. The hope was that the chimpanzee would come to imitate the sounds it heard and eventually use them in a meaningful way. Unfortunately, none of these studies provided any evidence that a chimpanzee could produce a meaningful utterance—either in English or Russian. As a result, a consensus developed among linguists, philosophers, and psychologists that—as clever as chimpanzees may be at solving difficult problems—they lack the competence to learn language.

During the 1960s and 1970s, independent groups of American psychologists established a new generation of projects whose common goal was to train an ape to use a natural or an artificial language. Apes are physically unable to produce the sounds of human speech, so chimpanzees, gorillas, and an orangutan were trained to learn visual languages. In some instances, the ape was taught American Sign Language (ASL)—a natural language of gestures and facial expressions used by hundreds of thousands deaf Americans. In other instances, the ape was taught languages whose words took the form of plastic symbols of different colors, geometric forms, and shapes.

Few recent scientific projects have caught the public imagination as strongly as did the second generation of attempts to teach apes to express themselves through language. Unfortunately, the widespread interest that this research program generated was followed quickly by a reaction that has bordered on benign neglect. The reasons for this abrupt succession of attitudes are numerous and complex. It is, however, not too much of an oversimplification to observe that two major factors were unrealistically ambitious goals and the uncritical acceptance of the purported meanings of the symbols used by the apes.

Sue Savage-Rumbaugh, the author of this volume, has more so than any other investigator of ape language steered clear of these pitfalls. Through careful and critical observations of what a chimpanzee does and does not know about a symbol, Savage-Rumbaugh and her associates have formulated a distinctively new approach to the study of an ape's linguistic capacity. That approach, which should be considered as a third generation effort to teach language skills to apes, is a healthy antidote to earlier excesses and is one that indicates much promise for future research on the linguistic competence of apes. Before turning to Savage-Rumbaugh's many positive contributions to our understanding of this complex issue, let us consider, if only briefly, the background from which her work emerged.

The program of research on ape language that was begun during the late 1960s focused almost exclusively on evidence of grammatical competence. That concern was a reflection of the prevailing assumption of psycholinguists that human language makes use of two levels of structure: the word and the sentence. In contrast to the fixed character of various forms of animal communication (e.g., bird song and bee dances), the meaning of a word is arbitrary. It is, however, widely recognized that animals can learn to make arbitrary requests (e.g., a rat pressing a bar for food and a chimpanzee offering a blue plastic triangle to the experimenter in exchange for a piece of apple). Accordingly, some psychologists have blurred the distinction between words and the arbitrary responses that an animal can be conditioned to make. A second level of structure, one that subsumes the word, is generally regarded as the essential feature of human language. Psychologists, psycholinguists, and linguists are in general agreement that using a human language indicates knowledge of a grammar. How else can we account for a child's ultimate ability to create an indefinitely large number of meaningful sentences from a finite number of words?

The initial reports of the second generation of ape language studies presented tantalizing anecdotes that were interpreted as evidence of grammatical competence. Washoe, the infant protege of the Gardners, was reported to have signed combinations such as *more drink* as a request for more orange juice. When her trainer asked her to identify a nearby swan, she signed *water bird.* Premack noted that Sarah, his most accomplished chimp student, requested an apple by arranging her plastic symbols in the order: *Mary Give Sarah apple.* Understandably, such utterances drew much attention to recent projects that sought to establish linguistic competence in apes. If an ape could truly create a sentence, the gap between animal communication and human language would narrow dramatically.

The psycholinguist's emphasis of grammatical competence as the hallmark of human language and claims by the Gardners and by Premack (Gardner and Gardner 1969, Premack 1970, Premack 1976) that their chimpanzees were combining arbitrary symbols so as to produce particular meanings created high (but, as subsequent work revealed, unrealistic) expectations. Such expectations had an unfortunate effect on public opinion. Projects were evaluated almost exclusively on the basis of the sentences that their pongid subjects were said to have created. From my own experience while directing Project Nim (a project modeled after the one that the Gardners' began with Washoe), I recall our great preoccupation with Nim's combinations in ASL. The approval of a large grant to support Project Nim resulted largely from the recommendation of a site visit committee of the National Institutes of Health. Though skeptical of Nim's ability to sign before their visit, the committee, one of whose members was fluent in ASL, was thoroughly impressed by the combinations they saw Nim produce. By Nim's third birthday, I was sufficiently impressed with his combinations to conclude that they constituted the best evidence to date of an ape's grammatical competence.

When I subsequently discovered, through frame-by-frame analyses of videotapes of Nim signing with his teachers, that Nim's combinations were artifacts of his ability to imitate his teacher's signing, I concluded that apes could not create sentences (Terrace, et al. 1979). In a review of Premack's work, I argued that sequences such as *Mary give Sarah apple* were rote sequences consisting of three nonsense syllables whose meanings Sarah did not know *(Mary, give, and Sarah)* and one paired-associate *(apple)* whose meaning Sarah understood in a particular context. Accordingly, I saw no basis for regarding such a sequence as a sentence (Terrace 1979). A year later, Thompson and Church (1980) published a paper arguing that there was no basis for regarding as sentences the sequences produced by Lana (a chimpanzee trained by Duane Rumbaugh). Thompson and Church observed that Lana's utterances were rote sequences performed as responses to conditional discrimination problems posed by the experimenter. At about the same time, Savage-Rumbaugh, Rumbaugh, and Boysen (1980) concluded that the available data on symbol combinations by apes did not justify interpreting those combinations as sentences. Rounding out the picture, Premack recanted earlier claims that apes could generate grammatical sequences (Premack 1979). The only investigator who continued to argue that apes did have grammatical competence was Patterson (Patterson and Linden 1981). However, her claims regarding the gorillas Koko and Michael have yet to be substantiated in a scientific article.

Skepticism about an ape's grammatical competence had the unfortunate effect of stifling interest in other significant questions about linguistic competence concerning the use of individual symbols. Those questions were pursued with exceptional creativity and rigor by Sue Savage-Rumbaugh. While other projects (including my own) were seeking evidence that would support the view that an ape can combine symbols so as to create particular meanings, Savage-Rumbaugh and her associates performed many painstaking experiments that revealed what a chimpanzee can learn about language at the level of the symbol. As this book will reveal through many examples, Savage-Rumbaugh's program of research shows how other projects missed a wealth of important issues about symbol use by pursuing the unrealistically ambitious goals of demonstrating grammatical competence.

As human speakers and listeners, we all too easily take for granted knowledge of individual words. It is essential to recognize that such knowledge cannot be taken for granted when a chimpanzee uses them. Consider, for example, a chimpanzee who learns to request a wrench by pressing some arbitrary symbol. (Wrenches, as well as other tools, are of interest to chimpanzees because they can learn how to use them to open food sites.) It is hardly news that when a child wants to play with a wrench, she can request that tool by uttering its name. Likewise, when a child sees a wrench and simply wants to register that fact to another person, she looks or points in the direction of the wrench and says, "wrench." A child can also respond appropriately to another person's utterance whether that utterance functions as a request or as a perceptual statement. In the first instance, the child furnishes the wrench to the person who requested it; in the second, the child acknowledges that she has also perceived the wrench. Additional knowledge about the word "wrench" is revealed by the child's appreciation of the relationship between that word and the more generic term "tool." Thus, when asked whether a wrench is a tool, a food, or an animal, the child responds by identifying the relevant superordinate category.

Through careful analyses of the circumstances under which a chimpanzee was able to use a particular symbol, Savage-Rumbaugh and her associates showed that none of the previously mentioned functions of a word can be taken for granted. Learning to use a symbol to request some object in one context does not imply competence to request that object in another context or the knowledge that the symbol in question could function as a name for identifying an object. Savage-Rumbaugh also showed that the competence to request an object with a particular symbol does not imply comprehension of that symbol when another organism uses it to make the same kind of request—whether that organism is the teacher or another chimpanzee.

A good example of Savage-Rumbaugh's approach can be seen in a paradigm in which the chimpanzees Sherman and Austin learned to request tools from one another. In that paradigm, which, at first glance, seems deceptively simple, the chimpanzees are situated in separate rooms. Chimpanzee$_1$ is shown a site in which a small portion of food is hidden. A particular tool is needed to make the food accessible. Chimpanzee$_2$ has access to a tool box containing the needed tool. In order to obtain the tool, chimpanzee$_1$ must request it from chimpanzee$_2$ by using an appropriate symbol.

The physical separation of the two chimpanzees at the start of each trial ensures that symbol use is the only possible medium of communication between the two chimpanzees. Once $chimpanzee_2$ provides the tool, $chimpanzee_1$ retrieves the food and shares it with $chimpanzee_2$. It was only after teaching each chimpanzee how to respond to their teacher's request, how to regard another chimpanzee as an audience for a request (no simple matter), how to share food rewards, and so on, that the seemingly unitary skill of complying with a request for a particular object could be established. In passing, I should note that, as far as I am aware, the study in which Sherman and Austin learned how to request tools from one another symbolically is the only example of such communication between two nonhuman organisms.

In other studies, Savage-Rumbaugh confronted the difficult problem of determining the conditions under which a chimpanzee would announce spontaneously an action or a choice he is about to make. In one situation, Sherman and Austin participated in the following cooperative game. Each chimpanzee took turns selecting a food that they would share, by pushing a particular symbol on their computer console. The other chimpanzee would then fetch that item of food and share it with his partner. At one point, Sherman appeared to become impatient with Austin's slowness in requesting a food. Accordingly, Sherman usurped both roles: that of the requester and that of the provider. Under these circumstances, Sherman declared in advance which food he would select.

In another situation, both Sherman and Austin indicated their preferences for a particular object without being required to do so when one of the chimpanzee's teachers appeared in their room with a collection of playthings. Instead of waiting for the teacher to decide which game to play, or to require that one of the chimpanzees request a particular object, Sherman or Austin would point to the object of their choice and then press the corresponding symbol on their computer console. After pressing a particular symbol, Sherman and Austin would often point to the object again.

From these handful of examples, it should be apparent that the study of the use of individual symbols by an ape is a highly profitable enterprise. Such research not only avoids the all-or-none problems of interpretation encountered by studies which focused on grammatical competence, but also serves to redirect research concerning an ape's linguistic abilities toward directions in which negative results allow the investigator to rephrase a question so as to produce positive results. Research on the understanding of individual symbols is especially interesting, since it is at this level of intellectual development that the human and ape species appear to diverge.

Savage-Rumbaugh's research has also yielded important practical applications. Chapter 15 of this volume summarizes various ways in which techniques that were developed to teach a chimpanzee to master the various functions of a symbol also facilitated language development in retarded children manifesting varying degrees of language impairment. In many instances, these techniques have proved successful where other approaches have failed.

A casual perusal of any section of this fascinating volume will reveal the patient and exacting approach that was followed to instill symbol use in minds which, for reasons of phylogeny or ontogeny, cannot process language normally. The reader will also be stimulated by Savage-Rumbaugh's conceptual analysis of language development and her skill in translating those insights into experimental questions. The answers to those questions have clarified subtle similarities and differences between symbol use in human children and apes. Now that Savage-Rumbaugh has successfully redirected research on an ape's linguistic competence toward realistic but important goals, one cannot help but to anticipate new discoveries from future phases of her fascinating project. Those discoveries should add significantly to our understanding of the origin and nature of language.

New York City
October 10, 1985

H. S. Terrace

References

Gardner, B. T. and R. A. Gardner. 1969. Teaching sign language to a chimpanzee. *Science* 162:664–672.

Patterson, F. and E. Linden. 1981. *The Education of Koko.* New York: Holt, Rinehart and Winston.

Premack, D. 1970. A functional analysis of language. *Journal of the Experimental Analysis of Behavior* 4:107–125.

Premack, D. 1976. *Intelligence in Ape and Man.* Hillsdale, N.J.: Erlbaum.

Premack, D. 1979. Species of intelligence: Debate between Premack and Chomsky. *The Sciences* 19:6–23.

Savage-Rumbaugh, E. S., D. M. Rumbaugh, and S. Boysen. 1980. Do apes use language? *American Scientist* 68:49–61.

Terrace, H. S. 1979. Is problem solving language? A review of Premack's "Intelligence in Apes and Man." *Journal of the Experimental Analysis of Behavior* 31:161–175.

Terrace, H. S., L. A. Petitto, R. J. Sanders, and T. G. Bever. 1979. Can an ape create a sentence? *Science* 206:891–902.

Thompson, C. R. and R. M. Church. 1980. An explanation of the language of a chimpanzee. *Science* 208:313–314.

Preface

Apes are man's closest living relatives. For many people this fact is, 100 years after Darwin, still a bit disturbing. Having accepted, with some chagrin, their biological closeness, many individuals nonetheless find it difficult to contemplate the existence of similar behavioral and psychological processes between man and ape. Such feelings engender the ape language controversy with a peculiar set of problems. Many scientists are not certain that they wish to share a sense of mental identity with apes. Consequently they tend to erect impossible standards that apes must meet before they can be considered "language users." Moreover, it is implicitly suggested that all behavior which does not meet these standards is to be explained by basic conditioning principles. Such views serve to maintain the comfortable conception that all animals, apart from man, are simple, irrational creatures.

The striking physical and neuroanatomical similarities between man and other primates, particularly the apes, should lead to the formulation of questions regarding their mental and emotional similarities. This has not happened because most psychologists view man's mental life as determined primarily, if not entirely, by the presence of language. The absence of language in other animals, even apes, leads easily to the assumption that they also lack mental processes, emotions, etc., that fall on the same continuum as our own experiences. Were we to allow that apes are capable of language, we would also have to allow the existence of mental processes similar to our own. It is really this latter issue which many individuals so strongly resist. In fact, critics of ape language have commented that they would find the language capacities of apes much more believable if only apes did not say so many things that sounded so human-like!

Other critics take a very different stance. They maintain that most animals, and certainly all primates, possess a rich mental life similar to our own, even in the absence of language. They urge that we study and describe the behaviors of animals using a language framework that attributes thought and intentionality to the behavior of all animals, instead of a behavioristically oriented descriptive language.

Both viewpoints overlook the important fact that brain size and structure display wide variability among vertebrates. Within the primate family alone, there is a steady increase in brain size that corresponds well with the appearance of various new primate species. That is, the larger the brain, the more recent the appearance of a given primate form. Man, with the largest brain of all, has made the most recent appearance, and his brain is, in most major respects other than size, nearly identical to that of the chimpanzee. This fact alone should convince us that we share some important mental processes with the apes. It should also suggest that the similarities between ourselves and other primates diminish as we move backward in time within the primate order.

Quite a different point is raised by those who argue that the study of ape language skills lacks validity because apes don't exhibit similar behaviors in the field. These critics contend that if linguistic skills are instilled in captivity, they are bound to be an artifact of training, and consequently will reveal little about the "true" cognitive capacities of apes. Moreover, they maintain that such skills cannot possibly have any adaptive function and therefore are of little interest to evolutionary theory. Those who take such positions ignore the extreme influence of learned phenomenon for our own species. Man, in his "natural state" does not know how to read or write, yet with training, he can use these skills to engage in forms of communication otherwise unavailable to him. Upon hearing that a man can read and write, one does not ask "How can this be since men do not do so in their natural state?" Yet upon hearing that chimpanzees can use symbols to communicate, it is regularly asked "How can this be since chimpanzees do not do this in the wild?" The lack of symbol use "in the wild" is taken as reason to question the validity of the skill among apes who have been taught to use symbols. The spurious nature of this assertion becomes apparent when we realize that it would be foolish to use man's lack of reading and writing skills "in the natural condition" to question the validity of such capacities in individuals who have been taught to read and write.

There should be no debate over the issue of whether or not "crucial differences" exist between the language skills of man and ape. They surely do. Moreover, men and apes differ in many other ways—anatomically, socially, and cognitively. Nevertheless, because of our unquestioned genetic and evolutionary affinity, man is behaviorally and cognitively more similar to the ape than to any other living species. The real questions revolve not around the existence of such similarities, but rather that what it is about such similarities that lies behind our unwillingness to acknowledge them.

Whether the distinction between ape and human language is formulated in terms of the absence of ape syntax, the absence of metaphor, the absence of the ability to discuss the past, etc. matters little. There will always be linguistic differences between man and ape. Such differences are not what the research in this book is about. Rather, it is about the ways the language using behavior of apes can come to be similar to the language of one- and two-year-old children. It is also about the behavioral changes which take place in apes as this happens. We have not attempted to prove that apes are able to cross the arbitrarily defined "syntactical threshold"; we have attempted to determine what they can and cannot learn to do with symbols and how they move from simple responses to referential communicative symbol usage. We have also searched for ways of productively applying our findings to human beings who have difficulties acquiring language.

There are still those who would hold that until the ape matches the skill of the normal human child, its symbol usage should be given little credence. Normal children, it is argued, learn and use language for "the sheer joy of it" while apes learn because of contingencies set up by their teachers. It is unrealistic to attempt to assess the amount of "sheer joy" an ape finds in using symbols to communicate. However, as will be discussed in chapters to follow, apes are not the only individuals who must be taught language. Many human beings suffer from mental abnormalities which prevent the spontaneous acquisition of language and they too must be taught to use symbols for communicative ends. As they learn these skills, they do not become "normal children" in all other aspects of their behavior, nor does their language use display all the characteristics of language use in the normal child. Yet to deny that such individuals have made significant progress in their ability to communicate with others would be foolish. Handicapped persons with no viable communicative skills are much happier when they become able to request foods, objects, and activities they desire. These skills become even more valuable from a communicative standpoint when they can be executed with the referent absent. Even when such skills reach only the single word level, there is unanimous agreement among the teachers and parents of such handicapped persons that communication of a language-like nature is clearly taking place and that, as such, it permits a range of social interchanges previously not available to such people.

One does not tend to look at these handicapped persons with such a piercing and critical eye, asking at every point "is it really language?" and concluding that otherwise it must be a "conditioned response." Instead, we ask what can they accomplish with language that they were previously unable to do. Even to tell someone that you want a drink is no small thing, if before you remained thirsty until someone thought to bring you water. The ability to communicate such desires can result in significant behavioral changes, both in human beings and in apes and behavioral change is really what language is all about.

The sense of wonder, excitement, and awe that one gains from working very closely with chimpanzees can never be conveyed in any book. Nor can the shared sense of accomplishment that one experiences with them as they come to understand a communication system that lets them go beyond their normal sphere and become a little bit more human. Humans and apes are both social creatures, and they each come to be as they are because of shared social interactions with members of their own species. When the two species share this realm of social interaction intimately and for long periods of time, both species are malleable enough so that each becomes a little like the other. For mind, like language, is a social construct, erected between knowledgeable communicating participants who come to share a history of significant social interchanges. It is possible, if one looks beyond the slightly differently shaped face, to read the emotions of apes as easily and as accurately as one reads the emotions and feelings of other human beings. There are few feelings that apes do not share with us, except perhaps self-hatred. They certainly experience and express exuberance, joy, guilt, remorse, disdain, disbelief, awe, sadness, wonder, tenderness, loyalty, anger, distrust, and love. Someday, perhaps, we will be able to demonstrate the existence of such emotions at a neurological level. Until then, only those who live and interact with apes as closely as they do with members of their own species will be able to understand the immense depth of the behavioral similarities between ape and man.

Video Tapes

Video tapes were made of Sherman and Austin during all phases of the work reported in this book. Individuals who are interested in seeing the work reported here may obtain a video tape by writing to:

Sue Savage-Rumbaugh
Language Research Center
Yerkes Primate Center
Emory University
Atlanta, Georgia 30322

Many aspects of the research become much clearer when visual material is viewed. The tapes depict unedited sequences of all major aspects of the work. They are not broadcast quality. They were made for documentary purposes and many of the data in the book have been drawn from such tapes.

Acknowledgements

The writing and preparation of this book for publication and the research reported in it were supported by grants from the National Institutes of Health—*NICHOL*—06016 and RR-00165. Special appreciation is expressed to the entire staff of the Language Research Center, whose support roles have helped make this work and book possible. I also wish to thank Herb Terrace for the extensive and thought-provoking editorial help and guidance he extended to me throughout the preparation of this book. A number of people have read and commented on various portions of the manuscript during its preparation and their comments were much appreciated. They include Duane Rumbaugh, Jack Michael, Mary Ann Romski, Fred King, Patricia Greenfield, and Emil Menzel.

Illustration Credits

Photos in figures 3.1–3.4, 3.6–3.10, 3.12–3.14 by Elizabeth Rubert.
Photos in figure 5.1 by Frank Kiernan.
All other photos by the author.

Ape Language

Chapter 1

Humans, Apes, and Language

Man's Fascination with Apes

Great apes, particularly chimpanzees and their close relatives, have held a fascination for layman and scientist alike since they were first discovered sometime around 1700 (see figures 1.1 and 1.2). Of course, before this "discovery" by the Europeans, apes were well known to the natives of their indigenous regions, among whom they inspired a variety of myths and legends (for an excellent description of these, see Yerkes and Yerkes 1929).

The first serious attempt to establish the potential of the ape for scientific inquiry in this country was begun by Robert Yerkes in 1916. Yerkes viewed the ape as a life form which opened unparalleled avenues of scientific endeavor, because it filled the vast behavioral and biological gulf which had existed between man and monkey. At about the same time, in Germany, another scientist was struck by the unexplored potential of the ape. Köhler (1925) wrote, "Chimpanzees not only stand out against the rest of the animal world by several morphological . . . characteristics, but they also show a type of behavior which counts as specifically human" (pp. 256–266). Nearly 50 years later, Jane van Lawick-Goodall (1968), the first scientist to study, in depth, the social behaviors of apes in the wild, echoed the sentiments of Yerkes and Köhler.

The views of Yerkes, Köhler, and van Lawick-Goodall are not unique. Virtually everyone who has had extensive opportunity to observe and interact with chimpanzees comes to view them as somehow different from other primates and from other mammals.

Portions of this chapter have appeared in an *American Scientist* article entitled "Do Apes Use Language," 1980, and also in a *Journal of Experimental Psychology: General* article, "Can a Chimpanzee Make a Statement," vol. 112, 1983.

HISTORY OF QUADRUPEDS. 413

The MONKEY kind are removed ſtill farther, and are much leſs than the former. Their tails are generally longer than their bodies; and, although they ſit upon their poſteriors, they always move upon all four.—They are a lively, active race of animals, full of frolic and grimace, greatly addicted to thieving, and extremely fond of imitating human actions, but always with a miſchievous intention.

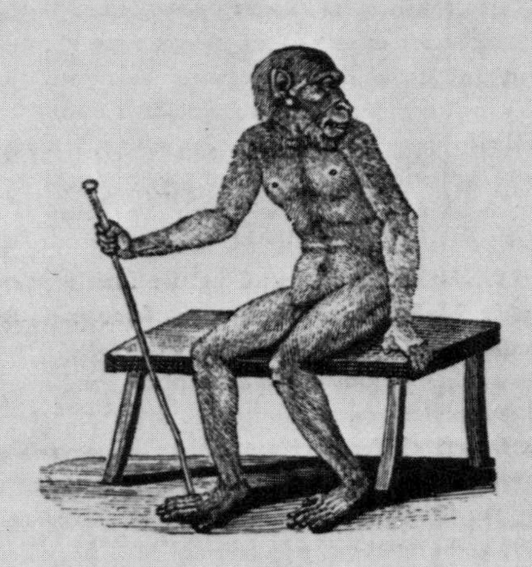

THE ORAN-OUTANG, OR WILD-MAN OF THE WOODS,

is the largeſt of all the Ape kind, and makes the neareſt approach to the human figure. One of this kind, diſſected by Dr Tyſon, has been very accurately deſcribed by him. The principal external differences pointed out

Figure 1.1. Wild Man of the Woods, from Thomas Bewick's *A General History of the Quadrupeds,* originally published in 1790. At this time it was generally believed that there was a single type of ape living in Africa, Madagascar, Borneo, and the East Indies. It was thought to be a creature which looked something like the above illustration and reportedly carried off human females and compelled them to stay with it. A few young apes had been kept in captivity even by this time and they were reported to be very fond of their keepers. Bewick asserts that while these creatures possess "the strongest affinity to the human form" (p. 450), nevertheless there exist "differences in the internal conformation such that the external similarities are only superficial and do not allow him to approach the nature of man" (p. 450). Thus it can be seen that Europeans were quick to discredit the seeming similarity between man and ape.

Figure 1.2. This drawing of one of the early apes to be kept in zoological gardens reveals the close affinity which soon appeared between apes and their keepers.

They seem—if the reader will permit the anthropomorphism—somewhat human. Apes are different from other animals. One feels this intuitively if given an opportunity to play with them, but the nature of this difference has been difficult to specify in scientific terms.

Apes' faces and movements appear expressive, even to the eye unaccustomed to the ape's physiognomy. We see happy apes, sad apes, tired apes, and puzzled apes on billboards, calendars, and T-shirts. The slogans under such pictures often comment upon the mood which we see reflected in the ape's face and posture. Our ability to understand and to empathize with apes on the basis of their emotional expressions is strikingly revealed in the movie *King Kong*. Guillerman, who designed Kong's face for the movie, spent many hours in zoos observing the expressions and movements of apes before he designed Kong. His intent was not to design a Kong that aped man but rather to design a Kong that was a true ape.

He did not worry about being anthropomorphic, or about whether the facial expressions which caused gorillas to look happy or sad to him actually reflected these emotions; he just presumed that they did, because of the common biological heritage of man and ape. He took it as a matter of common sense that ape and human facial expressions were similar enough that either species could readily perceive the moods and feelings of the other just by looking at one another. Furthermore, he knew that if audiences were to develop an empathy for King Kong, it would have to be because they understood how King Kong felt—and since King Kong did not speak, feelings would need to be communicated through facial expressions. The public would need to be able to see Kong's face clearly in order to infer anger, love, lust, hate, fear, and sorrow and to equate these emotions with similar ones in their own past experience.

Do apes really have these emotions? Do they display these emotions in a manner similar to our own species? Scientists have attempted to approach these questions more objectively than did Guillerman by labeling apes' expressions descriptively—e.g., open-mouth bared teeth face, bulged lips face, relaxed open-mouth face (see figure 1.3)—and correlating the occurrence of expressive behaviors with other behaviors. While objective analyses such as these lend more validity to our interpretation of an ape's play face or fearful grin, they cannot tell us whether the emotion the ape is experiencing when he produces such an expression is in any way similar to our own when we evidence such an expression. How can we tell whether the ape really feels like us, sees like us, thinks like us, and knows like us? These intriguing questions refer to complex neurological processes which at present we neither understand nor know how to measure. We are aware that as individuals we see, think, feel, and know, and we ascribe the same abilities to other human beings for two reasons: because they are members of the same species and share our biology, and because they talk about these abilities or ascribe them to themselves through other means.

Our Closest Living Kindred Species

We know that the great apes share our biological heritage to a remarkable degree. According to King and Wilson (1975) amino

Figure 1.3. Relaxed open mouth display in a young male *(Pan paniscus)* as he is laughing boisterously while being tickled.

acid sequencing, immunological, and electrophoretic methods of protein comparisons between man and chimpanzee all indicate that the average human polypeptide is more than 99 percent identical to its chimpanzee counterpart. This is an extraordinarily high degree of similarity—comparable with that found among sibling species who are virtually identical in terms of morphology. Man and chimpanzee, however, are not classified as sibling species, instead they are placed in different families.

There also exists a fundamental difference in the way scientists conceptualize their own behaviors and the way they conceptualize the behaviors of their closest living relatives—the apes. This fundamental distinction has no biological basis, but appears to be derived instead from a singular behavioral distinction—the ability of man to speak. The extreme importance of this enormous difference is aptly illustrated when we recall that, even as recently as the turn of the century, cases involving abuse of mentally retarded human beings were brought before the Society for Prevention of Cruelty to Animals, because this Society was considered to be the most appropriate body to hear cases regarding human individuals who had no speech. Thus we see that our concepts of human uniqueness are closely tied to our linguistic ability to express our ideas to one another.

The major distinctions between man and chimpanzee are behavioral, not anatomical. Furthermore, these behavioral distinctions pivot around two important skills, the ability to use objects in a complex fashion, and the ability to use language. It has also been asserted that man is also unique in other ways: for example, the ability to hunt, to draw, to make music, to transmit culture. These, however, are all outgrowths of the more basic skills of complex object usage and language. Indeed, Elizabeth Bates (1979) has argued that similar cognitive processes underly the emergence of both complex object use and language.

Apes and Language

It is no wonder then that Gardner and Gardner's (1969) initial assertion that they had succeeded in teaching a young chimpanzee, named Washoe, to use signs in much the same fashion that young human children use words produced a startled reaction among scientists and laymen alike. It is one thing to accept Darwin's thesis that man evolved to his present form from a primitive apelike creature. It is quite another to accept the idea that apes possessed, even in rudimentary form, that key ability which so clearly demarcated man's superiority over, and his freedom from, his apelike heritage.

Although others had tried to teach apes to talk, their efforts were unsuccessful (Hayes and Hayes 1951; Kellogg and Kellogg 1933). However, these approaches all had one thing in common: they attempted to teach apes to produce vocal sounds which approximated those of human beings. By contrast, the Gardners reasoned that perhaps the chimpanzee was not anatomically equipped to produce human speech [a fact later confirmed by Lieberman and his colleagues (Lieberman, Crelin, and Klatt 1972) in their comparative studies of the human and chimpanzee vocal tract].

Few, if any, animal learning studies have had the broad interdisciplinary impact which characterizes ape language research. The fields of linguistics (Brown 1973), psychology (Limber 1977), anthropology (Hill 1978), zoology (Griffin 1976), sociology (Meddin 1979) and neurobiology (Steklis and Raleigh 1979) have been affected by the claims that chimpanzees are capable of what was heretofore considered a peculiarly human capacity. Clearly, the presumption of human linguistic uniqueness is a cornerstone of many fields of scientific inquiry and, as Hill (1978:89) points out, the ape language research "may imply a paradigm shift, with Plato finally giving way to Darwin or perhaps an identity crisis for *Homo sapiens.*"

There have been three basic approaches to the teaching of language skills to apes, each of which has employed a different communication system and training philosophy (See Ristau and Robbins 1982 for a thorough review of ape-language work).

Gardner and Gardner, who worked initially with a chimpanzee named Washoe (and more recently with four other chimpanzees), chose to rear their subjects as much as possible like human children and to use American Sign Language of the Deaf (ASL) as the medium of communication. Work with Washoe began when she was a year old and continued for five years, at which time she was able to produce, in the proper context, 132 signs. She was also reported to have formed a variety of novel combinations, such as "Roger Washoe out" and "you me go out hurry," so that the Gardners concluded that Washoe compared favorably with children at Brown's Stage III and beyond (Gardner and Gardner 1969, 1971, 1972, 1973, 1974a, 1974b, 1975, 1978a, 1978b).

Following this initial work, Roger Fouts, a student of the Gardners, began teaching ASL to six chimpanzees at the University of Oklahoma. Fouts also continued to work intermittently with Washoe, using the same approach in teaching these apes signs that had been used with Washoe. He molded an ape's hands into the proper physical configuration with less and less precision and force until the animal placed its own hands in the proper position. Fouts, unlike the Gardners, permitted spoken English in the presence of the chimpanzee (Fouts 1972, 1973, 1974a, 1974b: Fouts, Chown, and Goodin 1976: Fouts, Chown, and Kimball 1976: Fouts, Mellgren, and Lemmon 1973). Shortly after Fouts began his work in

Oklahoma, Francine Patterson (1978) began a similar study with a gorilla named Koko. Like Fouts, she molded the gorilla's hands into signs, and she permitted spoken English. The findings of Fouts and Patterson lent corroborative support to the Gardners' conclusions regarding Washoe.

David Premack (1976a) taught the chimpanzee Sarah to select, under certain conditions, a variety of plastic chips that he glossed as English words. Sarah was typically required either to select one chip from a set of two or to arrange four or five chips in given sequences. Sarah's chip selection was considered by Premack to be analogous to spoken words. For example, Premack would place in front of Sarah a candy and three plastic chips. He called these chips food names such as "candy," "banana," and "orange." When Sarah selected the chip that Premack called "candy," she was said to have named and asked for the candy. Premack went on to perform similar chip-selection tests with more complicated sets of English words. One test, for instance, involved the selection of chips said to represent the plural indicator, in the presence of other chips glossed as "red yellow ?? color." Thus Sarah was said to be stating that red and yellow were colors.

Rumbaugh, et al. (1973) set out to develop a method of studying ape-language acquisition that would combine many aspects of the approaches taken by the previous researchers and would at the same time avoid the pitfalls of potentially cueing the ape. They used a mode of communication that would permit the gathering of data in a manner that was not dependent upon observer reliability. Unlike the "words" of other projects, those in this system were composed of a limited number of elements that were recombined in a variety of ways to produce geometric figures.

Lana, a female chimpanzee, was the first subject of this project. She was required to learn strings of lexigrams in order to manipulate food vendors and mechanical devices in her environment (e.g., *please machine give M&M* and *please machine make window open*) (Rumbaugh 1977). She came to realize that the machine had to be stocked by humans, and she readily transferred the use of such strings to the manipulation of her human teachers *(Please Tim give milk out room)* (Rumbaugh 1977). Lana executed these requests by depressing keys embossed with the geometric symbols and arranged on a large wall panel (see figure 1.4). The keys could be relocated to preclude simple positional learning.

Herbert Terrace began a signing project with the chimpanzee Nim, born in Oklahoma in 1973. He raised Nim as a human child and attempted to encourage communication regarding all aspects of daily life. Nim was taught to sign by a method modeled after the techniques of Gardner and Fouts. Terrace made the first permanent video record of the development of signing behavior across time, and it was this video record of Nim's progress which ultimately led Terrace and his co-workers to reverse their initially positive conclusions regarding the signing capacities of apes (Petitto and Seidenberg 1979; Seidenberg and Petitto 1981; Terrace 1979a).

Before Terrace's report, ape-language studies had received strong support from the scientific community. Many assertions regarding the capacities of apes were accepted with very little attention to methodology and the general atmosphere was one of enchantment

Figure 1.4. Lana at her keyboard. The horizontal array of projectors over the keyboard display the request she has just made—please machine make slide. This request causes one of a variety of slides to be projected for her to view.

about what apes would say or do next. Following the publication of Terrace's work, attitudes changed dramatically. No matter what was reported, and regardless of the care taken in the collection of the data, the general reaction became one of disinterest and incredulity. Both the initial overly ready acceptance—and the later thoughtless rejection—were unwarranted. The facts lie somewhere in between and they are very interesting in their own right. The purpose of this book, and the research reported herein, is to try to bring them to light.

Initial Assumptions of Ape-Language Researchers

The Gardners (1971), Premack (1976), Rumbaugh (1977), Patterson (1978), and Terrace (1979a) all began their work with the assumption that when an ape "correctly" used a symbol, it had some referent clearly in mind. These researchers were of the opinion that when a chimpanzee learned an associative connection between a displayed object and a particular behavior, the chimpanzee was, in fact, *naming*. Naming was viewed as a rather simple skill and thought to be readily accounted for by known principles of conditional discrimination learning. Thus, the real focus of these projects was the combinations of symbols that the apes produced and whether or not these combinations constituted evidence of syntactical competence.

The view that naming proper is a rather simple skill has not been supported by the recent work of developmental psychologists who have studied the preverbal child (Bates 1976; Greenfield 1973). What has emerged instead is a perspective which stresses the complex nature of the communication that is taking place between child and parent before the onset of one-word speech. Even at this early phase, human children engage in referential communicative exchanges that show many of the characteristics normally associated with language.

It is therefore important to ask whether chimpanzees also engage in nonverbal interchanges that have referential character before they are trained to use signs or geometric symbols. To answer this question, we must determine exactly what it is that the preverbal child is doing before he begins to speak. Recent work by Bates (1976) has focused on exactly this question and clearly delineates the behavioral communicative competencies of young children. By contrasting her findings with what is known about young apes, it is possible to compare the communicative skills of child and ape before the onset of language (which is spontaneous in the child and tutored in the ape).

Preverbal Children and Chimpanzees

Bates (1976) found that during the development of preverbal communicative competence, the human child passes through three stages: "from exhibiting self, to showing objects, to giving and pointing to objects" (p. 61). The initial stage, showing off, appears at about nine months of age and involves the voluntary and intentional repetition of an action which has previously been successful in evoking laughter or comment.

> For example, at 9;6 Carlotta is in her mother's arms and is drinking milk from a glass. When she has finished drinking, she looks around at the adults watching her and makes a comical noise with her mouth (referred to in some dialects as "the razzberries"). The adults laugh, and Carlotta repeats the activity several times, smiling and looking around in between. (p. 68)

Chimpanzees engage in similar "showing off" behaviors, to receive attention from both humans and other chimpanzees. The same behavior, "buzzing the lips" or "razzberries," in fact, is one which many chimpanzees use to gain attention. This would suggest that at a very primitive level both chimpanzees and children are capable of making a self-statement, and both are rewarded when they observe that their behavior has had an affect upon others.

Bates (1976) observes that this sort of "self showing" behavior is strictly social, that is, it does not involve reference to some third element in the situation. At 10;18, however, Bates (1976) finds that Carlotta now extends her arm to show an object to the adult, and thus an object (as opposed to Carlotta herself) becomes a tool for attracting adult attention. Bates (1976) notes that

> Within this new capacity to attract adult attention through use of a third element, we also found a succession of substages. Initially there is a very subtle passage from the showing off or exhibitionism of the previous stage, to showing. While Carlotta is playing with an object already in her hand (10;18), she extends her arm forward with the toy for the adult to see.

> Afterwards this showing becomes a complete and autonomous activity in itself. The child purposively seeks and takes an object in order to show it to the adult. . . . However, when showing first appears there is apparently no intention to give the object, as is confirmed several times when an adult tries to take it away. Carlotta will not let go of it, and may even pull her arm back to keep the toy. (p. 60)

Chimpanzees also advance to the substage of object showing, but not of giving. Object showing is often used to elicit a chase game between individuals. Chimpanzees deliberately search out interesting objects and show them to other chimpanzees and human companions. Attempts to take the object result either in avoidance behavior (which may be playful as in a chase game), or in aggressive behavior (in which case attempts to take the displayed object may result in a bite on the hand).

Showing off an object thus appears to be a primitive means of indicating possession. As such, it is a frequent point of confusion between naïve human experimenters and chimpanzees. When I was first introduced to Lucy, a five-year-old female chimpanzee who was taught American Sign Language (ASL) signs by Fouts (1972, 1973, 1974a, 1974b), she selected a plastic flower from her box of toys, carried it to me, pressed it to her nose, and then extended her arm toward me. I reached to take the flower, and she promptly bit me on my extended hand. (I continued to work with Lucy for several years and never again mistook showing for offering.)

Taking objects from chimpanzees, particularly objects to which they have drawn attention by showing, is frequently problematic. The observation that human children pass through a similar period of exhibiting without giving often goes unnoticed because the duration of this period in the normal human child is only a few months.

Following Bates (1976),

> The next substage occurs when giving becomes an autonomous means for establishing contact and interacting. . . . at 1;1;2 (one year, one month, two days) we have the confirmation that the child (Carlotta in this case) has differentiated showing from giving. She takes a wooden mask from a chair, crosses the room, smiling and looking at the observer, and drops the object in the observer's lap.
>
> At the same moment in which Carlotta is able to give objects, she also begins to point at objects while looking back at the adult for confirmation. (p. 61)

Here we find that the child makes a spontaneous transition from possessive showing to true giving and indicative pointing. Typically the chimpanzee falls short here. Although chimpanzees who are taught to sign have been observed to point at objects, these chimpanzees do not point at objects *and* look back to adults for confirmation. That is, they do not point at objects in order to show these objects to adults.

The noncommunicative type of point (pointing while looking only at the object of concern) also occurs in human children and precedes communicative, indicative gestures (pointing, showing, and looking back and forth between adult and indicated object) by three to four months. True communicative pointing appears just before the onset of the one-word stage.

Note that the chimpanzee lacks not only the onset of the one-word stage, but also the onset of *indicative communicative pointing.* Bates (1976) describes this behavior in Marta at 12 months, noting that she

> . . . shows, gives, and points to objects and awaits the appropriate adult response. If no response is forthcoming, she will repeat the behavior with great insistence, or possibly seek another audience. *At this stage she has yet to utter her first word* [emphasis is mine—S.S-R.], and shows no other signs of symbolic behavior. Hence, Marta has full control of the protodeclarative while she is still at sensorimotor stage 5. . . . Hence, the protodeclarative, for both subjects, (Marta and Carlotta) is a procedural rather than symbolic structure, beginning during the fifth sensorimotor stage. (p. 62)

Thus, we find a constellation of behaviors, all appearing spontaneously in the normal child before the onset of one-word speech, these involve:

1. Giving to others.
2. Indicative pointing for others (as opposed to pointing for self) (Werner and Kaplan, 1963).
3. Use of nonsense voluntary vocalization as an accompaniment to gesture to orient attention of others to indicated object.

None of these behaviors appears spontaneously in the chimpanzee. Nor is there any evidence to suggest that the chimpanzee comprehends the indicative or "referring" function of such behaviors when they are displayed to him by human experimenters.

Thus we see in the child the spontaneous emergence, before the onset of language, of a cognitive structure which employs and coordinates reference, cooperative giving, and joint orienting of attention in behavioral interactions. Therefore, in the case of chimpanzees, it is not *just* "words" or vocal speech that is absent; rather it is this entire cognitive complex which children use to orient, organize, and regulate cooperative referential behaviors between themselves and others.

The chimpanzee appears to be at a unique position within the phylogenetic evolution of communicative competency. Unlike most other animals, he is able to attribute intentionality to the general and the communicative behaviors of other apes (DeWaal 1982; Menzel 1974, 1975; Menzel and Halperin 1975; Premack and Woodruff 1978; Savage 1975; van Lawick-Goodall 1968). Also unlike most animals, the chimpanzee can separate *intended action*—as expressed by a communicative gesture—from the overt action of the gesture itself (Savage-Rumbaugh 1980).

Among group-living chimpanzees, such intentional gestural communication appears to be limited to situations in which food or reassurance is being requested. There are no reports of one chimpanzee gesturally requesting that another act upon a nonfood object in its behalf as is commonly seen among human parents and children. Chimpanzees do, by screaming and vocalizing, request assistance in fighting other animals and company in travel (Menzel 1974, 1975; Savage 1975; van Lawick-Goodall 1968), but such social solicitations lack the referential complexity which is seen in the nonverbal requests of young children. Giving of nonfood objects also appears to be nonexistent in chimpanzees, and the giving of food items is rare and limited to mother-infant pairs or the sharing of meat following a kill (McGrew 1975; Savage 1975; Teleki 1973; van Lawick-Goodall 1968).

Chimpanzees use gestures at the inter-individual level, but their gestures are limited to simple situations in which the lack of referential specificity is not a hindrance. Both the human child and the young chimpanzee learn to use the extended hand gesture as a means of requesting an object. Both learn to do so without any extensive training from older individuals. But the human child goes on to learn to specify particular objects through the use of symbols and is thus able, in complex situations, to specify one of a large number of outcomes.

The normal human child gains the capacity to refer to items in a specific representational manner before the onset of one-word speech. While the child does so without extensive intentional training from elders, the chimpanzee does not.

What happens when humans intervene and try to obtain behavior in the chimpanzee which is equivalent to that seen in the human child? The answer contains important implications for theory and research concerned with an ape's linguistic competence. Do humans simply provide a different and more complex model when they serve as conspecifics—thereby enabling the chimpanzee infant to spontaneously develop skills that he would not acquire when surrounded only with conspecifics as models? Is it true that *simply* by providing a human model and a nonvocal signal system, one creates the conditions sufficient for the spontaneous appearance of language in the chimpanzee?

In our view, the previous ape-language studies have not attempted to determine whether their subjects have moved from the use of nonreferential symbols to symbolic naming proper. It is not sufficient to simply test "labeling behavior" in chimpanzees because, as we shall show, chimpanzees, unlike humans, do not develop language of their own accord. This makes it necessary to show (not just assume) that a given ape is capable of using representational symbols before presuming to analyze his stage of linguistic development in terms of Brown's (1973) schema or before attempting to determine if his combinations reflect syntax.

"Names of Things"—the Central Issue

We agree with Nelson (1977) that the essence of human language is not found in syntax but rather in the fact that language permits "the translation of meanings, . . . knowledge of people, objects, events, and their relations into words: and the expression of these meanings to a social partner for some functional purpose. Reciprocally, it involves the interpretation of the meaning expressed by others to the child and the appropriate response to the functional purpose of the utterance" (pp. 567–568).

We suggest that every instance of referential symbolic communicative exchange, beginning with the one-word stage (once the child has moved beyond the pure performative use of words) is a complex phenomenon involving the following four components:

1. An arbitrary symbol which stands for, and can take the place of, a real object, event, person, action, or relationship.
2. Stored knowledge regarding the actions, objects, and relationships relating to that symbol. (This stored knowledge will not be identical for all symbol users, but the greater the degree of overlap, the more precise the communication.)
3. The intentional use of symbols to convey this stored knowledge about an object, event, person, action, or relationship to another individual who has similar real-world experiences and has related them to the same symbol system.
4. The appropriate decoding of, and response to, symbols by the recipients.

A word is an extraordinarily complex thing; in fact, it is not a thing at all. Rather it is an *activity* which occurs between two individuals. The activity of speaking a set of phonemes, which we term a word, in itself refers to the diversion of attention of both participants to something other than the actual activity of the word production in which they are engaging.

Words are typically spoken with the intent of evoking referents in the minds of the receivers, and receivers presume this intent on the part of the speaker. Words acquire the unusual ability to co-join the attention of 2 or more individuals upon items or actions which are not immediately present. They do so through interindividual interactions in which, initially, the referent is physically present and joint attention to that referent is accomplished by pointing or glancing. Vocal markers are overlaid once the joint object of reference is established. Later, these verbal markers alone can serve to refer to the referent even in its absence. At a more complex level, these verbal markers serve to mutually orient attention to relationships between referents and even to produce joint mental manipulations of these relationships in the absence of the referent.

We agree with Bates (1976) that "language is a tool, we use it to do things. . . . Words do not 'have' referents—speakers use words to refer or point to something existing in an hypostasized world" (pp. 1, 10). Consequently, the activity of referring with words is not really taught. The words themselves *are* taught. They are *used,* and as they are used they are learned. They initially aid the child in mediating and controlling his social interaction by allowing him to direct the orientation of others. Their use necessitates intentionality, that is, children come to use words with the intention of causing some behavior in the adult. As Bullowa (1979) has pointed out, "meaning and intent are fundamental to the definition of communication. . . . Relations between context and content, while variously conceived in detail, are universally recognized as important in understanding the developing system of interpersonal communication, which is itself seen as constructed through the joint efforts of mother and infant" (p. 5).

The position adopted here is that it is not possible to understand word acquisition as a conceptual process if attention is focused on the mere occurrence of words. Instead, we must focus our attention on the exchange of meaning between parent and child, or teacher and chimpanzee, and we must ask how this exchange is accomplished, elaborated upon, and refined—for the meanings of words, and the learning of words, are given life and substance only in these processes of exchange.

Chapter 2

In the Beginning Was the Request

Language at the "Request" Level

Although chimpanzees do not come equipped with all of the preverbal skills of a human child, they nevertheless show evidence of acquiring some rather sophisticated language skills (Gardner and Gardner 1971; Premack 1976a; Rumbaugh 1977; Terrace 1979c). How can this be unless they have either duped their teachers or, at the very least, understand some of what they are saying?

There is little doubt that chimpanzees can learn to use symbols to manipulate the behaviors of those around them—particularly to gain access to food, objects, and space that are dominated by others. Everyone who has worked in the field has found the apes quite competent in this regard (Gardner and Gardner 1971; Premack 1976a; Rumbaugh 1977; Terrace 1979c).

Both Lana and Washoe, for example, spontaneously began to use their learned symbol systems to control and regulate the behaviors of their human teachers to their own advantage. That is, instead of the human beings arranging the proper contingencies for a symbol (or group of symbols) to occur, the chimpanzees *began to create the contingencies* and then to use the symbols to bring about predictable situations. For example, Washoe began to go to the door and, spontaneously, to sign "out" (Gardner and Gardner 1971).

When Washoe initially learned to make this sign, it had been because she could not open the locked door herself. When she tried to go out, the teacher would stand by the door and sign "What you want?" and prompt Washoe to request opening the door by a query such as "Washoe want go play outside?" In such a case, the experimenter already knew that Washoe wanted to go outside because Washoe had been trying to open the door. When Washoe did sign after such encouragement by her teacher, the sign or signs she used (such as "out me," "Washoe out," etc.) did not convey *new* information to the teacher, but merely functioned as the occasion to open the door.

Thus, the teacher opened the door for Washoe not because she *then became aware* that Washoe wanted to go out, but simply because Washoe produced these signs in that setting. However, later Washoe began to spontaneously sign "out" *as* she approached the door, or even *before* she approached it. The teacher was then no longer setting the occasion for the sign to occur. The occasion was now set by Washoe's own desire to go outdoors and it occurred before any other nonverbal or contextual indicator that Washoe wanted to go out.

A similar sequence of events occurred during Lana's training, but in her case the lighting of a string of lexigram symbols (rather than a gesture) came to precede the experimenter's unlocking the door and allowing Lana to go out. For Lana, just as for Washoe, the use of symbols was at first simply overlaid on other more direct behaviors (like pushing on the door) which themselves revealed the chimpanzee's wishes. For Lana, just as for Washoe, these symbols then began to supplant and, in fact, precede the more overt non-symbolic behaviors.

At this point, for both Lana and Washoe, prelinguistic communicative intent can be said to have appeared in their use of the symbols, for symbols previously elicited only by the experimenter now occurred by the ape's own volition and began to be used as the sole means of communication.

The spontaneous requests by both Lana and Washoe were numerous and varied, but their emergence was almost surely shaped by three important factors. First, their modes of communication (signs and keyboard) were always available to them, just as a child's hand and voice are always ready to be employed. Second, their teachers behaved as though they attributed intent to behaviors where it had not necessarily existed originally. Third, because of the basic interindividual nature of the learning paradigm employed in the Lana and Washoe projects, it was made clear that the human's behavior depended on the chimpanzee's use of symbols. The chimpanzee's symbol selection was not merely judged correct or incorrect; it functioned to produce a behavioral and environmental change, mediated by the human interpreter.

By contrast, the chimpanzee with whom Premack (1976a) worked (Sarah) did not have all of her symbols readily available; the experimenters did not interpret her gestures or symbols as communications, but as correct or incorrect responses. And her symbol manipulation did not serve to alter the behavior of companions or their behaviors upon objects in the environment, but only to produce a reinforcement, or another trial. Thus Sarah did not begin to use her symbols to produce spontaneous requests.

The Influence of the Mother on Communication

These three factors (the availability of all symbols, the attribution of intent, and the interindividual nature of the communicative contingencies) are thought by many to be the elements which enable the infant to move from simple action-reaction communications to gestures. As a child reaches with open hand for an object far away, the mother believes the child is trying to tell her to hand him the object. The mother attributes communicative intent to the child's gesture *before* there is any clear behavioral indication (by glance and repeated gesture) that the child is, in fact, capable of intentional gestural communication (Lock 1978).

Lock (1978) suggests that it is the mother's premature interpretation which sets up behavioral contingencies that permit the child to infer a cause-effect relationship between his gesture and the mother's ensuing behavior. Once the child perceives this cause-effect relationship, he gradually gains control over his gesture and begins to watch his mother for the expected response. This monitoring and expectation serve to modify the gesture still further, rendering it more precise and decreasing the "pure action" function of the gesture while increasing the "communicative" function.

It is particularly noteworthy that this sort of premature interpretation of intentional gesture does not occur between chimpanzee mothers and infants (Savage 1975; van Lawick-Goodall 1968). In fact, just the opposite seems to occur. When the chimpanzee *(Pan troglodytes)* infant reaches toward an object that is too far away, the mother, if she notices this behavior at all, is more likely to remove the object than she is to hand it to the infant. She appears to interpret the infant's reaching not as a communicative gesture directed at her, but as a direct action. Chimpanzee mother-infant pairs *do* coordinate some interactions by means of gestures, but the uses of such gestures are limited to social exchanges (Plooij 1978; Savage 1975).

While a young chimpanzee may adopt a tickle or a grooming posture to initiate these activities, such behavior is quite different from gesturing toward an object while looking back at the mother and waiting. Social gestures communicate information about what one individual wants or does not want another *to do to him;* object-oriented gestures communicate information about what a second individual should *do to an object* for the first individual—a very different state of affairs. Such object-oriented gestures are preludes to cooperative object-oriented interindividual interaction schemas. They structure interindividual organized cooperative goal-directed behaviors, not just interindividual social behaviors.

For those who wonder why the chimpanzee does not develop language in the wild, the answer lies, then, in the absence of a caretaker who is able to attribute communicative intent *before its true onset.* Because chimpanzee mothers do not treat their babies' early actions as intentional communicative gestures, their infants do not have the opportunity to become effective communicators. A recipient of the action is needed who will perceive early actions *on objects* as incipient communications—and in so perceiving thereby produce the emergence of object-oriented communicative gestures. By withholding cooperation (such as opening the door) until a specific gesture has occurred, the human caretaker behaviorally lends meaning to the ape's gesture. By contrast, the chimpanzee mother would *not wait* for the gesture *before* opening the door; in fact, she would not open the door at all. (A chimpanzee mother who had herself been reared by human beings might behave quite differently, Fouts, Fouts, and Schonenfeld 1984).

When Humans Intervene

Human experimenters working with Washoe assigned intent to her early nonverbal gestures. Thus, Washoe's gestures received a very different response than would have been the case if she had been reared by her natural chimpanzee mother. This difference permitted, and in fact encouraged, in Washoe a perception of cause-effect relationships between gestures and the resulting objects presented to her and events she caused to happen. (The same is true of other human-reared apes such as Nim, Koko, Alley, etc.)

It is also true of Lana. When Lana lighted symbols, the human experimenters assumed that she did so with an intent to communicate. They also presumed that her nonverbal gestures were intentional. If she reached toward an M&M that was blocked by glass, the experimenters perceived the reach as a communicative gesture and, by pointing to the proper keys on the keyboard, showed Lana how to ask for the candies. When Lana touched these, she was given an M&M. Thus, the translation of action to symbol in Washoe's case was gestural, and in Lana's case it was from the action (or attempted action) to keyboarded request. In the process, Lana's reaching and pointing gestures took on a human appearance, but she did not develop the elaborate ritualized gestural repertoire displayed by Washoe. (Elaborate gestures were, in Lana's case, discouraged, since they were viewed at the time as a different communication system rather than an important link between action and symbol.)

It is therefore clear that chimpanzees can, given the proper human training environment, learn to use symbols to produce spontaneous requests, which appear to be reflections of internally motivated desires and needs. They can, in Skinner's (1957) terms, execute "mands."

What remains to be determined is whether or not they can use these learned symbol-object associations to engage in other linguistic functions which do not involve requesting. Can they use symbols that do more than reflect internal desires? Can they describe objects or events, or speculate about the nature of future events? These are more complex cognitive activities in that they involve a degree of distancing between symbol and symbolizer that is not found at the level of a simple request.

The Human Child: Language at the Request and the Descriptive Levels

In order to approach these questions, it is important to review what happens in human children as they learn to produce such behaviors. Keep in mind that we are dealing with a multilevel phenomenon. The child first learns to engage in symbol-object associative behaviors that are context-linked and may be described as an integral part of an interindividual behavioral routine. Thus a child may say "bye-bye" when people leave. Initially, this is not a request, it is simply a behavior that is linked to the routine of leaving. Later, the child may use "bye-bye" to precipitate the act of leaving and at this point "bye-bye" advances to a spontaneous request. Still later, when asked, "Where's Daddy?" the child may comment "bye-bye" to indicate where Daddy has gone. Such usage is an enormous leap forward; "bye-bye" is now being used in a way that gives information beyond that readily available from the routine or the context.

Earlier, when the child said "bye-bye" *while leaving,* it was a means by which the child participated in the leavetaking rituals characteristic of our species. Later, when the child used the word to initiate these rituals, the resulting events still involved the child as a participant. However, when a child says "bye-bye" to indicate that someone who was in the room has now departed, the word becomes divorced from the child's own immediate action in a most significant way.

Note that just before the emergence of word usage at this level, we also find the appearance of referential pointing, accompanied by an often indistinct vocalization. The child uses pointing and vocalization to draw the attention of others to objects and events within his environment. This suggests that the young human child has at its disposal a nonverbal skill—that of referential pointing—which can be combined with her intentional communicative capacities, and with previously learned symbol-object associations to produce the type of behavior commonly termed "naming." Furthermore, such naming activity does not appear to be limited to items that the child is interested in receiving, but is applied to all manner of items.

Braunwald (1978) provides us with an excellent description of the emergence of this phenomenon in her daughter, Laura, whose utterances were studied in detail during the period of transition from context-linked speech to representational speech. All such utterances, along with their nonlinguistic contexts, were recorded between 8 and 20 months of age.

During this prerepresentational period, Laura employed a number of single-syllable words in a wide variety of contexts. Braunwald notes that what was missing from Laura's utterances during this period was the arbitrary, culturally defined linguistic and social conventions for encoding objects and events by means of particular utterances. However, the use of utterances to form intentional requests was present. For example, Laura initially used the word "Ba" in all the following ways:

1. when she wanted milk, juice, or any liquid in a cup;
2. when she saw her mother carrying a milk bottle;
3. when she wanted to play with the milk bottle lid;
4. when she had finished drinking and her cup was empty;
5. when blowing bubbles in her drink;
6. when looking at a milk carton;
7. when drinking milk;
8. when spilling milk;
9. when milk was poured;
10. when mother put cups on the table;
11. when she wanted more to drink.

According to Braunwald, "Ba" seemed to refer to water, cup, juice, more, all gone, pour, drink, cup, bottle, bubble, carton, good taste, as well as milk.

Within just a few months, the words straw, blow, all gone, cold, cup, bottle, bubble, more, pour, spill, Laura, do, drink, and juice all appeared and were employed in contexts where before only "Ba" had been uttered. For example, when Laura spilled her milk at 13 months she said "Ba." At 16 months, when she finished drinking her milk she gave her empty cup to her mother and said, "Ba um good." At 19 months, in the same context, she said, "All gone."

Thus, even before Laura could produce a number of reasonably distinct words, she was using a single vocalization for the following communicative functions: to request food, to comment on her mother's actions, to request objects, to comment on her own actions and to comment on objects. Further, from Braunwald's description, it appears that from the very first use of "Ba," its occasion was determined by the child and not the mother. The mother did not withhold items until "Ba" was emitted—as human trainers have withheld items until the chimpanzee produces a sign or symbol. Presumably, Laura determined when to say "Ba" simply by listening to the speech of her parents and noting the contexts which they vocally marked. That Laura did *not* use "Ba" in a totally inappropriate context, such as while playing in the wading pool, suggests that even as speech first appears, the normal child has already learned an enormous amount of contextual information regarding the appropriate use of symbols.

Thus we find that at a very early age, the human child marks vocally a wide variety of situations, including social routines, with a single vocal element. This is done without instruction. Two things are clear from Braunwald's contextual description of the usages of "Ba." First, Laura used this utterance in many situations that were not associated with requests. Second, very few of these vocalizations can be attributed to attempts by her mother to elicit vocal labels. Indeed, Braunwald notes that attempts to get her daughter to label items generally were met with resistance. Laura was interested in such labeling activities only when she initiated them herself during social interaction routines.

By the time Laura was 19 months old, she was uttering phrases such as "Pretty nestor cup," "Laura spill milk," "Cold milk," "All gone," "Mama straw blow," "Pour juice," and "Laura do," all in contexts where her only previous utterances had been "Ba." These spontaneous utterances during feeding are quite different from those which have been reported for apes in a similar context. Terrace (1979b), for example, reports that Nim, while feeding, formed utterances such as "Eat me," "Drink Nim," "Eat drink," "Banana me," "Tea drink," "Grape eat," "Eat me Nim," and "Grape eat Nim."

Initial Use of Symbols in Child and Chimpanzee

All of Nim's utterances in the context of eating were essentially requests for food. By contrast, Laura commented on the attributes of eating utensils, her own actions of spilling, her mother's actions of pouring, and the state of having consumed all of her milk. Comments of this sort are not found in the linguistic productions of chimpanzees, who typically limit their use of learned symbols to requests. This is a dramatic and significant difference between use of symbols in children and in chimpanzees, and is obvious even at the earliest stages of language acquisition.

The vocabularies and symbols of Nim, Washoe, and Lana suggest that chimpanzees tend to learn symbols which correspond to very global requests, such as for contact, play, food, or change of location. However, as the human experimenters then attempt to teach symbols which denote *more specific attributes* of each of these situations (who is eating, what they are eating, the differences between eating and drinking, the act of transferring food from one individual to another, the pace of the transfer, etc.), the ape has difficulty in comprehending both the referents of such symbols and why their use results in anything more than the symbols he has already been taught to use when making requests. That is, combinations such as Nim milk, give milk, Laura give Nim milk, more milk, etc., all have the same result—the transfer of milk from a person to Nim. Consequently, the chimpanzee tends to link together all signs which have been appropriate in similar situations and to produce combinations such as "Give orange me give eat orange me eat orange give me eat orange give me you" (Terrace, 1979b).

In such combinations the specific referent of signs like "you," "me," or "give" does not need to be understood by the chimpanzee in order to select and use such signs. These "wild card" signs are nearly always correct and probably *any* occurrence of them would be encouraged. Thus one or more of these generally appropriate signs (you, me, more, give, the chimpanzee's own name) could be combined with other signs, thereby producing combinations which lack the true referents and specificity of children's combinations. The data on spontaneous requests from Project Nim support this view. Seventeen of Nim's 25 most frequent two-sign combinations contained one or more such signs.

We have seen that both chimpanzees and children are able to use symbols as performatives in complex social-interaction routines. When such routines become well learned or ritualized, both come to employ symbols with intent and to initiate desired routines. Children, however, move beyond this, reaching a point in which symbolic utterances begin to convey information that is not contextually linked to the child's immediate needs or desires. Children begin to specify attributes of the context which the apes show no interest in communicating.

Why is this? When we look at the gestural communication system of wild chimpanzees we see that it is also limited to contextually embedded social-interaction routines. A chimpanzee may gesture to seek reassurance, to be given permission to take a bit of food, and to initiate tickling, chasing, grooming, and sexual interactions. Such gestures can readily convey information about immediate desires within a given context. They cannot, however, convey information regarding objects removed in space and time, nor specific attributive information regarding present objects.

When Washoe, Nim, or Lana request objects or activities from their human experimenters, they are engaging in a type of communication which appears naturally in the chimpanzee. When they use a symbol to request that another individual open a door, turn on a slide projector, or smoke a pipe, they have moved beyond the typical topic (food) of chimpanzee requests. However, requests of this sort are still functioning in a context-tied, performative fashion. The chimpanzees have learned to use gestures or symbols to achieve objectives other than food, because their experimenters have structured their environments in such a way as to make the attainment of these objects dependent or contingent upon the use of gestures or symbols.

Washoe, Lana, and other subjects of ape language studies do not appear to have moved far beyond the natural communicative level of wild chimpanzees. Although apes have come to employ symbols in many situations (tickling, grooming, hugging, etc.), they can often convey similar information without tutoring (Savage 1975), the symbols merely seem to function as additional, ritualized, communicative symbols, just as certain versions of play faces or grooming postures become initiators of ritualized interaction in wild chimpanzees.

The attempts of the human instructor to encourage the chimpanzee to gain *specificity of referent* within a situation have resulted in primitive symbol stringing, as the chimpanzee learns more symbols for what, to him, is the same general contextual situation. Accordingly, chimpanzees form multisymbol utterances such as "You me sweet drink gimme," while children form utterances such as "Johnny's mother poured me some Kool-Aid." In the chimpanzee's case, all of these symbols are simply used as indicators that food is desired. However, in the child's case a variety of aspects of past occurrences are readily expressed. Thus, it appears that the chimpanzee is inclined to remain at the level of communication with which it was naturally endowed—namely, an ability to indicate, in a general fashion, that he desires another to perform an action upon him or for him when there exists a sole, unambiguous referent (as in the case when one chimpanzee has meat and the other has none).

Receptive Competence: The Overlooked Half of Language in Apes

Because the chimpanzee has yet to surpass the request level (using symbols to express his immediate desire), two other features of symbol usage commonly seen in human children are also absent in the ape. The first of these is receptive specificity. Human children often evidence receptive comprehension of words before the onset of productive competence. They can, for example, search among a set of objects and give the requested one, even before they are able to name the objects that they are searching through. Such skills have not been tested in chimpanzees, and there is no *a priori* reason to believe that such skills exist in apes unless they are taught. As a case in point, videotape analyses of Nim's responses to his teacher's signs suggest that he had little, if any, receptive comprehension of the sign for "apple" even though for two years he had been producing this sign accurately upon being shown an apple (Savage-Rumbaugh and Sevcik 1984).

Responding with cooperation and comprehension to the symbolic requests of others is as important as producing symbols, because the function of symbolic communication is to structure and coordinate behavior. Consequently, if symbols produced by others are not understood and responded to in a reliable, predictable fashion, their communications will serve no purpose.

Since apes acquire symbols by learning to produce them when they desire an item or event, and since the human teachers control all access to these desired items and events, it would perhaps be quite surprising if the chimpanzee did show adequate receptive competence. How could the chimpanzee, in fact, be the one to give a specific food when the teacher requested it, or open the lock for the teacher when the teacher requested it, unless he were in fact given access to the foods, the keys, etc.

The simple giving of an item in response to a symbolic request requires that the ape move beyond the ritualized performance of executing given symbols for particular goals. It requires that the ape attend to, and coordinate its behaviors toward objects with the symbolically expressed wishes of others. This means not only a conceptual orientation alien to the ape's natural level of communication, but an alien social orientation as well. It is not only symbol recognition which is involved, but an elaborate object-giving complex that must be developed within a social framework alien to the chimpanzee in its natural state.

Thus receptive comprehension entails a role reversal that requires a number of different skills that were not needed during the acquisition of the original symbol. For example, suppose a chimpanzee learns the symbols for banana, orange, apple, and drink; and that she can make these appropriately (when she wants these foods) by searching through her repertoire of known symbols and selecting the correct one. The inverse of this skill requires that the chimpanzee look over a number of foods, select the one that someone else suggests (by means of a symbol) and give it to the recipient. In this example, the set that must be searched differs with the role; it is the set of known food symbols when the chimpanzee is *producing* the symbol, but it is the set of real and available foods when the chimpanzee is *responding* to the symbol.

Additionally, the form of the response is quite different; in the instance of production, it is the formation, or choice of, a symbol. In the case of the receptive response it is the selection of a real object. Even more important is that the *consequences* of these two roles differ drastically. When the chimpanzee produces a symbol he gets to eat a favorite food. When the chimpanzee responds to the symbol production of another, he must give away a favorite food. From the chimpanzee's standpoint the consequences are surely vastly different.

This high level of receptive competence has generally not been expected of chimpanzees. However, a receptive competence which does not involve role reversal apparently does exist following the initial acquisition of the symbol as a communicative request, and it is probably this type of receptive competence which led the Gardners (Gardner and Gardner 1971) to conclude that Washoe comprehended far more signs than she could produce.

Receptive competence of this simpler sort simply requires anticipating the consequences that will follow symbol usage by another individual. Thus if the teacher always signs "tickle" before she tickles the chimpanzee, "out," before they go outdoors, etc., the chimpanzee comes to expect these consequences; and when he sees the experimenter sign "tickle" he assumes a play posture, when he sees the experimenter sign "out" he runs to the door, etc. This type of receptive competence requires neither role reversal nor the acquisition of new skills above and beyond those required by the original conditional discrimination.

Turn-Taking: Another Major Area of Competency Overlooked in Apes

Related to the apes' deficiency in role-reversal skills is their noted absence of turn-taking and the lack of coordination of symbolic requests with the nonverbal attention-orienting behaviors (Terrace 1979b). By contrast, human infants begin, between 8 and 12 months of age, to gain eye contact and attention before uttering a symbolic request (Schaffer, Collis, and Parsons 1977). Likewise, the mother uses similar methods to gain the child's attention before she utters a request.

Between mother and infant the general communicative format is as follows: (a) coordination of gestures and glances so as to mutually distinguish relevant contextual aspects of the situation; (b) utterance of verbal request; and (c) response to utterance by altered cooperative behavior or by rejection of verbal utterance. Neither the child's utterances nor the mother's utterances interrupt the other. Instead, attempts are made to gain attention before the utterance, as though to enhance and highlight the fact that an important signal is about to be given, and to establish the expectancy that the utterance of this signal is to meet with an alteration of behavior.

In the case of communication with chimpanzees, the human experimenter alone uses these attention-getting behaviors before a symbol is given. These nonlinguistic conversational regulators reflect knowledge of the roles of communicator and recipient. That they do not occur among chimpanzees suggests that the chimpanzee all too often perceives her role as that of performing an expected behavior, not that of conveying information.

New Questions That Need To Be Asked of Apes

My aim in comparing symbol use by children and apes is to cast the study of language acquisition in apes in a new light. At the very least, it should be clear why it is not fruitful to pursue such questions as "Do apes have language?" and "What do they have to say to human beings?" Instead, we must ask (a) how the processes of symbolic and nonsymbolic communication between two organisms differ; (b) whether the symbolic system appears *de nouveau* simply by learning which words "go with" what things, or whether it somehow emerges out of the nonverbal system; (c) what actually occurs when a creature who does not engage in symbolic communication with conspecifics is taught to use symbols with members of another species; (d) whether it is possible to determine objectively what these symbols actually represent to chimpanzees; and (e) why turn-taking is important.

These and other new questions will help address, in a more rigorous manner than was true of earlier studies, the nature and significance of differences between the communicative systems of animals and man. It can no longer suffice to view them as distinctly separate systems (Chomsky 1968) or as one system which follows simple laws of learning (Skinner 1957). The truth, it appears, is far more complex.

Verbal behavior emerges from and with nonverbal behavior, and as it does, it provides for a new means of coordinating interindividual object-oriented behaviors. We cannot completely understand the verbal system without going back to its roots, which are in the nonverbal system. The verbal system is mapped onto the nonverbal system through interactions that are interindividual in process and nature. Words are not learned by individuals; rather, interindividual interactions come to be coordinated through the use of words.

Language learning is not an individual accomplishment, but an interindividual *process,* and it presupposes mutual inferences about intentionality. This issue lies at the heart of that form of behavior we call language.

Chapter 3

The Project and the System

The "Animal Model Project"—Its Scope and Focus

The conceptual issues just presented have emerged from, and have been shaped by, the ape-language training program, which is the topic of the remainder of the book. Most books and articles which have described language skills in apes have begun simply by detailing what the ape was taught and what the ape then did. The reader with no firsthand knowledge of apes has been forced to accept the author's linguistic translations of the ape's behaviors with little feel for what actually was taking place as "words" came into being. Consequently, camps of believers and nonbelievers have arisen. This is unfortunate, since the issue of language acquisition by apes is clearly not an all or none phenomenon.

We hope that the perspective offered in the earlier chapters will have made apparent the conceptual complexity of these issues; we also hope that, as we describe this project, the reader will try to understand what it is that the chimpanzees Sherman and Austin do and how they came to do it without allowing previous biases about the "nature of language" to get in the way. For our part we will attempt to describe behaviors as accurately as possible instead of quickly translating button presses into "words."

This research program began in 1975 with the goal of using apes to help develop language-training techniques that could be used to teach language to mentally retarded persons who had failed to learn any language skills (even single words) by other available methods. Because of this goal, our program has been labeled the Animal Model Project and, in this respect, it differs from other ape-language projects, whose main focus has been to find the upper limits of language acquisition in apes.

The Animal Model Project inherited Lana's legacy and the Yerkes keyboard computer-based system, but the concept, design, and execution of the Animal Model Project has been quite different from the original Lana project. The keyboard communication system has been modified extensively to permit training techniques that were not possible with Lana. The Animal Model Project is, in a sense, a third-generation ape-language effort, and its primary goal has been to elucidate the processes of language acquisition in the ape and to determine the extent to which these processes are similar to—or different from—the phenomenon of spontaneous language acquisition found in normal children.

Chimpanzees were selected as subjects for the Animal Model Project specifically because apes do not acquire language on their own. Even when the ape is trained in a visual modality—a modality that compensates for the anatomical limitations of the chimpanzee's vocal apparatus—it still needs to be tutored in acquiring symbols (Fouts 1972; Rumbaugh 1977). The kinds of things apes must be taught should help provide insights into the assimilatory and conceptual processes which are occurring so rapidly and spontaneously in the normal human child. Because in the ape we are presented with a creature who has many of the prerequisite skills for symbolization, but lacks language, we can learn a great deal about the phenomenon of language itself as we attempt to produce it in apes. The common chimpanzee *(Pan troglodytes)* does not readily draw the relationship between symbol and referent; it must be helped extensively to attend to these relationships. By finding ways to manipulate the environment to facilitate attention to certain related events, we can learn how to help it to single out and attend to symbolic relationships, and thus enable it to formulate and comprehend communicative schemas that would otherwise be beyond its comprehension.

While we view the chimpanzee as an important animal model we are, nevertheless, constantly aware that many of the things which our chimpanzees do may have no counterpart in the behavioral repertoire of the retarded human being. For example, in contrast to many retarded children, the chimpanzee *(Pan troglodytes)* subjects in our project show no inclination at all to vocalize as they light symbols or point to objects. Even after years of training, when they have acquired sizable vocabularies they remain vocally silent.

Food barks, which occasionally accompany symbolic requests, constitute the one exception.

We have found it more profitable to begin the study of language acquisition with the chimpanzee rather than retarded individuals because of the tremendous variability in the retarded population. Rarely are two cases of retardation alike in the picture of language deficits which they present. For this reason, previous research projects investigating language acquisition processes in this population have tended to be case-specific (Rice 1980). Consequently, no general model of language acquisition in retarded individuals has yet emerged. On the other hand, our chimpanzee subjects present similar sets of deficiencies and competencies with regard to communicative skills. Accordingly, knowledge gained with one individual can reliably be applied to another.

We will describe a model of language acquisition in apes that we have developed during the last decade. This model is of sufficient generality to permit it to be applied to a wide range of language problems in the mentally retarded. (See, for example, summaries of the Child Project at the Language Research Center: Romski, Sevcik, and Rumbaugh 1985; Romski, Sevcik, and Joyner 1984; Romski, White, and Savage-Rumbaugh 1982.)[1]

Introducing Sherman and Austin

Sherman and Austin, two male *Pan troglodytes* (Figures 3.1 to 3.3), are the principal subjects of this book. They were both born at the Yerkes Regional Primate Research Center and, though assigned to the Language Project, remain the property of the Yerkes Center. Sherman was born in 1973 and Austin in 1974. Sherman spent the first 1½ years of life with his mother before being removed so that she could be bred again. Austin was with his mother for only two months and was removed after having failed to thrive. He was believed to have had difficulty digesting her milk, and has displayed continuous and frequent allergic reactions to a number of complex proteins.

Both Austin and Sherman were assigned to the language project in 1975 when they were 1½ and 2½ years old, respectively. Sherman was approximately half again as large as Austin at this time and has retained some of his size advantage even now at 13 years of

Figure 3.1. Sherman, 1978.

Figure 3.2. Austin, 1978.

Figure 3.3. Austin (left) hugs Sherman (right) who smiles.

age. A large number of personality differences were evident between Sherman and Austin in 1975, and these characteristics have—like their size difference—remained exceptionally stable through adolescence. Sherman has always been the dominant chimp of the pair, because of his size and also his rough and tumble, blustery personality. Wherever the social action is, Sherman is in the middle of it. It is impossible not to pay attention to him nearly all of the time because he is always doing something and doing it very fast. Austin, by contrast, is demure and "laid back." He must often be coaxed into the group activities and care must be taken to ensure that he is not so upstaged by Sherman that he simply refuses to participate.

Sherman is rarely careful with anything. Cups, blankets, and toys in his presence generally take a pounding, not because he tears them up on purpose, but because he uses them up. Austin is very careful with most objects. He carefully stacks extra glasses and places them out of the way; when things come apart, he tries—though ineptly—to put them back together. If Austin does break something, it's usually deliberate: he is attempting to destroy that object because he doesn't like it. Austin always destroys representations of infant chimpanzees and humans. Whatever the form of these representations (dolls, photographs, or video images), he always wants to bite them and becomes agitated when not allowed to do this.

Austin, unlike Sherman, also displays a rather common trait of young chimpanzees separated from their mothers during the first six months of life: he rocks rhythmically from side to side while holding a blanket whenever he is distressed. However, in spite of his smaller size and quieter disposition, Austin is less fearful than Sherman. While Sherman may be upset for days when something unusual happens (such as when the fire alarm accidentally goes off near him, or a stranger goes up onto the roof to check the heating system), Austin deals with these events by displaying. Once, he threw a large stick at the fire alarm while Sherman cowered in the corner. Sherman, however, displays ferociously when events are not out of the ordinary.

When Austin does display fears, they often seem to be of rather unusual things. For example, he became fearful of eating food from a particular box after he once found small metal filings accidentally stuck to a piece of food in that box. He also became afraid of entering a test room when the teacher insisted on remaining outside the door to run a control test in which she wanted to avoid any possibility of cueing Austin. Presumably, Austin figured that if the teacher were not going to enter that room, he was not either. Sherman has never displayed fears of this sort.

Sherman prefers to communicate anything he can nonverbally. If he wants you to tickle, he approaches you with a large playface, rolls headfirst into your lap (even if he is twice the size of your lap), pulls one of your hands into his mouth, and starts wiggling in five different directions at once. He is capable of making it virtually impossible not to play with him. Austin has preferred to use the keyboard to request tickling bouts ever since he first learned that it was possible to do so. He says "Tickle" and then quietly waits for you to begin tickling him. If he does not feel quite at ease with a teacher, he will say "Chase" first, because in a chase game the teacher does not get too close to Austin until he has had a little time for interaction.

These personality differences apply not just to the use of "tickle" and "chase", but to virtually all uses of the keyboard. If Sherman can accomplish what he wants without using the keyboard, he will. Austin does not like to have to negotiate things without the keyboard. Once he learns a way to get what he wants with the keyboard, he prefers to keep using it rather than to try to make his wishes known in some other manner.

After eight years of daily language training with Austin and Sherman, it is not possible to say which is truly more intelligent. There are, to be sure, learning differences, but they do not make for simple generalizations. Tasks requiring close observation are learned more readily by Austin, those requiring active participation are learned more readily by Sherman. Once learned, both display virtually identical competence and accuracy. If a blind test is given to Sherman and he misses two, then predictably Austin will also miss two, though they will be different items. We have endeavored at every point since their assignment to the language project to keep their environment and training identical, yet their individual styles persist and continue to exert an impact on their acquisition processes and language use.

Two other young chimpanzees were initially part of the Animal Model Project and will be mentioned from time to time in the remainder of the book. Erika, a wild-born female was caught and sold as a pet at three months of age. She was kept by her original owners until she was two years of age and was donated to the Yerkes Center with the provision that she not be used in biomedical research. For this reason, she was assigned to the Language Project (when she was approximately 2½ years of age). Kenton was born at the Yerkes Center in 1974. He was removed from his mother at birth and was placed in the Yerkes nursery until he was 14 months of age. At that time he was assigned to the Language Project. Both Erika and Kenton were in language training for 18 months. We discontinued work with them because the effort with four animals was too great for our staff of four and our overall progress was compromised by attempting to work with such a large population. However, Kenton and Erika were acquiring skills similar to Sherman's and Austin's, and there was no reason to suspect that their progress would not have continued to be comparable had we continued to work with them.

At times throughout the book we will also mention Lana. She was born at the Yerkes lab and was the first pilot animal in the Language Research Project. Since her training history has been detailed extensively elsewhere (Rumbaugh 1977) we will discuss her only when it proves relevant in analyzing differences between the performance of Lana and that of Sherman and Austin.

Rearing Environment

Sherman and Austin were reared in an enriched social environment which included interactions with other apes and human caretakers during nearly all of their waking hours. They participated in all laboratory activities, from answering the phone, to cleaning floors, to the preparation of food (see figures 3.4–3.14). Periods between training or testing were spent tickling, watching TV, going for walks outdoors, building nests, drawing, painting, etc. They were worked with every day, seven days a week.

Strong emotional bonds developed between Sherman and Austin and many of their teachers, including myself. Twelve different

Figure 3.4. Sherman using screwdriver to remove bark from guide track so that door can be closed.

Figure 3.5. Sherman helps put food away in refrigerator.

Figure 3.6. Austin helps clean a clogged drain.

Figure 3.7. Austin, with wig, admires himself in the mirror.

Figure 3.8. Sherman uses the phone. He produced a breathy "Agh" repeatedly in attempts to talk on it.

Figure 3.9. Austin attempts to repair a broken lock.

Figure 3.10. Austin helps put kitchen utensils away.

Figure 3.11. Sherman teaches us how to climb trees.

Figure 3.12. Sherman blows gingerly on a hot marshmallow he has just roasted.

Figure 3.13. Sherman carefully tastes the still hot tidbit.

Figure 3.14. Sherman and Austin use portable keyboard to request foods outdoors on a "cookout."

teachers worked with them during the period covered by this book, although the major part of their care and training was accomplished by four people: Sally Boysen, Janet Lawson, Liz Rubert, and myself. Social interactions with these four teachers grew to be far more important than food rewards, and two of these teachers continued to work with Sherman and Austin as adults.

Their rearing environments were modeled as much as possible after those created for Washoe by the Gardners. The Gardners' view of language as first and foremost a social behavior was adopted as the guiding philosophy of the Animal Model Project. Language was integrated into their daily lives in as many ways as possible once a request vocabulary of ten symbols was acquired.

The Computer-Based System

The communication system developed and refined in the Animal Model Project has a number of components, each of which will be described in detail. First and foremost, the system is a graphic one—a symbol, once produced, is not transitory as is a signed gesture or spoken word. When a symbol is touched on the keyboard it lights up and remains lit until it is deliberately dimmed by the experimenter. The act of lighting a symbol is treated as the equivalent of uttering a word or producing a gesture. As each symbol is lit a unique tone or series of tones also sounds. The symbol also appears on the projectors above the keyboard and on TV monitors located at other positions in the laboratory. Thus symbol production is a significant and noticeable event. It results in lights and noises (unique to each symbol) and commands attention, so that if the receiver is not looking at the keyboard at the time a symbol is produced, he is nevertheless aware that symbols are being used because of the accompanying tones. The receiver can then look at the keyboard and see what was said, since the keyboard will be displaying, in a brightly lit manner, the specific symbol or symbols which have just been touched.

The keyboard is an active graphic system that elicits a child's or chimpanzee's attention in a way that pointing to a symbol painted on a board or placing a magnetic chip upon a piece of metal cannot.[2] The symbols are also highly discriminable. This permits the teacher-experimenter to respond to each symbol in a consistent way without having to guess which symbol the chimpanzee is producing.

The saliency of symbol production when using the keyboard communication system contributes significantly to its efficacy as a language acquisition device. Perhaps the greatest single problem in teaching language to apes or mentally retarded persons is the difficulty of developing consistent responses which can be discriminated easily by others.

Signs are often inadequate because the fine motor movements which are needed to produce a large number of signs are simply not within the capacity of the ape or the retarded person. Additionally, before the subject knows which signs to produce, or even why he is required to produce signs at all, he tends to produce signs which are not very discriminable. This places a considerable interpretive burden on the receiver. Because it is often difficult to determine which sign a subject is producing, one tends to give the beginner the benefit of the doubt and interprets a questionable sign as correct. Unfortunately, this serves only to further reinforce sloppy and indiscriminable signing.

When using communication boards that require a pointing response, the subject is also often unclear when making the response and may waver between several alternatives. By being unclear, the subject can force the teacher to interpret, and again "sloppy" responses become inadvertently reinforced. By contrast, the electronic keyboard leaves no doubt as to which response was made. Consequently, teachers can respond reliably in a consistent and predictable fashion. When the subject is attempting to learn what these symbols mean (in the sense of how their use is treated by other individuals), it is most important that the social environment provide both the chimpanzee and the human retardate with a consistent response.

The actual symbols placed on the keyboard are composed of nine distinct elements. Following Rumbaugh (1977), we refer to these symbols as "lexigrams." The elements of each lexigram are shown in figure 3.15, along with Sherman's and Austin's present vocabulary. The assignment of particular symbols to particular words is completely random. Because there are only nine possible elements to combine, each word is made up of similar subcomponents. These nine elements can be reversed left to right with no change in their configuration, and all but one element can be inverted with no configurational changes. Therefore, while the particular pattern of a lexigram is important, the orientation of the pattern is not.

Only nine elements were used because of the technological limits of the original IEE projection system (Rumbaugh 1977). However, any number of symbols and any form of symbol can be placed on the faces of the keys themselves. We have recently developed the capacity to project keyboard symbols on TV monitors, and thus no longer need to be limited to the original nine elements. Consequently, any sort of symbol (Bliss, English type, etc.) can now be placed on the keyboard and can be projected on any TV monitor.[3]

Unlike Lana's keyboard, the present models do not require the subject to push directly on the symbol itself. Instead, the subject's touch activates a conductance-sensitive plate located under each symbol. This causes the symbol to light automatically with no pressure. Any symbol can occupy any location on the keyboard at any time. Indeed, its location can be changed by software on every

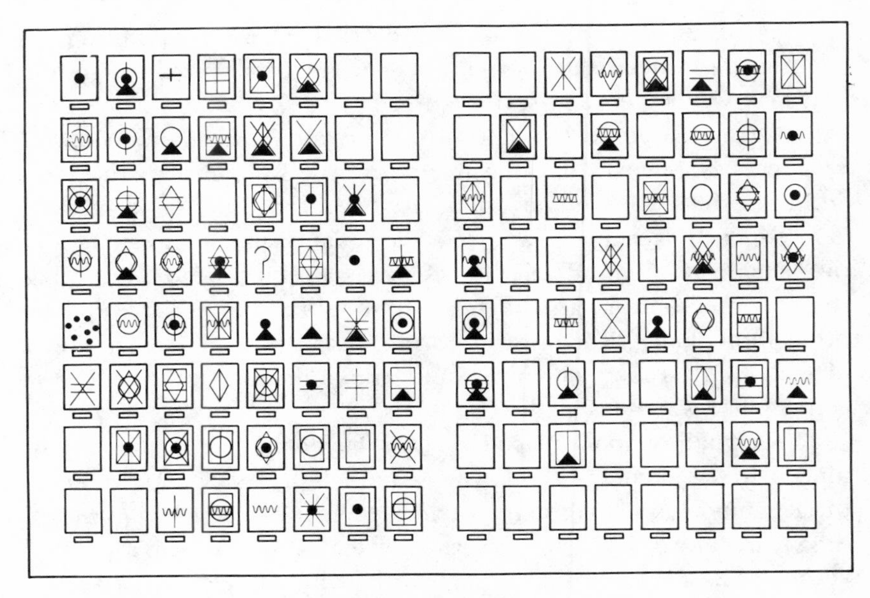

Figure 3.15. Sherman and Austin's keyboard, circa 1983.

trial should a subject come to focus unduly upon a particular position.

Any key can be rendered inactive should a subject tend to develop a tendency to always select that key before choosing the correct one. This is a common strategy, since a subject is rewarded indirectly for the incorrect symbol if he always produces it before producing the correct one. By turning off keys temporarily we can prevent the occurrence of this erroneous type of symbol stringing during the initial stages of training. Since the subjects are not allowed to develop a training history of stringing symbols together when they do not know which single symbol to employ, their later acquisition of meaningful symbol combinations is enhanced considerably.

All of the keys on the board remain dimly lit to signal that they are active and available for use. Any key that is touched becomes brighter than the others. As each symbol is touched, it appears on the projectors above the keyboard. The left-right order of the symbols on the projectors reflects the sequence in which they were touched. Once all of the symbols in any given message have been lit, touching the plate just below a solid yellow key (generally referred to as the "period" key) causes all of the keys which have *not* been lit during that message to darken. Touching the period key also brightens the message keys. As a result the most recent message stands out clearly from the rest of the symbols on the board and becomes highly visible to both listener and receiver.

This is particularly important during the introduction of new symbols, because the teacher and subject have an immediately available record of what has been said. Thus, if the teacher has asked for a specific item (such as a key) and the chimpanzee gives her something else (such as a blanket) she does not have to repeat her original request. Instead, she can simply gesture to the lighted keyboard to refer to what she has said, and thereby indicate that it is what she still wants. By contrast, when attempting to communicate in sign language, repetition of the particular sign is necessary. Once the sign is made the teacher can only refer to what she has just signed by repeating herself. There is no reason to believe, in such cases, that the subject understands that repeated signs are the same as the previous utterances instead of new ones.

The keyboard display is also used to help the subject understand production errors. Suppose, for example, that the subject is asked to identify a jar of jelly which the teacher holds up and that the subject responds "Peanut butter." Since "Peanut butter" will remain illuminated, the teacher can refer to the subject's response by pointing to the lighted symbol and thereby convey that the response was incorrect. It is not possible to do this when using rapidly fading systems such as sign or speech. With subjects such as severely retarded individuals and apes, one cannot presume that they can recall their previous utterances or that they are attending to them at the time the teacher attempts to refer to them as being incorrect. With our computer-linked graphic system, the teachers can use the display to refer to the immediately preceding utterance of the subject, while the subject can also emphasize what he has just said by referring gesturally to the display.

Following every keyboard message, the symbols employed in that message remain lighted until the teacher records the name of the individual who lit the symbols, whether or not the symbols used were the correct symbols, and the semantic nature of the message (for example a request as opposed to the name of an item). This is accomplished through the use of a small hand-held numeric keyboard device which transmits utterance codes along with the actual utterances. Our permanent records of each session include all of the utterances and information regarding their correctness (given the context) and the purpose of the utterance. Although the rate of exchange of information is slowed down by the recording process, it does not appear to hinder training. Once a teacher becomes familiar with the coding system, codes can be entered almost as rapidly as the communication takes place.

The coding device can also be used to turn off the entire keyboard if a single button is pressed. This feature is used frequently when the teacher wants to focus the subject's attention upon some item before requesting that the keyboard be used to name it. This prevents a premature response and ensures that the teacher has the subject's attention to the topic on hand before he starts selecting symbols.

The opportunity to focus the subject's attention has proved very helpful during training with Sherman and Austin. Both easily learned to use the coding device to turn the keyboard off and on, and they do not hesitate to turn the keyboard on when the teacher has turned it off should they wish to say something. This kind of instruction cannot be done with signs, since the subject can emit them whether or not he is attending to the topic of interest.

The Value of Computerized Data Collection

At the end of each language session it is possible to request a number of different types of data summaries. These include:

1. A transcription of all utterances which occurred during a given session, both by the teachers and the subjects. Each utterance is marked correct or incorrect (this coding having been made when the utterance occurred). The time that each utterance occurred is also indicated on the transcript next to the utterance (see figure 3.16).
2. A transcript of all utterances made during the session by a particular subject or teacher.
3. A summary of all occurrences of a particular symbol. This summary reports when a given symbol was used and which other symbols, if any, were used in conjunction with it.

Figure 3.16. Daily transcript showing coding schemes and typical "conversation" in 1983.

Excerpt From a Day's Printout

Austin: TALK TALK .OK+ 10:44:24
(The OK+ indicates that given the context, this utterance was correct. The specific context was described in writing on the data sheet as being that of Austin using the keyboard to say this to himself.)

Janet: YES CHASE .OK+ 10:44:51
(Again, the OK+ means that Janet's utterance is correct, given the context. The context was that of Austin using a gesture, hand clap, to request the activity of chasing, and Janet using the keyboard to reply to the request.)

Janet: QUESTION CHASE OTHER-ROOM .OK+ 10:45:10
(Instead of immediately beginning to chase Austin, Janet asks him if he would like to chase in a different place, and specifies an adjacent room where they could go and engage in a game of chase without disturbing Sherman who is working nearby with another teacher.)

Austin: PHONE .OK+ 10:45:43
(Austin specifies a different location for chasing, namely the hallway which runs along the outside of all the training rooms. The phone is located in this hallway, along with a refrigerator and it is the chimpanzees' favorite area to chase in, although they are not allowed to go there to chase unless they are being very cooperative.)

Janet: QUESTION CHASE OTHER-ROOM .OK+ 10:45:50
(The teacher does not immediately understand why Austin has responded PHONE to her original question and so she repeats it.)

Austin: GIVE AUSTIN REFRIGERATOR CHASE .OK+ 10:45:57
(Austin now attempts more clearly to tell the teacher what he is attempting to communicate by specifying that he wants to go out into the area where the refrigerator is located and chase. Since a lexigram has not been assigned to the hall area, the words PHONE and REFRIGERATOR are the two best ways Austin has of indicating the area to which he wants to go. It is interesting that when his teacher does not understand the one word utterance PHONE, he uses a different term to indicate the area and he tells her who should go there and what he wants to do there.)

Janet: OTHER-ROOM .OK+ 10:46:48

Janet: QUESTION CHASE OTHER-ROOM .OK+ 10:46:48
(Apparently, Janet is willing to chase in only one location.)

Figure 3.16. *(continued)*

Austin: GIVE OTHER-ROOM .OK+ 10:46:55
(Austin gives in.)

Janet: YES, GO OTHER-ROOM .OK+ 10:47:02
(Austin and Janet go to the other room and chase for approximately one minute, then return.)

Janet: GET FOOD REFRIGERATOR .OK+ 10:48:12
(Janet is anxious to start the day's training session which will entail getting food from the refrigerator and placing it in a location where Austin will need a tool to obtain it. Austin must then use the keyboard to request the tool. This work will be described in detail in chapter 9.)

Austin: GIVE AUSTIN OTHER-ROOM .OK+ 10:48:24
(Even though the teacher was not asking Austin if he wanted food, but was instead telling him what they were going to do, Austin asserts what he would like. He had a good time chasing and is not ready to eat. He wants to go back to the other room and continue chasing. A resumption of the previous activity is implied in his statement and is apparent in his behavior.)

Janet: YES AUSTIN GO OTHER-ROOM .OK+ 10:48:33
(By her answer, Janet indicates that yes, Austin can go to the other room if he wants, however she does not include her own name and does not accompany Austin, hence the chasing game cannot be resumed. Janet adopts this strategy because she does not want to chase with Austin anymore. Austin will often chase for over an hour if all of his requests are complied with and this is not what the teacher wants to do at this point.)

Janet: GO REFRIGERATOR .OK+ 10:48:52
(When Austin realizes that Janet is not going to accompany him into the other room to chase, he decides not to go himself. Janet then tells him what they are going to do.)

Austin: REFRIGERATOR .OK+ 10:48:56
(Austin agrees to the activity which Janet specifies. Note he does not mention chase now as he knows Janet is no longer going to engage in a game of chase.)

Janet: YES GO REFRIGERATOR .OK+ 10:49:05
(They go.)

4. A list of each different utterance used by a given individual for a specified period of time.
5. A summary of the occurrences of any behaviors which have been given special codes (such as novel or spontaneous usages).

These summaries make it easy to chart daily progress, to spot errors and confusions which may appear among old skills as new symbols or tasks are introduced, and to evaluate the stability of each animal's performance with each symbol. Indeed, teachers use daily the types of summaries described above to review what has been accomplished, and to spot more precisely where a given subject is having difficulty. Ready access to transcripts of previous sessions allows teachers to reflect on what is happening in a way that is simply not possible when one is interacting in real time with a subject. Other communication systems (signs, plastic chips) do not provide the teacher with ready access to transcriptions and summaries of the subject's progress.

Mothers may not need such transcriptions to help them teach language to a normal child, but the same is not true of chimpanzees and retarded children. Such transcripts often help the teacher to better understand the subject's difficulties and permit ready conference and discussion with other teachers who may not have been present at that session.

Elimination of Background Color Cue

In the original system of Yerkish grammar that was used with Lana, different parts of speech, and in some cases different categorical classes, were assigned a specific background color (Names of animates were purple, names of objects orange, foods red, etc.). These distinctions have not been used with Sherman and Austin. Words of all types have been assigned a variety of background colors, and colors have been changed for many symbols. Most of the food and tool names initially had a red background, however once these symbols were well learned, the chimpanzees could identify them on backgrounds of any color, including black. They could also recognize black symbols on a white background. In general, differential background color served only as a means of adding visual saliency and pattern to what otherwise proved to be a rather monotonous display of rather homogeneous symbols. Recently though, we have eliminated all background color from Sherman and Austin's symbols so that all symbols are white on black background. This has not affected their performances.

The Tonal Auditory Feedback Component

In addition to using normal speech in the presence of our chimpanzees, we have devised a specific sequence of electronic tones to accompany each symbol. These tones allow both teacher and subject to draw attention to the fact that a key is being used. The specific pattern of tones is created by randomly assigning a tone to each phoneme of the English referent for the lexigram. However no stress value is given to different syllables.

Some of the human teachers have learned to identify 20 to 30 tonal patterns, but many other teachers cannot tell which symbol has been lit on the basis of tone alone. Neither can the chimpanzees. Yet the tones are a valuable part of the system. They broaden its use by allowing a speaker to get another's attention simply by pressing a key. However, since their precise patterns are not learned by the chimpanzees, their system remains essentially visual with an auditory onset cue.

We originally chose a tonal system because we knew that the chimpanzees were exhibiting little promise with spoken words and they do use variations in pitch in their natural vocal repertoire. We added the tonal system after three years of training with the visual system, which may account for the difficulty both teachers and apes show with the tonal system. We have attempted to specifically train with the tonal system, without any visual symbol present for a number of months. Under these conditions the human teachers made relatively rapid progress and began to discriminate tones and even to identify new ones on trial 1, presumably drawing on the relationship between the tones and English phonemes that was designed into the system. The chimpanzees, however, made no progress at all, and their visual skills during this period became somewhat disrupted.

Since the introduction of the tonal cue, some changes have in fact been observed in the chimpanzees' behavior which attest to the value of such cues. Both chimpanzees and teachers now use the keyboard regardless of whether or not they have the immediate visual attention of the other. Before the introduction of an auditory cue, one chimpanzee was often saying something while the other was not looking at the keyboard, hence the message would go ignored. The chimpanzee whose message was ignored would often interpret this as a lack of willingness to respond on the part of the other chimpanzee and hence he too would cease to cooperate. The chimpanzees will also light keys even if the teacher is in another room and expect her to come and see what they have said. The coordination of symbolic communication between chimpanzees (see chapter 7) has been significantly enhanced, since each chimpanzee now attends to the auditory cue as a signal that a symbol has been lit and turns to see what symbol that is.

The Ape's Inability To Comprehend Spoken Speech

One of the greatest disadvantages of a nonvocal modality is the absence of an auditory cue. When using vocal speech, the child does not need to be looking at the mother (as is the case with signs) or at a keyboard (as is the case with graphic systems). Instead, he can be engaged in all manner of activities and object manipulation while the mother provides a running commentary on what he is doing and the objects he may be manipulating. Not only do vocalizations allow the mother to encode the child's activity without drawing his attention away from that activity, but they also allow her, through the use of intonation, to convey her feelings about what it is that has engaged the child's attention (Newson 1978).

By contrast, a child communicating with signs or a graphic system must divert attention from what he is doing in order to produce a symbol. Likewise, the teacher must divert the child's attention to get the child to observe which symbol the teacher is using in order to comment on what the child is doing. Once a child has ceased activity, he may not realize that the observer is referring to a behavior the child has abandoned. Deaf children of normal intelligence seem able to cope with these problems and to develop some language skills in a nonvocal modality, even when they are not provided with language models who employ a modality they can process (Goldin-Meadow and Feldman 1977). It is, indeed, much more difficult for apes and retarded children to deal with the separation of reference from action.

We have attempted to minimize this problem by using spoken English around the chimps. Apes appear to exhibit a striking sensitivity to speech patterns and intonation. Indeed, they give every appearance of comprehending, in context, a wide variety of spoken words if they have been raised in a human environment with extensive exposure to human speech (Fouts, Chowin, and Goodin 1976; Patterson 1978). By speaking to the chimpanzees while we use the keyboard (and also when we are not near it), we are able to let them know that we are monitoring their behavior and what we feel about that behavior. This is true even if they are not looking at us. Since chimpanzees engage in eye contact less frequently than human beings, a vocal channel of communication is the only effective one when they are not looking at us.

In situations that are rich in contextual cues, Sherman and Austin appear to understand a wide variety of spoken requests. They respond readily to utterances such as "Sit down," "Go over there," "Turn on the water," "Close the refrigerator door," "That's very good," "Don't go out the door," "Yes, you can go outdoors," "That's a very good job," "Can you stick out your tongue," etc. However, all such utterances are delivered in a social context in which the participants share a common focus of attention and in which the utterances are accompanied by gestures and facial expressions. "Sit down" for example, may be uttered just after the chimpanzee has stood up. Additionally, the tone of the utterance (or the teacher's facial expression) may convey disapproval. Furthermore the teacher may accompany the utterance with a pointing gesture to indicate where she wants the chimpanzee to sit. Accordingly, the chimpanzee will be able to conclude relatively quickly what behavior is expected even if he cannot understand the spoken words. Likewise, a phrase such as "Turn on the water" may be uttered both when the chimpanzee and the teacher are standing by the sink and when the teacher holds an empty glass directly under the faucet. "Don't go out the door" is likely to be uttered just as the chimpanzee is headed out the door, and the intonation of the phrase (and possible threats of adverse consequences) can readily convey the seriousness of the sender's message.

This ability to understand utterances in context has led many people to conclude that the chimpanzee can understand spoken English. None of the work which claims that apes or other animals are capable of understanding human speech actually tested this capacity using a large array of objects with multiple trials and controls for cueing (Fouts, Chown, and Goodin, 1976; Kellogg and Kellogg 1933; Patterson 1978; Warden and Warner 1928). Indeed, considering the extremely limited nature of the research in this area, it is quite surprising that many scholars so readily accept the fact that primates and other animals comprehend human speech, particularly when they equate the comprehension of speech with the "capacity for language" (Passingham 1982).

Attempts to determine whether or not Sherman and Austin can understand spoken English in the absence of contextual or nonverbal cues have demonstrated repeatedly that they cannot, even though they have been reared with constant exposure to speech. Figure 3.17 shows the results of a test comparing their receptive skills for spoken English with those for lexigrams. The test for comprehension of lexigrams and words of spoken English has been administered many times in our lab by different teachers, with blind controls, and with different items. In administering this test the teacher placed 16 items which Sherman and Austin had learned to name on a table in front of them. She then asked them to hand the items one at a time using either spoken English or lexigrams. (Both types of trials were randomly interspersed.) Once an item was handed to the teacher she thanked them and rewarded them if they were correct. When they were wrong, she told them that they were incorrect and gave them no reward. Both correct and incorrect items were returned to the table after each trial.

	LEXIGRAMS		SPOKEN ENGLISH	
SHERMAN	Correct Item Given	16	Correct Item Given	2
	Incorrect Item Given	0	Incorrect Item Given	14
AUSTIN	Correct Item Given	16	Correct Item Given	1
	Incorrect Item Given	0	Incorrect Item Given	15

Figure 3.17. Results of comprehension tests of spoken English versus lexigrams.

The items we have used on the test are very familiar to the chimpanzees. They have heard these items named thousands of times in a wide variety of circumstances, as we use English with them at all times. The results of these tests are unequivocal and highly reliable. The chimpanzees can readily look at the items on the table and find the correct one to give when the request is made with a lexigram. However, when it is made with a spoken English word, the chimpanzees vacillate and look to the teacher for more information. They will frequently gesture toward the keyboard to encourage the teacher to use a lexigram symbol to tell them which item to give. If forced to guess, they will select an item close to them and hand it to the teacher very tentatively. Often, after making errors on a number of such trials, they will refuse to continue unless the teacher gives them more information, either by pointing, or using lexigrams.

Even though Sherman and Austin cannot understand English words independent of contextual cues, the use of spoken English is effective in getting the chimpanzee to do what we ask. For this reason we use it as part of our efforts to communicate with them. Spoken English allows us to convey all manner of affect to Sherman and Austin. Even though they cannot comprehend our spoken words they do understand the affect conveyed by our intonation. They learn to respond appropriately to our affect with no special training. Indeed, how we feel about their behavior helps shape the chimpanzees' own feelings regarding their activities.

Intonation and affect in speech patterns are believed to be right hemisphere phenomena in human beings (Heilman, Scholes, and Watson 1975; Tucker, Watson, and Heilman 1977). It is possible, for instance, for patients with right hemisphere lesions to comprehend speech content while being unable to determine affect. At present we have no means of determining in which hemisphere the perception of affect is located in the ape. Nevertheless, the differential sensitivity of the ape to affect versus content suggests that the gap between ape and man with respect to the processing of speech may be smaller than is commonly assumed. During daily interaction with apes, one feels subjectively that their affective comprehension of speech is well on par with one's own and it is only the processing of symbolic content that is lacking. However, throughout the animal kingdom, it is common to find that species which exhibit peculiar patterns of communication also exhibit specific decoding capacities that complement the encoding capacities in virtually a one to one fashion (Zoloth et al. 1979).

It seems likely that chimpanzees possess the conceptual apparatus to understand some human speech but lack sufficient decoding apparatus to process phonemes and phonemic transitions. Tallal (1978) has presented evidence which suggests that a similar state of affairs may be obtained in some cases of childhood aphasia. The chimpanzees' ability to respond to the affective component of human speech most likely reflects a similar use of affect in chimpanzee and human vocalizations. It suggests that, as human beings have evolved, the affective component of vocal signaling has changed significantly less than has the arbitrary symbolic component.

Chapter 4

Apes Who Don't Learn What You Try To Teach Them

Asking the Wrong Question

The initial research questions addressed by the Animal Model Project were raised against the backdrop of the prevailing view that apes could learn symbols quite readily. Our primary interest revolved around the issue of how a child comes to categorize the world linguistically (Rosch and Lloyd 1978). Within the field of child language emphasis was on the child as a "processor." A "processor" was defined as an individual who brought "a variety of capacities and strategies to the task of making sense out of the stimuli, objects, and events of the world" (ibid.). We were hopeful that we might be able to shine some light upon the chimpanzee as "processor" by controlling the input of the "buzzing, bloomin' confusion" as William James termed the onslaught of sensory stimulation which faces every child. Since we could differentially control the linguistic input which various chimpanzees received, we believed it possible to begin to determine how different sorts of linguistic inputs led to different processing strategies.

We designed a study to teach one group of chimpanzees that each different object in the world was represented by one symbol only (for instance "box," "bowl," or "cup"). Other chimpanzees were taught that many objects (which appeared to be different) all shared the same symbol, for example, red, blue, green. Since these chimpanzees would have no linguistic experiences apart from this training (in contrast to normal human children who are exposed both to specific names and categorical names from birth), we reasoned that these chimpanzees would be forced to group their symbolic worlds very differently from the beginning. Would this affect their perceptual worlds? Would both groups be equally adept at learning their first symbols? Could apes who learned that each object had a specific name go on to learn that one symbol could, on some occasions, be used to designate a number of different objects?

In retrospect, these were premature questions. We had not stopped to ask why a chimpanzee who had had no previous language training would know that symbols encoded anything, much less specific names or colors. We anticipated very little difficulty in instructing these initial lexigrams for colors and names. It seemed simply a matter of pairing the items and the lexigrams enough times, and using procedures that could efficiently correct errors. Such training would readily produce—or so we thought—the associations *cum* symbols that we desired.

We began training these associations by sitting in front of the keyboard with a chimpanzee. An object would be shown to the chimpanzee who was then encouraged to light a symbol. If the chimpanzee illuminated the correct symbol he would receive food and social praise. If not, another trial would follow. Initially, only one key was available, thus the chimpanzee did not have to make a choice: he needed only to learn to respond to the presentation of an object by pressing a key. Once this response was occurring reliably, we introduced a brief period of errorless training with the second symbol, then we randomly presented both objects while both symbols were available. The chimpanzees began to have difficulty at this point, but with continued training, they all improved. We then introduced the third symbol.

All of the chimpanzees showed significant performance deficits and concomitant attentional decrements with the introduction of the third symbol. Once the third item was introduced, all four chimpanzees consistently made errors on the second and third items. It seemed as if they had been following a two-factor strategy: when the teacher holds up object A, hit key A; when the teacher does anything else, select key B. With the third item came a profound hesitancy to select any symbol, accompanied by a focusing on the teacher instead of the keyboard. The chimpanzees now frequently attempted to elicit help by waving their hands just above the keys while steadily keeping their eyes on the teacher. At times, they even lighted the keys without looking at the symbols on them, looking instead only at the teacher (see figures 4.1–4.5). After four months of daily training for several hours per day, none of the chimpanzees had learned the six associations we intended to teach them. Even the best animal was doing no better than 80 percent correct on three items, and his performance was not stable.

What is most noteworthy is that this kind of symbol training is very similar to the training other apes had received. An object was held up and the chimpanzees were encouraged to select the symbol which had been paired with the object previously. Washoe and Lana had both seemingly learned such object-symbol associations with ease. In addition, Asano et al. (1982) had more recently taught three young chimpanzees object and color names using a keyboard-lexigram system virtually identical to our own. These chimpanzees required eight to nine months of daily training and performed at accuracy levels greater than 90 percent.

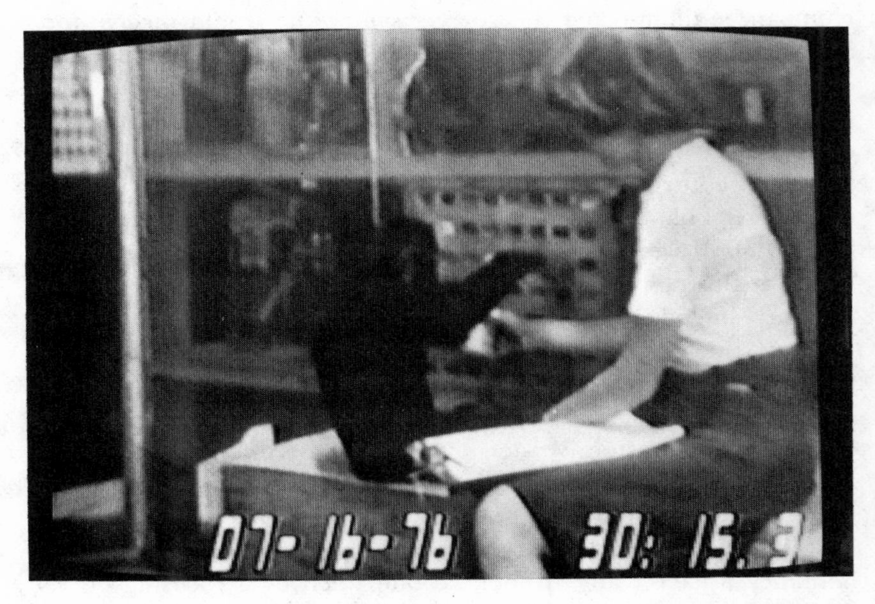

Figure 4.1. The teacher shows Sherman an object and Sherman reaches to the board to light a symbol.

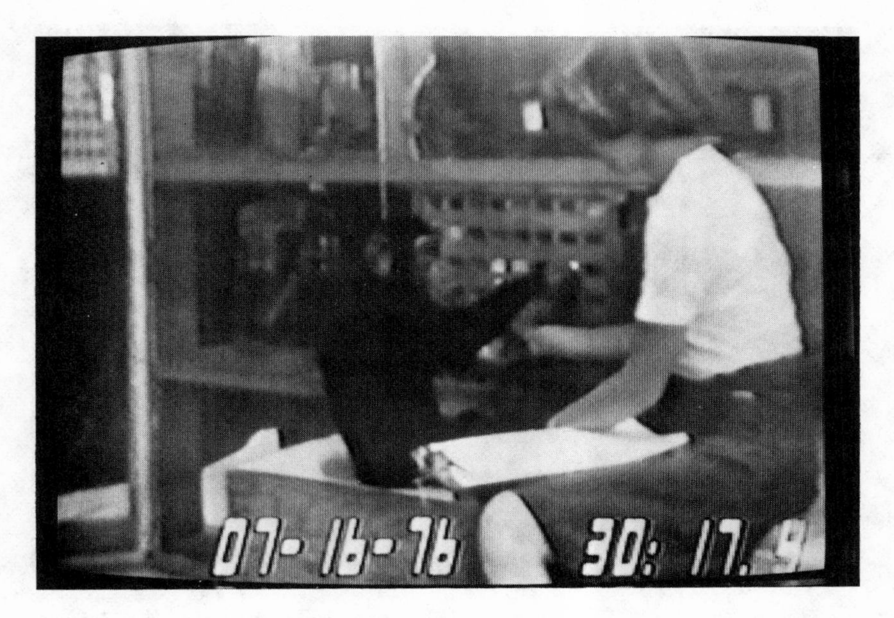

Figure 4.2. As the symbol lights (note the brightened area on the projectors above the keyboard) Sherman looks toward the teacher with a pout face to determine if he has selected the correct symbol.

Early Lessons from Video Tapes

Why then, with four months of daily training, had Sherman, Austin, Erika, and Kenton not learned at least a few simple associations? It did not seem reasonable to conclude that, of four randomly assigned chimpanzees, we should have somehow obtained four of the least intelligent apes at the Yerkes Center.

The answer emerged from analyses of videotapes of our training sessions. These showed that there was only one aspect of our training situation which seemed to be salient from the chimpanzee's perspective. This was the teacher's decision to give, or not to give, the food in her possession after a key was lit. Instead of attending to the relationship between the object the teacher held up and the symbol they then lit, our chimpanzees were attending to the relationship between the symbol they lit and the ensuing event—i.e., whether or not the teacher gave food to them. They treated the object display as simply a discriminative cue for the onset of the

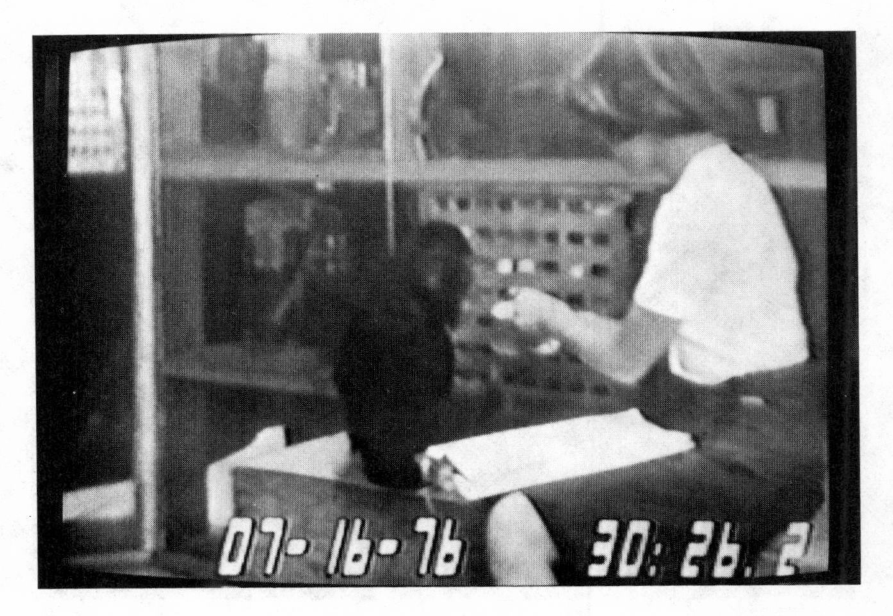

Figure 4.3. The teacher points to the object to show Sherman what he was supposed to label and then to the symbol which he should have selected. As he realizes that he is not going to be rewarded on this trial Sherman's expression begins to change to one of distress.

behavior of symbol selection, while they essentially ignored the type of object the teacher displayed. Their behaviors suggested that they thought the teacher's decision to call a response "correct" might have something to do with *how* they lit the symbol, how they treated the teacher, etc. as opposed to *which* symbol they lit.

In brief, it could be said that the teacher was attempting to cause the chimpanzee to select certain symbols by "showing" him certain objects. Meanwhile, the chimpanzee was attempting to cause the teacher to give food by hitting certain keys; and when this was unsuccessful, the chimpanzee began to look for cues or attempted to manipulate the experimenter's mood. Because the teachers had little comprehension of why the chimpanzees seemed to do well at some times and poorly at others, many superstitious behaviors appeared in their manner of presenting objects. For example, the teachers became very particular about just how an object was to be held, and they engaged in numerous means to satisfy themselves that the chimpanzees had looked at all the alternatives before

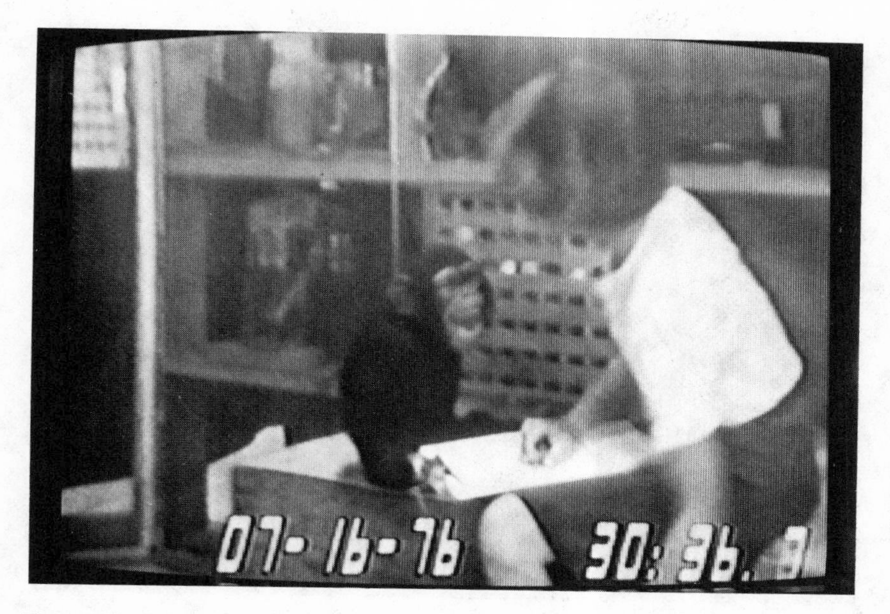

Figure 4.4. The teacher puts the object down, which clearly signals to Sherman that the trial has ended and that he has not received food. He registers considerable emotional distress, and from the orientation of his visual regard, it is apparent he is directing this emotional communication to the teacher.

responding. In retrospect, it was evident that the chimpanzees were seeking ways of controlling the food-giving behaviors of the teacher and the teacher was seeking ways of controlling the chimpanzee's attention to the keyboard and the stimulus objects. However, since the chimpanzee's attention was entirely focused on the teacher's decision to give or not to give food, neither party in the transaction made progress in their attempts to shape the behavior of the other. The appearance of superstitious behaviors in the repertoires of both parties (teachers and chimpanzees) was the result.

Thus the chimpanzee clearly appears to view itself as a stimulus for the teacher's ensuing behavior. It is probably for this reason that request-based tasks have been more effective than naming tasks as an initial means of symbol introduction with chimpanzees (Rumbaugh and Gill 1976a). The chimpanzee wants to know what it is that causes the teacher to give him food, to tickle, or to provide trips outside. Thus the chimpanzee appears to be *predisposed* to

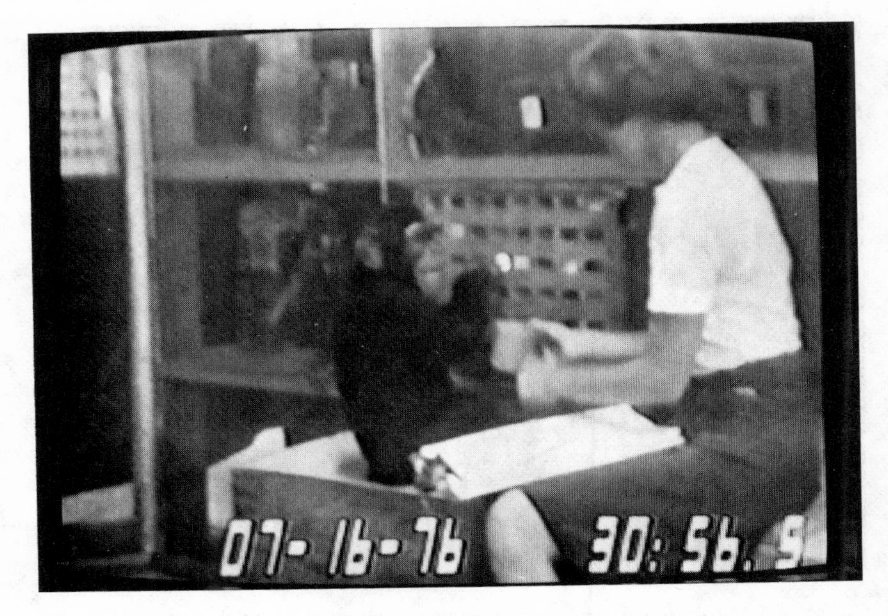

Figure 4.5. The teacher again selects the same object and shows it to Sherman, who now selects another symbol. It is evident from his facial expression and orientation that he is focusing upon the teacher and again waiting for her judgment. Even though the teacher has just pointed back and forth between this object and the correct symbol (figure 4.3), Sherman still does not select the appropriate symbol when given a second chance with the same object. This suggests that he did not understand that the purpose of the teacher's previous behavior of pointing back and forth between the symbol and the object was to show him the correct symbol for that object. These photos were taken in the midst of many sessions of similar attempts to use pointing to try to convey to the chimpanzees that particular symbols were to be associated with particular objects.

attend to the relationship between what he does and what effect his actions have on the environment, both animate and inanimate. The chimpanzee does not appear to be equally predisposed to attend to the more complex three-part relationship between what the teacher shows him, what symbol he selects, and whether or not he receives food given that he selected symbol A_1 when shown object A. It is suggested that it is, in part, the "common outcome" of receiving the same reward for different responses that causes the chimpanzee difficulty in this sort of "word learning" task.

The Importance of Differential Outcomes

Lana's early training did not start with the complex task of selecting certain symbols when shown various objects and colors. In fact, she did not even begin such complex training until well into her third year of using the keyboard. Her earliest training emphasized cause–effect relationships between the symbols she touched and the ensuing salient changes in her environment. Lighting particular keys caused the vending devices outside her room (visible through the Lexan wall) to vend special foods. Other symbols caused different vending devices to vend favorite drinks. Still other symbols opened windows or turned on movies or music. In fact, nearly every key that Lana selected initially caused something salient and interesting to happen. That is, every symbol produced a differential outcome. Moreover, no specific behavior on the part of the experimenter necessarily preceded Lana's key selection. She could open her window, look at slides, ask for M&Ms, etc. at her discretion. Her response was not contingent upon the experimenter's doing or showing anything. That is, the "onset" cue was not under experimental control. (Later on, it became possible to simply point to any device and ask Lana to name or operate that device, but such teacher-initiated behaviors were not part of her initial training.)

Since this initial work with Lana, a number of investigators (Brodigan & Peterson 1976; Peterson & Trapold 1980; Peterson 1984; Trapold 1970) have begun to look at the effect of differential response outcomes in a number of other animal species. In the typical differential outcomes procedure, the animal must learn to discriminate between two environmental stimuli (generally colors or geometric patterns). One group of animals received different outcomes for responding correctly to each stimulus (food or water, food or tone, grain or peas, etc.). For this group, one stimulus is always associated with one outcome and the other stimulus is always associated with the other outcome. The animals in a corresponding control group receive the same outcome (only food, only water, etc.) for responding correctly to either stimulus.

Without exception, animals in the differential outcomes groups learn much more rapidly than animals in the similar outcomes groups. Furthermore, the differential outcomes groups achieve and maintain far higher accuracy levels. Even more important, differential outcomes groups are able to continue to perform accurately when delays are inserted into the task. Peterson, Wheeler, and Trapold (1980) demonstrated that their differential outcomes animals could handle a 10-second delay, while the introduction of a similar delay reduced the performance of the common outcomes group to chance levels.

The differential versus similar outcomes procedure used by Trapold and Peterson is, in many respects, like the request versus naming procedures used in ape-language studies. That is, when an ape "requests" an item, he receives that item, and each different request results in a different outcome ("tickle" results in social play, "movie" results in pictures on the wall, "M&M" results in a vending device dropping an M&M in front of the chimpanzee, etc.). However, when the ape "names" an item he may receive food and social praise, but this outcome is similar for all named items. In many cases, ape-language investigators do not distinguish between naming and requesting. Instead, they assume that the ape is "naming" something that he wants. In this view, no behavioral or procedural difference exists between naming and requesting. However, when such a distinction is made at the procedural level (Savage-Rumbaugh, Rumbaugh, and Boysen 1978a), it becomes clear that learning is far more rapid when a differential-outcomes or request-based procedure is employed.

The difficulty in using differential-outcomes or request-based procedures to train symbol use does not lie in any technical aspects of the procedures themselves, as these are relatively straightforward. Rather, the difficulty is theoretical. Does the animal really have an "expectancy?" Does it possess some representation of the anticipated consequences of its actions? Does a chimpanzee know that when it selects the symbol it will be tickled? A further theoretical problem lies in the "message" aspect of symbol usage. Even if we are to grant that a chimpanzee is capable of anticipating the consequences of its actions, can we grant that it is using the symbol as a communicative device to request that others engage in particular actions? To do so requires that the ape be capable of conceiving of itself as an agent who can communicate causal messages—that is, an agent who is capable not only of producing predictable outcomes, but also of asking others to do the same and of specifying the precise nature of those events.

Learning To Ask Better Questions

We attempted to address some of these issues by designing studies that broke them down into smaller questions. We began by altering radically the nature of the training procedure after our failure in obtaining object and color naming behavior in Sherman, Austin, Erika, and Kenton.

We made the keyboards of two of the chimpanzees (Sherman and Kenton), operate as large vending devices. Whenever they selected any of four symbols (juice, M&M, sweet potato, or chow), the keyboard delivered that food to them immediately. Each symbol produced one and only one sort of food, and when that food ran out its symbol became inactive. Consequently, Sherman and Kenton could not make "errors" since every key they selected always produced food. Whether or not they selected the key which they "intended" and received the food which they "anticipated" could be determined only by indirect means.

One approach we used was to offer all four foods in a free choice situation each day. All the foods were placed in front of them and they were allowed to eat any food they chose, in any order they chose. We simply recorded the order in which they freely selected these different types and then compared the order of food selection in the free consumption setting with the order of request in the keyboard setting.

One would assume that if a chimpanzee were requesting foods in order of preference, he would ask for foods in the same order he chose to consume them when they were freely available. Sherman and Kenton's keys changed location each trial, so if they wanted to request M&Ms, they could *not* do so simply by selecting the same key over and over. Instead they would have to hunt on each trial for the M&M symbol. Not only did they have to distinguish this symbol from the three other food symbols, they also had to discriminate it from four other irrelevant symbols also located on the keyboard. Thus, in order to ask repeatedly for a favored food, such as M&M, the chimpanzees had to perform a complex search task on each trial. By contrast, if they did not care which food they received on a given trial, they could simply hit any food key and receive something.

Given the complexity of the search task, it seemed that Sherman and Kenton did anticipate which food they would receive: otherwise why should they bother to search for a particular food symbol? It would be much easier to select any food symbol. Moreover, their order of selection on the free preference test corresponded rather closely with the order in which they chose to obtain these foods when using their keyboard.

The two other chimpanzees, Erika and Austin, were trained differently. In their case, the teacher stood outside the room and held up one of the four foods listed above. If they selected the symbol which corresponded to the food being shown to them by the teacher, she "dispensed" the food by dropping it into the vending tube which deposited the food near the chimpanzees' sitting bench. If they "requested" another food, the displayed food would be removed and another trial would follow.

Thus, while Sherman and Kenton could request any food they wanted, Erika and Austin had to limit their requests to the single food item which was displayed by their teacher on any given trial. Yet for both groups there were differential outcomes associated with the selection of each different symbol. The main distinction was that the experimenter determined the constraints placed upon the onset of the response for Erika and Austin, but not for Sherman and Kenton.

The outcomes of these two different training procedures were dramatic. The chimpanzees who were allowed to request any food they chose, when they chose, learned rapidly to distinguish the four food symbols from each other and from the irrelevant symbols. Although these chimpanzees could not make a true "erroneous response," since there was no measure of correspondence between the item displayed by a teacher and their response, they could make a response which did not agree with their free choice response, or they could select an irrelevant key. In order to permit a between-groups comparison we counted as errors (for Sherman and Kenton) all responses to irrelevant symbols, and all symbol choices that did not match the free preference test for that day. Errors were scored for Erika and Austin on trials when they selected a symbol not assigned to the displayed food. Our acquisition criterion was 90 percent or higher across two successive sessions, with a minimum of 60 trials per session, and a maximum of 100.

Table 4.1 compares the performance of these two groups on this task in terms of trials to criterion on each of the four food symbols.[1] Table 4.1 also presents the acquisition data in minutes to criterion and contrasts it with Fouts' (1973) data on initial acquisition of signs.

As Table 4.1 illustrates, both groups required almost an equal amount of time to acquire their first symbol. During the acquisition of this symbol, the task was basically identical for both groups, since only one food was available and the chimpanzee's job was to discriminate the symbol which resulted in the receipt of food from

Table 4.1. Trials to Criterion*

	M&M (4 irrelevants)	*M&M & Sweet Potato*	*M&M, Sweet Potato, & Chow*	*M&M, Sweet Potato, Chow, & Juice*
Self-determined request(Kenton & Sherman)	$\bar{X}$ = 879	$\bar{X}$ = 152	$\bar{X}$ = 270	$\bar{X}$ = 133
Teacher-determined request	$\bar{X}$ = 823	$\bar{X}$ = 450**	Did not learn	Did not learn
	Minutes to Criterion***			
	1 word	*2 words*	*3 words*	*4 words*
Self-determined request	176 min.	30 min.	54 min.	26. min.
Teacher-determined request	165 min.	90 min.	Did not learn	Did not learn
Mixed request format; Behavior is ASL hand motion (Fouts, 1973)	163 min.	28 min.	118 min.	23 min.

* 90% correct or better across two consecutive sessions.

** Only 1 chimpanzee (Austin) reached criterion.

*** Our data are converted to minutes in order to make it more comparable with Fouts' (1973) data; however, even when converted to time, there are still important criterional differences, since Fouts used 5 consecutive correct responses and we used 90% or better across two sessions. Had we used Fouts' criterion, our data would suggest spuriously short acquisition times since our chimpanzees did not have to learn to *form* a sign, only to select a particular symbol.

the four other irrelevant symbols. At this point, Erika and Austin did not have to respond differentially to the items which the teacher displayed on different trials, since the teacher held up only one item, M&Ms.

The similarity of the groups at the one-symbol level suggests that they were closely matched in competence. The differential results beyond the one-symbol level appear to show that as long as the chimpanzees can say what they choose to say (as opposed to what the teacher would have them say), learning is fairly rapid.

It could be argued, however, that allowing the chimpanzee to determine which symbol to select, and when, was not really the major reason behind the extreme differences in learning rate. A simpler explanation might reside in interpreting Sherman and Kenton's performance as evidence for the existence of a learned hierarchy of lexigram preferences. That is, they could have simply learned "look for the first symbol you learned, if you don't see it, look for second symbol," etc. Such a strategy seemed unlikely, however, since the order of symbol training (M&M, sweet potato, chow, juice) did not correspond with the typical order of daily preference (M&M, juice, sweet potato, chow). It was also observed that when the chimpanzees received chow as part of their basic daily diet, they would typically request only two or three pieces of chow and it was always the least preferred food. However, if we eliminated chow from their basic daily ration (and replaced it with something else) then chow became the most preferred food. Thus, as food preference was manipulated, keyboard performance altered correspondingly. This indicated that a hierarchical response strategy could not account for the differences between groups.

From the Lexigram as Stimulus to the World as Context

Learning symbols involves much more than simply asking for M&Ms until someone turns off that symbol, then asking for juice until someone turns off that symbol, etc. At some point the language learner must become sensitive to the contexts in which utterances occur. Apparently this point occurs relatively early in normal children. Even before a child reaches the one-word stage, she clearly demonstrates a sensitivity to whether or not the caretaker is within hearing or viewing range and makes sounds and facial expressions only when the caretaker is available as an audience. If the caretaker is not available but is nearby, the young child will seek his presence before emitting such behaviors. Consider another commonly observed example. Children often appear to take a minor injury such as a bump on the head or a scratched knee well in stride if no parent is nearby. However, as soon as the parent appears they may burst into tears. Children also tend to limit their early food requests, at the one-word level, to situations when a caretaker is present.

Like human children, Lana became sensitive to the contexts in which utterances should occur. She astutely and reliably attended to the state of the vending devices located just outside her room. When she saw M&Ms in the vending device, she asked for them: when it was empty she stopped asking. The same was true for all the other foods and drinks which she knew. Lana was sensitive not only to whether her vending devices were filled, but also to whether someone were there who might replenish them. When this happened, Lana immediately stopped requesting food of the machine with her stock sentences such as "Please machine give M&M." Instead, Lana turned her attention to the experimenter, tapped on the wall till he or she looked at her, and then—while maintaining eye contact—asked "Question you put food into machine?" Lana never made such "experimenter-appropriate" requests while the experimenter was absent.

To determine whether or not Sherman and Kenton would likewise be sensitive to which food was present in the vending devices, we moved their devices from behind the keyboard into full view, as were Lana's vendors. As soon as their keyboard was turned on, Sherman and Kenton began asking for M&Ms even though their dispenser was obviously empty. While they watched, we placed sweet potatoes in the dispenser, yet Sherman and Kenton continued to request M&Ms. Unlike Lana, Sherman and Kenton's requests were not sensitive to the external context. Since their previous training had not required them to attend to the state of a dispenser, perhaps we should not have anticipated that they would do so. Just the same, we did not expect that they would *repeatedly* select the M&M key given that the absence of M&M was coupled with the visible presence of other desirable foods. We thought that after a few M&M requests they would stop, note which food was clearly available, and ask for it. They did not. They continued to select the M&M key even when they were shown (by the teacher) that lighting the sweet potato symbol activated the dispenser and produced sweet potatoes. Since they quickly ate the pieces of sweet potato during this demonstration, it was reasonable to presume that they wanted them. Indeed, they looked directly at the pieces of sweet potato and food barked excitedly as they selected the M&M symbol. Yet instead of switching to the sweet potato symbol when no food came out, they continued to select M&Ms for many trials. In fact they selected "M&M" more times than it had taken them to learn the difference between the M&M and the sweet potato symbol in the earlier free request task.

Apparently, their previous training had enabled Sherman and Kenton to encode food desires in a primitive way but it had not enabled them to take into account important contextual cues as did Lana. At this point they employed symbols only in a primitive cause-effect manner—just as simple actions like pushing, shoving, or biting can be used to affect another directly—rather than to communicate. That is to say, they had learned what symbols did for them (in terms of obtaining M&Ms, etc.) but they had not learned any contextual cues regarding the appropriateness of symbol production.

Such contextual cues are generally referred to as "conditional discriminations." Yet they are actually more complicated than typical conditional discriminations and should be thought of as third-order conditional tasks. A first-order conditional discrimination (CD) would require that a response be made whenever a key was lighted. A second-order CD would require that a response be made whenever the key was lighted and food was present. A third-order CD would require that a response be made whenever a key was lighted, food was present, and the food was being displayed by the experimenter. In addition, as multiple foods are used across trials the subject is being required to learn multiple CD's simultaneously.

These contextual cues also differ from those typically presented to other animals in CD tasks in that they occur in a complex visual-auditory environment. The chimpanzees are not confined to a test chamber with limited visual-auditory input. Rather, their environment is filled with diverse and changing visual and auditory information. They must somehow separate out the particular "cues" to be treated as *the* conditional stimulus from all of the other visual and auditory stimuli in their surroundings.

Conclusions Regarding the Difficulties of "Rote Associations"

How did Sherman, Kenton, Austin, and Erika finally learn to be attentive to the relationships between the teacher's behaviors, their symbol selection behaviors and the ensuing consequences? Chapter 5 describes how this sort of learning eventually took place. For the moment, though, it is important to use the "failures" described above as an object lesson for animal language experiments as a whole. An obvious implication of these failures is that associations *between particular symbols and particular objects* are not readily acquired by apes. For those who have viewed ape-language skills as little more than the appearance of particular signs in the presence of particular stimuli (Seidenberg & Petitto 1981; Terrace et al. 1979), we must raise the question—how did the apes learn all of these associations to begin with and are such associations really simple?

Our failure to obtain symbol-object "associations" with Erika and Austin in one context, and the relative ease with which we seemingly obtained these "associations" in Sherman and Kenton in a slightly different context, strongly suggests that not all item-symbol "associations" are as similar as we might initially presume. The ability to learn associations between what a subject desires and the means of obtaining it appears to differ significantly from the ability to learn associations between the item a teacher holds up and a particular symbol. The task presented to Sherman and Kenton was a first-order CD; the "stimulus" for selection of one symbol as opposed to another was differential internal motivation. By contrast, the task presented to Erika and Austin was a third-order CD, with the stimulus presented by the experimenter. Both groups experienced differential outcomes.

What does this difference mean with regard to the acquisition of "words" or "names"? Does either sort of training lead to the learning of "names"? Gardner and Gardner (1978b) have asserted that chimpanzees using the lexigram keyboard system are not "naming" but are instead executing rote associations. They note that the naming tests reported for Lana by Rumbaugh and Gill

(1976) follow "hundreds, sometimes thousands, of trials on the same small set of keys and stimulus arrays, and none of these tests can have placed any great strain on the memory capacity of the chimpanzee" (p. 562). By contrast, chimpanzees who have learned to produce an ASL gesture are described as truly "naming," although the Gardners do not report the number of training trials required to teach reliable gestures.

The Gardners are correct in stating that the memory capacity of the chimpanzees can handle a large number of rote tasks. The problem, however, lies in determining precisely what a "rote task" consists of. It certainly does not consist simply of exposure to numerous training trials. Erika received over 2500 training trials on the symbols M&M and sweet potato. Yet, she could not reliably name either of these foods when they were displayed by the experimenter. If "rote training" could accomplish the inculcation of associative "naming" behavior, it should have done so in Erika's case.

The fact that such "rote" training did not produce associative learning emphasizes the importance of analyzing such tasks more carefully. Apparently chimpanzees can readily learn associations between symbol production and desirable events such as the receipt of food or different types of food, a game of tickle, a trip outdoors, etc. Moreover, if producing such symbols always leads to the receipt of these desired items or events, the chimpanzee seems to acquire "vocabulary" rapidly—particularly as long as he can initiate his requests at will, in response to internal cues. However, when the occasions for the occurrence of a given symbol are set by others, the chimpanzee does not learn symbols nearly as quickly, if at all.

Why Wasn't Cueing More Helpful When We Needed It?

The difficulty of learning two symbols, even with thousands of training trials, speaks to the issue of cueing. It has been assumed that, with prolonged training, the opportunity for the chimpanzee to learn inadvertent cues increases markedly. In fact, some critics (Umiker-Sebeok and Sebeok 1981) have categorically rejected all ape-language studies, asserting that, because apes must be trained by people, any and all training results in "tainted" performance.

It is important to note that during most of the training described in the previous sections a human teacher was seated with the chimpanzee by the keyboard. Even when objects were shown from outside the room by one teacher, a second teacher still sat inside the room next to the keyboard—often with the chimpanzee on her lap to comfort him. Such companionship was essential, as the chimpanzees were very young during this initial training and were desperately afraid of being locked in a room by themselves.

The constant presence of a human companion clearly produced a situation in which inadvertent cueing could take place. In addition, only a few stimulus objects and symbols were being used at this time because of the limited capacities of the animals. Such conditions are often considered ideal for the development of inadvertent cueing. Yet there is no evidence that any cueing occurred. If it had, the chimpanzees surely would have been able to perform appropriately, at minimum, the few symbol-object associations we were attempting to teach them.

Cueing was ineffective because the chimpanzees were not even attending to the teacher's overt and obvious behaviors designed to help them, much less to subtle behaviors of which the teachers were unaware. When chimpanzees are not able to attend to obvious object showing behaviors it becomes unrealistic to suggest that the teacher's unwitting gestures and glances are cueing the chimpanzee with regard to which key to select (cf. Umiker-Sebeok and Sebeok 1981). It is also often overlooked that ape-language tasks are not dealing with the "onset" and "offset" of a single response, as in the case of a "counting" horse who needs only to learn to start hoof-stamping on an overt cue and terminate it with a covert cue. Rather we are dealing with a complex visual search task which includes multiple alternatives that are relocated each trial and in which the "correct" alternative is to be linked to an *obvious* external cue—the display of a particular object. Thus subtle cues such as a raised brow, a change in tone of voice, etc., cannot possibly provide the ape with enough information to produce the correct symbol—should such cues even occur inadvertently or intentionally.

The role of cueing as a factor in our subjects' performance can also be ruled out on other grounds. Only if the chimpanzees ran their fingers up and down the rows of keys waiting for the teacher to nod her head when they were over the correct key could cueing lead to a correct response without the chimpanzee's having learned the proper symbol-object association. This is because the symbols are so close together that it is not possible to use glance alone (advertently or inadvertently), to guide the chimpanzee in symbol selection. At best, a teacher's glance could focus the chimpanzee's attention on the proper area of the board, but even so the chimpanzee's selection would still have to be made from among a number of symbols. Since it is behaviorally obvious if a chimpanzee is running his finger across the keyboard waiting to be told which key to select, this sort of cueing cannot occur without the direct knowledge and intentional participation of the teacher.

Early Conclusions

In summary, it can be seen that we had learned three important things from this work. The learning of symbol-object associations could not necessarily be produced simply through sheer practice. Cueing was not only *not* a factor to be worried about, it was ineffective even as an intended means of teaching. And most important, "symbol-object associations" were not a single class of responses. There were many different sorts of associations that chimpanzees could learn and basic reasons why some types of tasks produced far more rapid learning than others. The previous view that a "name" was just a "rotely learned" symbol was clearly inadequate. It was necessary to specify much more precisely the conditions under which the "symbolic" response occurred. The ability to use a symbol such as M&Ms "to get M&Ms" could mean many different things that had not been differentiated in previous reports.

Table 4.2 details some of the different types of training settings which could lead to the production of a gesture or selection of lexigram. It is important to remember that even though, for purposes of clarity, table 4.2 lists alternatives for a single gesture or lexigram, the training tasks typically involve alternation between a number of these responses. As table 4.2 reveals, for the first four types of situations, there is no ready means of determining whether, in fact, the symbol "juice" was selected in response to a specific desire for juice as opposed to other edibles. If the chimpanzee were given a banana (or other food) when he asked for juice—and refused to consume anything but juice—then there would be strong evidence that the symbol "juice" was being used to procure that specific item. However, since hungry apes will as readily consume bananas as juice, this sort of data is difficult to gather. Consequently, tasks such as V and VI of table 4.2 must be used to assert that the chimpanzee "knows" juice. Not only are these two tasks quite different from the others, they are quite different from each other with regard to outcome.

When the learning parameters are broken down, as in table 4.2, it is clear that generalization can occur in any or all paradigms. Thus a chimpanzee might respond by selecting "juice" regardless of whether the liquid is grape juice, fruit punch, apple juice, or bananas. However, such generalization must not be taken as evidence of true "naming," since generalization itself need not be correlated with a capacity to produce the response "juice" under all of the settings listed in table 4.2. For example, Sherman and Kenton called many small chocolates "M&Ms," yet they were basically performing only at level I of table 4.2.

If we look at the reports of "vocabulary" acquisition for apes such as Washoe, Koko, and Nim in terms of the model outlined in table 4.2, it is obvious that symbol-object associations could function at any of these levels and still be classified as learned symbols. The criterion for inclusion as vocabulary was simply the repeated (15 occasions) occurrence of the symbol in what seemed to be a relatively interpretable context. As table 4.2 shows, measuring the mere occurrence of a symbol such as "juice" reveals almost nothing about the contingencies surrounding the use of that symbol.

Are "Associations" the Same Things as Names?

Even if it is granted that Washoe, Nim, Koko, and others are capable of learning item-object associations at level VI in table 4.2, we must still ask whether such associations are equivalent, at a functional level, to "names" as we know and use them in human language.

Does the ability to produce symbols A_1, B_1, C_1, when shown objects A, B, and C mean that the chimpanzee has a representation of these behaviors which is elicited by the showing of objects?[2] In a limited sense, the answer is, yes. That is, in order to do well on a set of such item-symbol associations, the subject must be able to recall and produce the correct symbol when shown each item. Such recall requires that the subject not emit similar behaviors (e.g.,

Table 4.2.

	Internal Stimulus	*External Stimulus*	*Response*	*Event*	*General Description of Behavior*	*Verification that juice is desired*	*Sufficient for inclusion as Vocabulary item*
I.	General hunger	None	"Juice"	Receives juice	Asking for juice	Drinks juice	No
II.	General hunger	Presence of juice	"Juice"	Receives juice	Asking for juice	Drinks juice	No
III.	General hunger	Presence of experimenter	"Juice"	Receives juice	Asking for juice	Drinks juice	No
IV.	General hunger	Presence of juice plus experimenter	"Juice"	Receives juice	Asking for juice	Drinks juice	No
V.	General hunger	Presence of juice + experimenter, + the experimenter shows or points to juice and then to other items	"Juice"	Receives juice	Asking for juice	Drinks juice	Partial

VI.	General hunger	Presence of juice + experimenter, + experimenter shows or points to juice as opposed to some other item	"Juice"	Receives social praise or food reward other than juice	Naming juice	Not appropriate	Partial
VII.	Desire for juice	None	"Juice"	Receives permission to go to food larder and retrieve juice	Asking for juice	Selection of juice from among other foods, followed by consumption	Yes

lighting a lexigram that looks very similar, or moving the hands through a similar sequence of actions), but that they recall and selectively execute particular behaviors. Can the chimpanzees recall such behaviors apart from actually acting them out? Can they, for example, recall that the symbol "juice" goes with juice, even if they do not actually make the gesture or light the key on the board? If so, one would have to conclude that they have some sort of representation of juice without actually engaging in the act of producing the symbol. Additionally, the ability to negotiate a delay between presentation of the object and selection of the symbol would suggest that some symbolic representation of the object is present at the covert level. In later chapters, the use of symbols to represent items and events removed in space and time will receive major focus. It is, after all, not the associations themselves, but the representational power that these associations called "words" lend to our communications that is important. The ability to "name" things in view is rather mundane; the ability to communicate about things out of sight unquestionably endows upon its users an enormous adaptive advantage.

Chapter 5

Talking to Teachers Instead of Machines

Errorless Training: Does It Work?

As we learned in the previous chapter, symbol-object associations are both misleadingly simple in theory and difficult to teach. This finding is not typical of ape-language studies. For example, Gardner and Gardner (1971), note that:

> The acquisition of individual signs is the aspect of this project that is most clearly related to the paradigm of S-R reinforcement theory. This paradigm, which had a strong influence on the tactics that we used for teaching individual signs, serves as a convenient point of departure for the description of specific teaching methods. According to the theory, particular responses are made in particular stimulus situations. When a response is followed promptly by reward, there is an increment in the probability that that particular response will be repeated by the subject when that particular stimulus situation is repeated. Thus, the first step in training is to have the subject make the to-be-learned response in an appropriate situation. The to-be-learned response need not appear in its final form at the beginning of training. If some approximation of the response, even a poor approximation, is made in the appropriate situation, then the probability of the approximation can be raised by reward. After that, selected variants of the approximation can be rewarded, and by a series of approximations the final form of the to-be-learned response can be achieved. This procedure of rewarding successive approximations has come to be known as shaping. (p. 129)

According to Gardner and Gardner (1971) this procedure worked quite well and Washoe's vocabulary "grew in length with each passing month at an accelerating rate" (p. 140).

Premack (1971, 1976a) similarly views symbol-object associations as being acquired with relative ease by the chimpanzee, although he differs from the Gardners in explicitly specifying that reinforcement is not necessary for this to occur, claiming that language and evolutionary considerations

> imply a rather different acquisitional model than the ones urged either by Skinner (1957), on the one hand, or by Chomsky (1965), on the other . . . any acquisitional model must take into account two key facts: reinforcement plays no direct role in learning, . . . and not only language but all basic human skills can be and have been acquired on an observational basis. (Premack 1976a, p. 4)

In place of reinforcement, Premack (1971, 1976a) substitutes "errorless training" which is, in effect, Guthrian (Guthrie, 1952) association by contiguity revisited. According to Premack (1971):

> the basic procedure for teaching words to a naive organism is extremely simple (p. 195). . . . For example, we may start with the fruits that are offered. The set of possible objects is defined by offering different fruits on different trials, each time with a corresponding change in the language element. When the fruit is banana, the plastic chip is one kind, when apple of a different kind, and when orange still a third kind. On each trial the chimp's task is the same: place the piece of plastic, which is beside the fruit, on the board before receiving fruit.
>
> Two tests will show whether or not the subject has formed an association between members of the object class and of the corresponding language class. Trials on which the chimp is given two would-be words but only one piece of fruit will determine whether it can match the word with the fruit. (pp. 193–194)

The procedures used by Gardner and Gardner (1971) differ from those used by Premack (1976a) in that Premack does not deal with shaping (because this is not an important issue with a plastic-chip symbol system) and he does not stress the role of reinforcement. The methods used by the Gardners and by Premack are similar in that they both stress the importance of producing or acting on the symbol in the presence of the stimulus in order to achieve object-item associations. Additionally, both procedures advocate repeated practice in a setting where the subject does not have to determine which response to make, just when to make it (i.e., when the stimulus is shown). Washoe was given repeated trials on the same sign as it was being learned and shaped, and Premack's chimpanzees were given similar repeated trials on one item in his errorless training program.

Since our earlier training had not involved massed repeated trials on a single item, we decided at this point to test this approach. We had not done so previously since it appeared that when we did repeat trials (even 5 or 10 times), the chimpanzees simply lit the same key over and over, and then another. They also did not seem to attend to the stimulus which they were supposed to learn to pair with that symbol. They simply sought out one symbol and responded to it repeatedly. It appeared, in fact, that such massed errorless trials hindered symbol acquisition far more than they helped. However, this conclusion ran at odds with those of others in the field and so we decided to attempt to try their approach more systematically.

We designed a study to test two things about symbol acquisition from this perspective: (1) would repeated associations between symbols and their referents—given as a series of errorless training trials—lead to the acquisition of object symbol associations as readily as Premack (1971, 1976a) claimed, and (2) would these associations, if learned, permit the subject to use symbols to "stand for" objects as opposed to being simple paired-associates of objects that required the presence of the object for their occurrence?

Eight chain or "stock sentences" of the form ABC (food name) period or ABD (liquid name) period were introduced in an errorless training paradigm. The English glosses for the ABC and ABD chain were "Please machine give" and "Please machine pour." The eight foods and liquids were beancake, banana, chow, orange, orange drink, water, milk, and Coke. We placed two food vending devices just outside the clear Lexan walls of the training room, one on either side of the animals' keyboard. In the vending device on the right we placed slices of orange or beancake, in the vending device on the left we placed slices of chow or banana.

Each of the eight chains was taught individually, using an errorless training procedure in which only the keys necessary to form a given chain (plus irrelevant distractor keys) were lit on any given trial. As the keys needed to produce each chain were activated, the corresponding food was placed in a vendor in view of the chimpanzee. For example, when chow was loaded into the vendor the symbols for "Please machine give chow" would be activated on the keyboard and when the chimpanzee touched them (in that order), chow would be dispensed. The chimpanzees did need to discriminate the appropriate chain symbols from the similar irrelevant symbols. All symbols moved randomly each day through a total of 24 positions to prevent positional responding. Training continued on each sentence until all sentences were mastered easily and reliably without error, regardless of key location. The level of mastery required was 100 percent on any given sentence across five days. During this training the subject was never required to choose between any two food names. A single food was loaded into a dispenser and that food name, plus the keys for the ABC (food name) chain, and the irrelevant keys were illuminated.

Learning the Contextual Appropriateness of Requests

During this errorless training, we also attempted to sensitize the chimpanzees to the ineffectiveness of requests for food when the dispenser was empty. Since Premack's chimpanzees did not have their symbols available at all times, but only when the trainer presented a trial, Premack never encountered the problem of chimpanzees repeatedly selecting symbols when no food was present.[1]

Our subjects could light symbols any time they chose. If, for example, they requested bananas before the teacher had loaded them into the machine, the dispenser turned as though to push out a piece of banana. Since none was available, none was dispensed. Additionally, once all the bananas in the machine had been requested, continued requests activated an empty dispenser. Thus the chimpanzees were required to attend not only to the consequences of their actions (whether they received food or not) but also to the antecedents—that is, whether food was loaded into the machine and available to be vended.

Sensitivity to the state of the dispenser was produced simply by allowing subjects to execute chains such as "Please machine give x," where x was the name of a food as many times as they wished until they learned that these chains produced food only when x was visible in the dispenser. Attention to the state of the dispenser was facilitated by the experimenters, who used gestures and chimp-like vocalizations to engage the animals' interest and orientation as the dispenser was being loaded.

It required a rather large number of trials for the chimpanzees to cause dispensers to operate only when there was food in them

(Sherman, 257; Austin, 1119, Erika, 2113; Kenton, 1593). When a chimpanzee no longer would request a given food after the last piece had been delivered, and would not begin requesting it again until additional pieces were placed into the machine, we considered the animal had demonstrated a knowledge of the dispenser's state. We were surprised at the large number of trials required for the chimpanzees to learn that requests for a specific food were conditional upon the presence of that food in the vending device. It appeared that Lana learned this rather easily, although, no record had been made of the number of trials which Lana required.

It would seem reasonable to ask whether Washoe's requests were conditional upon the presence of food. When Washoe produced food signs in the absence of a given food, the Gardners inferred that she was asking them to retrieve that food, commenting about having eaten previously, or simply letting them know that she was thinking of it. Could the same be said to be true of Sherman, Austin, Erika, and Kenton, as they busily pressed keys without looking to see whether the food they were specifying was even present? Since we could not verify such assumptions, we viewed them as premature. Moreover, since our chimpanzees looked toward the food slot after lighting symbols, they appeared to expect that the machine could vend food whether the dispenser was filled or not. They did not appear to be asking for an "abstract" food.

Errorless Training as Evaluated with Choice Tests

Errorless training on each chain plus food name was continued until each chimpanzee reached our criterion; this amounted to more than 1750 trials per food name for each animal. The large number of errors made by our animals could not be attributed to sequencing problems, since by this point in training they no longer made any sequence errors. Nor could they be attributed to lapses of attention. Our chimpanzees were clearly attentive to the individual foods themselves as each piece of food was shown them and then placed into the dispenser. Furthermore, because we had required that the chimpanzees both chain their responses and determine relevant from irrelevant symbols, we were certain that the chimpanzees could adequately discriminate these symbols at a high level of accuracy.

It is difficult to compare the number of massed trials which these chimpanzees received with those given to Washoe, since Washoe's data are presented as a ratio consisting of the total number of correct responses divided by the total number of correct responses, incorrect responses, and prompts for a given session (Fouts 1972). Still, it is not likely that Washoe received more trials than these animals, since Gardner and Gardner (1969) and Fouts (1972) both stress that Washoe did not require a large number of trials to acquire a sign. Premack (1976a) reports giving a total of 148 errorless trials with words and nonwords in a paradigm that is comparable to the one we used. Thus, as far as we can tell, we had given the methods of errorless training a fair test.

To determine the effectiveness of this errorless training procedure, we administered a series of choice trials similar to those described by Premack (1976a):

> Choice trials differed from errorless ones only in that two words were given on each trial, in the presence of one piece of fruit. To receive the piece of fruit, the subject had to place the correct word on the board. If it placed the incorrect word on the board, it was told "No," was not given the fruit, and was either advanced to the next trial or, in case of an anxious subject, allowed to correct and then advance to the next trial. (pp. 57–58)

During our tests, elements of two combinations were lit. For example, the symbols "Please," "machine," "give," "orange," and "banana" (plus the irrelevant symbols) would be lit and either oranges or bananas would be loaded into the dispenser. The subject had to note which food was placed into the dispenser and use the appropriate food symbol, in the correct chain, to request it.

In contrast to Premack (1976a), we did not find that errorless training produced symbol learning. That is, it did not enable the subject to choose the appropriate symbol given the foods which were displayed in the dispensers. Table 5.1 shows the results of a 100 trial choice test for our four chimpanzees and the results of similar tests given to two of Premack's chimpanzees, Elizabeth and Peony. Clearly, our chimpanzees did not perform as did Premack's animals. They tended, with the exception of Erika, to be able to ask correctly for only one food. That is they always asked for the same food (beancake in the case of Sherman and Kenton and banana in the case of Austin) and consequently were correct about half of the time. Erika was not correct on any trial since she tended to produce the following sequence on every trial "Please machine give banana beancake period." Sometimes she inverted the order of the two food names but she always included them both. (This is a strategy which Premack's structured system of symbol presentation and withdrawal precluded.)

If repeatedly pairing the object (in this case a food) with the symbol which was to be associated with that food produced "words" at best, or even symbol-object associations at worst, then at this point, our chimpanzees should have acquired words. Furthermore, they had been rewarded for each correct response by receiving the food they requested with their combinations. Each combination had been associated with a differential outcome, and each combination produced consequences which the chimpanzees attended to, as evidenced by the astute visual regard they paid to the food as it dropped out of the dispenser and the rapidity with which they then consumed it.

Why should our results regarding the effectiveness of errorless training be so different from those of other investigators?

1. All of the Gardners' blind tests were given after signs had been used for some time. During this period of use, Washoe was required to choose between a number of different signs in order

Table 5.1. 100 trial choice test on banana-beancake pairing with either food randomly loaded into the dispenser.

Chimpanzee	*% correct*
Sherman	50
Erika	0
Austin	35
Kenton	48
Peony*	81
Elizabeth*	79

* Premack (1976a) reports this data on page 78. It represents choice performance across 4 different pairs. His figures are somewhat puzzling, however, since on page 77 Premack gives the actual number of total trials and incorrect trials from which he presumably computed the above percentages and these figures produce percentages of 60% correct for Peony and 50% correct for Elizabeth. Premack does conclude on the basis of this data that "both animals finally mastered all the desired associations" (p. 77).

to execute the correct response. Thus none of the Gardners' test results actually reflect the effect of errorless trials, coupled with shaping, on a single given sign. They all reflect the extensive additional training which followed initial acquisition of signs.

2. Premack's structured situation inhibited many of the errors which might have been made, and thus they may not accurately measure the subjects' knowledge, or lack thereof.
3. Premack's choice trials were not administered with blind controls; consequently inadvertent cueing cannot be ruled out.

We are not asserting that our chimpanzees did not learn anything during these thousands of errorless training trials. They learned to discriminate lexigrams, to sequence lexigrams, and to use lexigrams to obtain food when a dispenser was loaded. What they did not learn were precisely those things which errorless training *does not require* but is purported to produce: that is, how to choose one symbol when one type of food is shown, and another when a different type of food is shown.

Symbol-Item Associations Require Multiple Alternatives

What our errorless training results suggest is that item-symbol associations are not small units of learning in and of themselves that can be acquired by repetitive practice. Rather, if we want chimpanzees to be able to choose between one symbol or another under certain external constraints, then that choice itself must be part of the conditions of acquisition; otherwise this ability does not appear. Yet recall from chapter 4 that when we first attempted to require the chimpanzees to choose between two symbols when shown two different objects, they attended to the consequences of their actions and not their antecedents (i.e., the displayed stimulus object).

The single-antecedent condition which we *had* successfully taught the chimpanzees was to attend to the state of the dispenser. This was an onset-offset type of discrimination: when the dispenser is full, light the symbols; when it is empty, stop lighting the symbols. It took the chimpanzees a large number of trials just to become sensitive to this simple contextual cue.

One might be tempted to view this as a reflection of meager intelligence. Yet other events made it seem more likely to be an attention related problem. Once the chimpanzees understood this antecedent condition, they *quickly* began to anticipate even earlier antecedent events in the chain. For example, they became attentive almost immediately to the events which produced a loaded dispenser—that is, the experimenters' behavior of placing food in the dispenser. They soon began to attempt to influence the occurrence of this antecedent condition by gesturing to the experimenter, to the bowls of food from which we took the cut-up food pieces, and to the dispensers. By so gesturing *to direct the experimenter's attention,* they demonstrated that they understood the relationship between the bowls of food, the desired location of the food, and the agent whose behavior moved the food from one location to another.

They also understood that they could affect the behavior of this agent by communicative gestures. These were unritualized but elaborate gestures (hand shaking while looking at the food, hand extension toward the experimenter of the food, etc.) and they certainly were not taught. The spontaneous appearance of these appropriate communicative gestures contrasts sharply with the chimpanzees' seeming lack of ability to pair a graphic symbol with a food or to figure out when to operate a dispenser. It suggests that our chimpanzees were more attuned to learning to control the social nonverbal communicative realm about them than they were to learning certain object-symbol connections simply because these connections produced food.

It might also be said to indicate that a gestural mode of communication is easier for apes to acquire than a graphic mode. In response, it must be noted that the chimpanzees *did not* spontaneously begin to produce gestures which *represented* various foods. Rather, their gestures were more appropriately termed "primitive attention directing devices." By drawing the experimenter's attention to the bowl of food and then waving their hands, they let it be known that they expected the experimenter to notice the bowl of food and put some of it in the dispenser. Such gestures were very effective when the items or actions of interest were present and could be referred to by such generalized primitive attention getting devices. However, such gestures became ineffective without contextual support. It is in this situation that graphic symbols have the greatest advantage, as will be seen in later chapters.

Multiple-Alternative Training Paradigms

Having concluded that the chimpanzees were not going to learn symbol-object associations solely by means of errorless training trials, we decided to require that they make a choice between sets of alternatives and sets of lexigrams. We lit the ABC chain and two food symbols, banana and beancake. Whenever the chimpanzees selected "Please machine give beancake," the dispenser on the right turned. Whenever they selected "Please machine give banana," the dispenser on the left turned. Five to ten pieces of food were loaded in either dispenser, but food was placed only in one dispenser at a time.

The chimpanzee's task was thus to ask for banana when the left vending device was loaded with bananas, and to ask for beancake when the right vending device was loaded with beancake. This would seem to be a fairly simple task, particularly since three of the chimpanzees had demonstrated that they could correctly request at least one food with both symbols available on the prior choice test. Furthermore, as long as only two foods were being used, dispenser location was redundant with food type.

During the initial phases of training, the chimpanzees performed as they had on the choice test. That is, they would concentrate on just one symbol, for example, banana. When the banana dispenser became empty, the chimpanzees watched closely as the beancake dispenser was loaded and then began to light "Please machine give *banana*." As they did so, they would look expectantly at the beancake dispenser, holding their hands under the vending slot as though anticipating that the big chunks of fresh beancake were going to drop into their grasp at any moment. When this did not happen, and the dispenser simply sat there silently, doing nothing, they quickly repeated "Please machine give *banana*," and again nothing happened. After several more failed attempts, the chimpanzees engaged in such behaviors as banging the Lexan in front of the dispenser or pacing. Eventually, they would return to the keyboard and try again, often at this point using both food names in the chain "Please machine give beancake-banana." This chain was designated invalid and operated neither dispenser.

These sorts of behaviors continued to persist until the chimpanzees finally began to notice that erroneous requests *not only produced no food, but that in fact they caused the other dispenser to operate.* Thus, if the chimpanzee were lighting "Please machine give banana" while the beancake dispenser was full of beancake and he happened to take his eyes off the beancake long enough to notice that the empty banana dispenser turned when he executed this chain, his whole demeanor immediately underwent a dramatic change. Typically the chimpanzee would first stare at the banana dispenser in astonishment, then repeat the chain "Please machine give banana" several times while looking back and forth between both dispensers to see which one turned as a result of what he had done. When he satisfied himself that each time he said "Please machine give banana," the dispenser on the left turned, he then looked at the dispenser on the right and said "Please machine give beancake."

The Effect of Monitoring Differential Outcomes of Symbol Selection

The immediate and striking shift in demeanor and correct responding which followed the simple act of noting that a given sentence caused the wrong dispenser to turn warrants special attention. Once this occurred it was only a matter of hours (180–200 trials) before the two older chimpanzees (Erika and Sherman) always selected the banana lexigram when bananas were loaded in the left dispenser and always selected the beancake lexigram when beancake was loaded in the right dispenser.

However, the two younger animals, Austin and Kenton, continued to fixate on the dispenser which was filled with food, never noticing that the effect of their chain was to cause the other dispenser to operate. Since attending to the *effect* of an *incorrect chain* had been so pivotal in Erika and Sherman's behavior, we attempted, by pointing and vocalizing, to draw Austin and Kenton's attention to the action of the empty dispenser when they lit an incorrect sequence. This was to no avail. They were not interested in looking away from the food toward an empty dispenser and did not visually follow our pointing gestures. If they did look, it was after the other dispenser had already turned and they saw no effect of their incorrect sequence.

To compensate for this tendency to visually fixate on the filled dispenser, we added an auditory beeping signal to each dispenser.

Now, whenever the banana dispenser turned, it emitted a high noise and whenever the beancake dispenser turned, it emitted a low noise. The chimpanzees learned these auditory associations in a few trials (the reader should note that these associations are between stimuli and not between a stimulus and a choice behavior on the part of the chimpanzee) as evidenced by rapid glances in the direction of the appropriate device when either noise was heard. These noises, which served to indicate that a particular dispenser was operating rapidly drew Austin and Kenton's attention to the effects of their sentences and they then easily solved the task, as had Sherman and Erika.

All four chimpanzees now attended carefully to two antecedent conditions (dispenser on the right loaded with beancake and dispenser on the left loaded with banana). They also responded differentially and appropriately to these two conditions. Such responding, at close to 100 percent accuracy, appeared in a few hours. Recall that we had attempted (unsuccessfully) to obtain such differential responding by training these animals for months at a time to select one symbol when shown one food and another symbol when shown another food. Why now, in a few hours, were all four chimpanzees learning what we had previously failed to teach them?

Chimpanzees as Cause and Effect Analyzers

The answer is simple and revealing. Whenever a choice was required between banana and beancake, that choice produced an outcome, regardless of whether or not a reward followed. That is, a chain terminated by either banana or beancake operated a dispenser. Every symbol had an effect that was stable if not desirable. In prior training, outcomes had not been predictable, because they were right-wrong judgments made by the experimenter. Thus at times lighting food symbol A produced a judgment of correct (if E were holding up food A) and a reward, and at other times lighting food symbol A produced a judgment of incorrect (if E were holding up food B) and no reward. Nothing consistent happened when choosing A versus B. Since outcomes were unpredictable, the chimpanzees could not focus on antecedent conditions, for they were continually trying to determine outcomes. The rapidity with which they focused on antecedent conditions, and learned them, once outcomes were predictable even though not always rewarding, was remarkable. It suggests strongly that the chimpanzee is attempting to perform a cause-effect analysis of each situation, and until he can discern a regular pattern of effect he continues to produce changes in his own behavior, or in the social behaviors of other animals rather than to look for conditionally related changes in the environment. We encountered this principle again and again (as will the reader throughout the remainder of this book). It proved so pervasive that, in fact, no successful training strategy could be designed which did not take it into account.

Once the chimpanzees did learn to attend to the contents of the food dispensers, what they had learned could be characterized as follows:

1. If condition A (bananas in left dispenser) is present, light the sequence "Please machine give banana."
2. If condition B (beancake in right dispenser) is present, light the sequence "Please machine give beancake."
3. If neither condition is present, get the teacher to do something about it.

The behaviors of item 3 were quite elaborate and were not taught. Whenever a dispenser was empty the chimpanzees looked directly at the teacher and requested gesturally that the teacher do something about the situation by waving his hands in large circles. If all of the prepared food had been consumed and the teacher's bowl of food were empty, the chimpanzee would look and gesture, with more hand waving, toward the refrigerator and then back toward the teacher to suggest that the teacher go and refill the food bowls. Gestural behaviors of this sort did not appear slowly, in a trial and error fashion suggestive of shaping. Instead, they appeared suddenly, and as the chimpanzees learned how to obtain food reliably from the machine they became very interested in getting the teachers to put food reliably into the machine. Their gestures served this purpose quite well; we understood what they wanted and we complied. The more we complied, the more frequently they asked us to do things. They even attempted gesturally to convey which food was to be loaded into the dispensers by waving their hands toward the bowl which contained the food that they wanted to have loaded on that trial. However, we often had to deny this request since it interfered with the experimental need to load the foods on a random basis.[2]

Our goal, of course, was not to teach chimpanzees how to operate a number of different vending devices in order to obtain food, but to teach them that symbols could be used communicatively to refer to or represent things both present and absent. In order to achieve that goal it was necessary to teach a greater variety of symbols. We also hoped to go one step further and, at some point, teach our chimpanzees to use symbols to communicate ideas and information to each other that could not be conveyed by their primitive attention-directing gestures.

Unconfounding Food Type with Dispenser

Once the chimpanzees could request banana and beancake from separate dispensers, we removed one of the dispensers and alternately loaded pieces of banana or beancake in the remaining one. If the chimpanzees requested banana when bananas were in the dispenser, they got a piece of banana. If they requested beancake when bananas were in the dispenser, nothing happened. We thus removed the redundant cue of food type plus location, and focused attention solely on food type. Additionally, we removed the consequence of an incorrect response. When the chimpanzee selected the incorrect symbols, nothing happened. The empty dispenser did not turn. The older chimpanzees now handled this change adroitly. Although their performance dropped a little at first, they did not revert to their earlier behaviors of chaining both food names in a "banana-beancake" request, nor did they evidence frustration and cease responding. When incorrect, they altered their response on the following trial. Given only two food alternatives, they were always correct on trial two. Their performance steadily increased until they could, on trial one, accurately request either food that was placed in the dispenser. Other foods were then rapidly added to their vocabulary.

When we alternated between putting beancake and bananas in the same dispenser, the two younger animals experienced more difficulty than the older animals. Errors of chaining both food names together reappeared. If the dispenser sat idly on trials when the wrong food symbol had been chosen, Austin and Kenton seemed

Figure 5.1. Having just requested and consumed a piece of orange (with the orange lexigram still illuminated on the keyboard), Sherman carefully checks the dispenser before selecting the next lexigram. The food about to be vended, if the correct lexigram is selected, is beancake.

unable to sort out why their symbols worked on some occasions but not on others. Frustration responses and guessing behavior, in which the chimpanzees lighted keys without even looking at the symbols, reappeared.

We then made a number of changes to reinstate the previously successful component of "separate consequences" for each response and also broadened the scope of possible outcomes. First we went back to placing bananas and beancake in separate dispensers. We then arranged for each to provide one or two other kinds of food. The dispenser on the right was loaded with beancake or oranges, while the one on the left was loaded with bananas or chow (see Fig. 5.1). Finally, we introduced two new dispensers. Each would dispense one of two liquids if the sequence "Please machine pour (liquid name)," were executed and if the vending device had been filled with the named liquid. Each liquid vending device was also given a particular tone. This meant that if oranges were loaded into the left dispenser and the chimpanzee requested any of the other three foods, the empty dispenser on the right beeped and turned. If the chimpanzee chained together two food names nothing happened. Under these conditions the chaining together of separate food names rapidly dropped out. The younger chimpanzees' attitudes altered and their performance rapidly improved.

For these younger animals it seemed important to make the task more complex without violating their previous expectancies. Presumably, they had attended more closely to the positional cue of dispenser location than the older chimpanzees and when this was eliminated they could not adjust. However, when this cue was kept intact (and rendered irrelevant),[3] they were able to adjust without having to negate what they had previously learned. We soon became able to eliminate these separate dispensers with their tonal cues and put all of the different foods into a single dispenser with no auditory cue.

Delayed Effect of Errorless Training

The previous extensive errorless practice with each of these individual foods and drinks seemed at this point to pay off, for we were able to reintroduce these foods and drinks rapidly to the chimpanzees. As they were reintroduced, each additional food symbol remained active, thus presenting increasing possibilities for errors among food symbols. In one month, the chimpanzees moved from appropriately requesting just two foods (banana and beancake) to requesting eight different foods and drinks from a common dispenser. Accuracy rates dipped with the addition of each food name, but rose again rapidly to over 95 percent and remained there until another food name was introduced. At this point the performance differences between the older and younger animals ceased to exist. The chimpanzees now watched to see which of these eight foods or drinks were loaded into their dispensers and requested each food correctly on trial one.

All four chimpanzee subjects demonstrated such keen attention to food type and symbol selection at this point that we decided to reintroduce the three foods that had given them such difficulty earlier. At this time, they had not seen the three symbols M&M,

juice, or sweet potato, for over six months nor had they ever used any of these symbols in a chain.

By this point, all of the chimpanzees would happily work alone inside a room (with the door open) for 15 to 20 minutes. Accordingly, it was not difficult to administer their first major test with appropriate blind controls. The teacher sat outside the room and randomly loaded the different foods and drinks into the dispenser. All of the previously mastered food symbols were activated along with the ABC and ABD chains, some new irrelevant symbols, and the three symbols: juice, M&M, and sweet potato. When the chimpanzees were eagerly engaged in requesting the eight foods and drinks they knew, the teacher placed a new food (M&M, juice, or sweet potato) in the dispenser. In order to perform appropriately during this test, the chimpanzees had to integrate their newly acquired concepts of the dispenser's function, one-to-one correspondence between symbols, foods, and the use of chains with symbols that had never previously been adequately mastered. All four chimpanzees did remarkably well on this test. As shown in table 5.2, Erika made no errors at all. The results of this test were most encouraging, for they revealed that these chimpanzees had considerable ability to integrate past experiences with new ones in a productive way without additional training. The test also revealed that the *same* symbols that had previously resulted in failure were now used easily in response to the appropriate external cues. Moreover, not only could these symbols be used appropriately, they also were not used inappropriately, or linked together as "M&M-juice." Since the chimpanzees had experienced *no* further training with these symbols beyond their previous failures, they clearly had learned more in the previous task than their performance revealed.

Even though the chimpanzees now demonstrated clear and reliable food-symbol associations, many readers may find themselves tempted to conclude that this is still basically a mechanistic sort of performance with little relevance to language. It could be argued (and has been) that these responses are tied to a computer, and bear little resemblance to what passes for symbolic communication among members of our own species, even very young ones. Food that comes out of a complex vending machine cannot make for a social transaction.

Table 5.2. Test of chimpanzees' ability to use M&M, sweet potato, and juice lexigrams in a completely new context where stimulus-dependent choice and stimulus-dependent combinations are required, but have not been taught.

Chimpanzee	*Food or drink*	*Errors before correct response*	*Errors after correct response*
Erika	Juice	0	0
	M&M	0	0
	Sweet potato	0	0
Austin	Juice	0	0
	M&M	21	0
	Sweet potato	6	0
Sherman	Juice	2	0
	M&M	6	0
	Sweet potato	16	0
Kenton	Juice	28	0
	M&M	3	0
	Sweet potato	6	0

Readers of such persuasion might recall that one of the advantages of using a machine to teach the *initial* skill of associating specific symbols with specific foods (and thereby developing a concomitant concept of one-to-one correspondence), resides in the reliability and predictability of the machine. It does not inadvertently frown or change its visage if the chimpanzee makes an error. Further, it cannot be perceived as an arbitrary and emotive judge who whimsically decides to give or not to give a chimpanzee food. Young human infants with large cranial capacities may have little trouble sorting out the vagaries and foibles of their caretakers' responses when learning to communicate naturally. Common chimpanzees, on the other hand, seem to need more consistency in their worlds, at least at the start, in order to develop a similar sorting process.

Another reason to start symbol training with a machine has to do with the fact that it does not take a young chimpanzee long to realize that he can communicate his desires for food to his human companions through gestures, glances, and whimpers. He is therefore not eager to learn something so difficult as a symbol when he can get his message across so directly and easily through the use of primitive attention-directing gestures. It is hard for the chimpanzee to refrain from blaming a teacher who will not give him the food he wants simply because he did not touch the right key. A machine, however, knows nothing of the gestures, glances, and whimpers of chimpanzees. If it does not dispense food, there is only one thing the chimpanzee can do: search for a way to make it operate properly. He certainly cannot change its mind by pleading with it. Moreover, it is apparent that at a very fundamental level, the chimpanzee knows this. In this sense, the machine forces the chimpanzee to analyze and change its own behavior, and to avoid attributing failure to the arbitrary decisions of other animates.

The central question is not really whether the chimpanzees' initial dispenser training was too mechanistic. It is, rather, whether the chimpanzees could now learn to use these associations not just to control a machine, but to control the behaviors of other apes or their teachers in complex social situations. Furthermore, would these associations allow them to communicate things that go beyond the communicative capacities of primitive attention-directing devices and the dependency of such behaviors upon the external context?

From Operating Machines to Controlling People

During one training session, instead of gesturing for the experimenter to go to the refrigerator and take some food out when the bowls became empty, Erika lit the symbol "orange," and stared directly at the refrigerator. Noting this, the experimenter selected orange slices from the refrigerator and loaded them into the dispenser. As soon as Erika had finished requesting and consuming the orange slices, she again looked toward the refrigerator and this time said "banana." Again the experimenter fulfilled the request by selecting bananas from the refrigerator and placing them in the dispenser.

Was Erika using her symbols to request that we select a particular food from the refrigerator? It seemed so; but we had no means of clearly determining whether that was indeed what she was doing. When we tried to let her go to the refrigerator after such a spontaneous request, to see if she selected the food that she had encoded, she became highly excited and distracted. When we opened the refrigerator she would grab the first large piece of food that caught her eye and hurry away to eat it. If we attempted to test her comprehension of the food which she had spontaneously requested by getting her the wrong food (for example bringing bananas instead of oranges) she seemed perfectly happy with the bananas and would ask for them straight away once they were loaded into the dispenser. Yet Erika no longer limited her requests for specific foods to times when all of the food dispensers were empty.

Erika's use of food symbols when no food was loaded into the machine contrasted with her earlier insensitivity to the state of the dispenser. Previously, she hadn't even looked at the dispenser before she began hitting keys. After responding, she looked expectantly to the slot where food was dispensed, obviously waiting for the empty machine to provide food. Now, she swiftly checked the dispenser and upon finding it empty solicited the experimenter's attention. Having gained this, she lighted a symbol and directed the experimenter's gaze, by glance and gesture, toward the refrigerator. She paid no attention to the machine's vending slot, showing no expectancy that her symbol use would cause anything to come out. She did continue to watch the person and wait for the response.

Even though we could not determine unequivocally that Erika was using her symbols to ask us for a specific food that she had in mind, it did seem clear that she was now using symbols where in the past she had only used gestures. The gestures conveyed a general sort of information (get food out of the refrigerator, put it in the dispenser) in a highly structured and ritualized context where the teacher's role was clearly defined and known to both parties. The gestures did not tell the teacher which food, but they did serve as an encouragement from Erika to go ahead and carry out the actions as expected. Erika's use of a specific symbol, however—intentional or not—clearly functioned to orient the experimenter's attention quickly and directly toward a certain food. Erika's failure to ever request chow or water (low preference items) in this context suggests that her behavior was not capricious.

We noted that Erika's tendency to communicate spontaneously with food symbols was most prevalent when the experimenter was slow in executing some well known part of the routine, such as going to the refrigerator to refill an empty bowl. We therefore intentionally acted slowly in similar situations with the other chimpanzees. We found that they also would occasionally light specific food symbols when there was no food in the dispenser and accompany these symbols with gestures which clarified their intent. As with Erika, such requests were limited to favorite foods and were, at this point, still sporadic.

Once it became clear that the chimpanzees would try to use symbols to communicate with us in the circumstances just described, we decided to ask whether they would do so in other situations. We sat beside them at the keyboard and held up one bowl of food at a time to see if they could simply use their symbols to ask us for the food we showed them. They could! In fact, it was as easy for them to use their keyboard to ask the experimenter for food as it was to request food from the machine. They would ask for any food we held up and if they should happen to make an error they would readily correct themselves. They also spontaneously used pointing gestures to show us which food they wished to have held up as we sat next to them with eight separate bowls. They appreciated our compliance with their gestural selection, but if we chose not to comply, they readily asked for the food we selected with no decrement in their overall accuracy of 95 percent or better.

Symbols for "Teacher" and "Machine"

Since the chimpanzees requested food so readily, either from the teacher or from the machine, we decided to attempt to get them to encode this animate-inanimate distinction with the chains "Please machine give [or pour] [food name]." versus "Please teacher's name give [or pour] [food name]." Although the chimpanzees evidenced keen awareness of the food's location (in the machine or in the teacher's hand) they had difficulty inserting the relevant terms appropriately into their well learned ABC-D chains. If they made an error (for example, requesting "Please *machine* give beancake." when they should have said "Please *Sue* give beancake"), they often attempted to correct it by inserting a different food name in place of beancake (e.g., "Please machine give *banana.*"). Less frequently, they would omit some portion of the chain (usually the newest portion—"machine" or the teacher's name), or invert chain elements. At this point there were several hundred different errors the chimpanzees could potentially make—a number so large that it was no longer possible for them to sort through all the alternatives in order to determine what was the correct or "expected response." When the chimpanzees failed after three or four attempts, they would stop responding unless the teacher showed them the correct sequence. Hence, we became hesitant to continue requiring chains of lexigrams.

Chains had been used initially simply because similar chains had always been required of Lana and it was believed that they could serve to familiarize the chimpanzees with simple syntactical forms even if the chimpanzees did not initially comprehend their function. At this point, however, we decided to abandon chains entirely, since they clearly compounded the types of errors the chimpanzees could make.

From this point forward the teachers accepted any and all relevant communications that they could understand or intuit. They did so in the same spirit in which a human mother accepts questionable utterances from her child. If, for example, a chimpanzee said "Coke" and looked at the Coke which we were drinking, we did not ask "? What Sherman want." or provide a model for what the chimpanzee was to say such as "? Sue give Coke to Sherman." (as was done with Washoe, Nim, and Lana in attempts to get them to specify agent, action, and object). Instead we provided a model through answers such as "Yes, Sue give Sherman Coke." and then transferred the Coke.

At times, during specific training tasks, we did try using some of the above questions in an attempt to get the chimpanzees to encode agents or objects. These attempts were always to no avail as long as we structured them and we decided *what* the chimpanzee should encode instead of allowing the chimpanzee to encode what he saw fit. Later, we were to learn that the chimpanzees could and would encode a variety of things if we did *not* prod them with questions or insist they model their symbol use after ours (Greenfield and Savage-Rumbaugh 1984). The key was to allow them to encode those elements of a situation which were salient for them. Often this required manipulating the situation so that things not normally noticed by the chimpanzees would become critical elements to communicate to others.

Chapter 6

The Functions of Symbols: Requesting, Naming, Comprehending

A Language of Requests

Once our chimpanzees learned to use specific symbols to ask for specific foods, their rate of progress in mastering new symbols improved markedly. It was also gratifying to observe that they required no additional training to use symbols with their teachers as readily as they had come to do so with the computer. Pleased with their progress, we introduced some nonfood requests such as "tickle" and "out." These requests were learned rapidly. We therefore felt confident that, at this stage of their training, our chimpanzees were able to acquire different kinds of symbols without returning to "computer-exact" responses.

Yet we were still encountering some distinct training difficulties. For one, the chimpanzees were not learning symbols for things that they had little or no interest in requesting. They always wanted to be tickled, and they always wanted to go out, thus these symbols could and would be employed repeatedly. However, the chimpanzees did not want to play with the same objects over and over and consequently requests for particular objects (such as a blanket or ball) were not readily elicited or easily recalled across sessions without prompting. It seemed that each day they needed to be reminded of the names of these objects. Once reminded, they would use them accurately for that session, but two days later they would forget. We did not encounter any such forgetting of food names or "tickle" and "out," even if these symbols were not used for months. A second problem lay in the request-oriented nature of their use of symbols. As long as they wanted something, they could reliably recall which symbol to use to request it, but they did not use any symbol for purposes other than requesting things. It could be said that they had only one "speech act" (Searle, 1969), and they were completely dependent on human beings for the communicative success of that speech act. Without their teachers to fulfill their requests, their symbol usage could not go far, since they showed little inclination to attend to or to fulfill each other's requests.

As has been discussed in previous chapters, knowing how to use a symbol such as "banana" in order to get someone to give you a banana is not equivalent to knowing that "banana" stands for or represents banana. Let us put aside, for the moment, the philosophical issue of exactly what it means for an innocuous scribble to stand for or represent something to which it bears no resemblance. It is still clear that something about a symbol demands that the user be able to divorce the symbol from its referent and that the user should not expect that every use of a symbol will produce the immediate appearance of the referent.

Searle (1969) pointed out that human speakers put symbols to many different uses (requests, commands, questions, statements) and that they can use these same symbols to accomplish all the different communicative functions which they are capable, as human beings, of conceptualizing. A chimpanzee who is capable, as human beings are, of using a given symbol for only one communicative function cannot be said to have an understanding of the symbol apart from its immediate benefit to him. Given that state of affairs, there is no basis for arguing that he knows that such a symbol stands for or represents a thing. That is, he may know how to use "M&M" or "Coke" to get these foods, without knowing why it is that the production of these symbols produces these foods. Only the human recipient of the chimpanzee's communication knows that different symbols stand for different foods, and only the human recipient will search through a number of foods at the request of another individual (in this case, a chimpanzee) and find the requested food.

The Other Half of the Speech Act: The Listener

Why is this? The chimpanzees had not been taught anything regarding the role of a listener, nor were they spontaneously inclined to behave as competent listeners. For a speech act such as a "request" to be effective, there must be a *listener* who can decode the symbolic behavior and respond to it in a reliable and cooperative manner. To do so, the listener must be capable of more than simple requests. That is, if the listener, like the speaker, expects every utterance of M&M to be associated with getting to eat an M&M, then when another individual says M&M, the listener will not respond as though this speech act is a request for the giving of M&Ms, but will simply view it as an occasion for eating M&Ms should they become available.

The development of cooperative decoding and responding on the part of the listener is what lends the speech act meaning; it is the constitution of rules, but rules of a very special kind because they are interindividual rules (Searle 1969). No single party can execute such rules, only two individuals interacting in a speaker-listener episode can do so.

The Emergence of Naming out of Requesting

Because the chimpanzees would need to be able to divorce their symbol usage from the immediate receipt of food if they were to engage in speaker-listener episodes, it was decided to produce that separation by asking the chimpanzee to name foods without giving them the food to consume immediately thereafter.[1]

We began by asking Sherman and Austin to name foods which we held up as we asked at the keyboard, "?What this." We did not expect that they would initially understand the question simply because we bothered to ask it using the keyboard. We tried to make use of nonverbal behaviors to differentiate this new question from the request situation in which we held a food out toward them with an offering gesture and asked "?Give." When "?What this." was used, we displayed the food with a showing gesture (as opposed to an offering gesture), and held it closer to our bodies to indicate our intent to retain possession of the to-be-named food. We showed the chimpanzees a second food which they could have if they named the displayed food correctly. Thus, while they would be rewarded for correctly naming a food, they would not get to consume that particular food. In so doing, they would have to maintain a one-to-one correspondence between a given food and a given symbol without the expectation that using symbol *A* would produce food *a*.

Could they do this? The answer is ambiguous. The chimpanzees began by selecting the correct symbol for each food which we held up. After having lit the symbol they then reached for the food with the clear expectancy that it would be given to them. When they were praised but given another food, they often appeared surprised and distracted. They continued to gesture toward the named food even after consuming their reward. When that food was put away, they seemed to assume that they had used the incorrect symbol to request it. They would often light another symbol even before the next trial and look directly at the food which had been presented on the previous trial (which the experimenter had placed to one side) as though attempting to "correct" their error by using another symbol. When we still did not give them that food they seemed even more upset. They were *not* frustrated because they were not getting to eat; the food reward they were given was always a desirable food—indeed, often more desirable than the food they were being asked to name. Rather they were upset because their expectancies were violated.

When, after the first few trials, it became evident that they were not going to get to eat the particular food which the experimenter held up, their performance began to disintegrate. They could not accurately select different symbols from trial to trial when shown different foods if the results of selecting these different symbols were always the same; that is, if they received social praise and a single food common to all trials.

In changing from a requesting to a naming procedure, we essentially moved from a paradigm which stressed differential outcomes to one which stressed a common outcome. However, the stimulus-response demands of the task remained unchanged. A given food was shown and the chimpanzees were to select the symbol they had always selected when shown that particular food. It would therefore appear that it was their expectancies which caused the performance decrement, not the stimulus-response link itself. This interpretation is bolstered by the fact that for the first 10 to 15 trials of this task, the chimpanzees' performance was quite high, nearly 100 percent. It was only with continued presentations that unfulfilled expectancies became more disruptive, and performance began to deteriorate.

In order to get things back on track, we returned to the original request paradigm. It took only three to four trials for the chimpanzees to realize that they were again going to get to eat the food we were holding out; their performance returned immediately to previous levels. When we returned to the naming task, we again observed high initial performance which gave way again after 15 to 20 trials to rapid deterioration. We went through this cycle for several weeks and made no progress. In this and in numerous other situations in which chimpanzees initially do well on a task, and then begin to experience difficulty, we believe that the problem is not one of a lack of ability on the part of the chimpanzee, but of differing expectancies between experimenter and subject.

Plastic Foods: The Appearance of Differential Outcomes

Since the distinction between naming and requesting was, in essence, a procedural and theoretical distinction we recognized but the chimpanzees did not, we decided to attempt to make the distinction more concrete from the chimpanzee's point of view. One means of so doing was to make each piece of the to-be-named food quite obviously inedible. Toward this end, we coated pieces of food with a thick layer of plastic resin (something like what one finds in the windows outside many restaurants in Japan). Thus the chimpanzees could clearly see which food was being displayed on any given trial. In fact, we allowed them to touch, mouth, and smell these plastic-encased bits of food. From their initial inspection, it was clear that our chimpanzees definitely did not regard these pieces of food as something they would wish to eat. A single attempt to bite through the plastic usually convinced them of that.

The chimpanzees had no difficulty in recognizing which food was which. This was evident from the fact that the plastic encased foods which they sought to inspect most closely were the bananas and M&Ms, while they paid little attention to the carrots and chow.

It was our hope that, if the chimpanzees knew that they could not eat the piece of plastic food the teacher held up while asking "?What's that" (on the keyboard) then their expectancy that they would be allowed to consume the displayed food would not interfere with their ability to light the correct symbol when shown each food. Unfortunately, our expectations did not seem to be reasonable to Sherman and Austin. Again, they followed the pattern of doing well for the first few trials. Then they began to make numerous errors. When they received another food for having correctly named the plastic-encased food, they still reached for the plastic food as though they expected to receive it. When the experimenter handed it to them, they bit down on it as though to eat it. When this did not work they handed the piece of food back to the experimenter and pushed on the experimenter's hands as if to make a gestural request that the experimenter somehow open the food for them. When this did not work, they looked at the experimenter and looked toward the refrigerator as though they expected the experimenter to obtain for them a "good" piece of that food for them. When this also did not work, their performance deteriorated and they became uninterested in the task. Thus, even though they did not treat these plastic-coated foods as edible objects, they nevertheless seemed to expect that these foods would or could become edible or that the teacher would make them so if they lit the correct symbol.

Speculating that the chimpanzees simply perceived their teachers as individuals who were being extremely uncooperative, we decided to load these pieces of plastic-covered foods into the vending devices we used earlier. Recall that these vending devices had previously successfully counteracted the chimpanzees' view that we were arbitrarily deciding to give them food on some trials and not on others. By objectifying the situation once again, and letting the machine vend these plastic inedible foods (along with an edible reward food common to all trials), perhaps the chimpanzees would stop looking to the teacher to make the food edible, and would simply name it. In this paradigm we still posed the question, "?What that" prior to each trial, but did so from outside the room using the experimenters' keyboard. This caused the question to appear on the chimpanzees' projectors.

This time the vending devices did not seem to help. The chimpanzees acted as if they expected the teacher to do something about the fact that the food was not edible. When Sherman answered the first question correctly upon seeing M&Ms in the dispenser, the plastic-coated M&Ms and a piece of banana were vended to him. He quickly ate the banana, picked up the plastic-covered M&M and brought it out of the room where he showed it to the teacher, put it on the floor, stomped on it, and looked at the teacher. Sherman and Austin exhibit such object-showing behavior infrequently. Indeed, this was the first unequivocal instance of such behavior in Sherman. He acted similarly and also showed that he was quite unhappy with the plastic-covered foods for the next six trials. After that he refused to respond. Austin's reaction was not so dramatic. He simply dropped the plastic-covered food to the floor and looked at the teacher. He continued to respond, but his performance fell to a chance level after the first ten trials.

Procedural Fading: Turning Requesting Into Naming

These unsuccessful attempts to train Sherman and Austin to name foods led us to try a new approach. We began a fading procedure in which the teacher held up the to-be-named food and asked "?What that." When the chimpanzees answered correctly we gave them a small bite of the displayed food, then praised them profusely (patted them, hugged them, played with them). We then gave them a sizable portion of the reward food located outside the room by pressing a hand-held button that caused a vending device to operate.[2] By taking the food reward out of the room and placing it in the machine, we decreased the possibility that the chimpanzee would interpret his symbol selection as representative of the reward food.

This procedure allowed us to fade from one type of reward to another, while maintaining the multiple learned associations between specific foods and specific symbols.

The bites of the displayed or to-be-named food became smaller and smaller until finally they were inconsequential and the chimpanzees did not even bother to eat them. We then ceased even offering bites of the named food. During this fading training, we asked the chimpanzee to name only three foods: M&M, banana, and sweet potato.

This worked readily, quickly, and easily. It required only 102 trials before Sherman could name each of these three foods accurately across 30 consecutive trials without error. Austin required 201 trials to reach the same level of competence.

Spontaneous Generalization to New Symbols

Our success with M&M, sweet potato, and beancake spurred us on to see if Sherman and Austin could similarly name other foods. When the additional foods and drinks (chow, juice, Coke, milk, orange, orange drink, banana) were first presented to Sherman and Austin, both chimpanzees named them correctly on 21 of 21 trials with no additional "naming" training on these items.

Each of these items was presented at least three times in a random series of 100 trials that also contained the three previously trained foods. The test was administered with controls to prevent cueing. The experimenter stood outside the chimpanzees' room and held up the foods or drinks one at a time while asking "? What this." This question appeared on the chimpanzees' projectors. Although the chimpanzees could see the items, the experimenters stood behind a blind and could not see the chimpanzees either as they looked at the foods or drinks or as they selected the correct lexigram. Since the chimpanzees' lexigram responses appeared on the experimenters' projectors outside the room, the experimenters did know when and how the chimpanzees responded.

Sherman and Austin's high levels of performance on this test showed that they could use lexigrams learned in a request-based task to label foods without specifically being taught to do so, once they understood the contingencies of the labeling task. It is also of some interest that while Sherman and Austin required only 102 and 201 trials respectively to be able to label foods or drinks, Lana required 1600 trials to do the same thing. Lana was taught labeling skills using only two training items and fading techniques were not used. Like Sherman and Austin, she transferred her labeling skills to other items. She did not, however, transfer on trial one. Typically she required about ten training trials per item once she mastered the "naming paradigm."

Some Instructive Differences in Naming and Requesting as Revealed by Lana's and Sherman and Austin's Errors

Once Sherman and Austin successfully negotiated the labeling test, they had no more difficulty in alternating between naming and requesting tasks as long as familiar symbols were used in each instance. Switching between these two tasks soon became a regular part of our daily procedure. Additionally, errors made during request tasks were generally similar to errors made during naming tasks. Such errors generally would occur between the two newest foods or drinks in Sherman and Austin's vocabularies or between two foods or drinks that were similar in physical appearance (for example, grape juice and Coke; both are dark liquids).

Lana's errors did not follow a pattern similar to that shown by Sherman and Austin. After labeling a food correctly, she would often be unable to recall the correct lexigram when required to use it as a request. There were also foods she could request readily but had great difficulty naming. Thus, while her general performance on both tasks was relatively high, her pattern of errors differed distinctly from that shown by Sherman and Austin. In addition, virtually all of Lana's requests for foods were inserted into long strings such as "? You give beancake to Lana in room." Likewise, her naming responses were also emitted in chains such as "Beancake name of this that's in bowl." Because of our earlier decision to drop all chaining requirements, Sherman and Austin typically used a single word (for example, "beancake") whether naming or requesting. What differentiated names from requests in their case was the context and not the response string. Perhaps the similarity of the response requirement in labeling and requesting caused Sherman and Austin to draw parallels between these two different speech acts which Lana did not. We will return to this issue in chapter 11, where we will describe specific tests which elucidated how Lana's concept of a symbol and its function differed from those of Austin and Sherman.

Our Goal of Symbolic Communication Between Apes

We were encouraged with the progress that the chimpanzees had made in learning to distinguish the request and labeling functions of a symbol. However, we still had not achieved behaviors characteristic of true speech episodes in which a listener and a speaker use symbols to control and coordinate each other's behavior in meaningful rule-bound exchanges. Such exchanges are an essential feature of human languages (Searle 1969; Shotter 1978; Skinner 1957).

The execution of such exchanges requires, at minimum, three skills which Sherman and Austin lacked. First it requires that there be a listener as well as a speaker in every exchange. The listener must be able to decode the symbol used by the speaker and relate each symbol to its particular referent. This type of association, while similar to that required of speakers, runs in a different direction. A speaker must perceive an object and produce the proper symbol to refer to that object. However, a listener must, in the present instance, perceive the symbol and then pair an object with that symbol. In short, the speaker must form object→to→symbol associations while the listener must form symbol→to→object associations. It is not clear that such bidirectional associations are spontaneously formed by chimpanzees, although they have been readily credited with listener competencies on the basis of extremely limited evidence (Gardner and Gardner 1978b; Premack 1976a; Terrace 1979c).

Cooperation is a second skill needed for rulebound exchanges between a speaker and a listener. The listener must be predisposed to respond to the speaker's message, preferably in the manner that the speaker intended.

The third basic requirement is that of turn-taking. Not only must speakers be able to serve as listeners and listeners be able to serve as speakers, but the episodes of speaking and listening must be arranged and agreed to by the participants. Without the three minimum requirements listed above, it is not possible for two individuals to engage in a speech episode. More complex episodes with more elaborate requirements may occur, and generally do among humans, but less elaborate ones do not (Chalkley 1982).

The requirements outlined above are basically those one would expect to find in a competent listener. Human children typically become competent listeners before they become competent speakers (Benedict 1979). Additionally, they have no difficulty comprehending words which they regularly use in their own vocabularies. For that reason, relatively little attention has been given to the receptive capacities of apes. Indeed, virtually all of the data regarding the linguistic competencies of apes have dealt with the productive capacities of apes. It has also generally been assumed that if an ape can use a symbol correctly, it can also comprehend its use by others (Gardner and Gardner 1978b; Rumbaugh 1977; Terrace

1979c). Apart from data presented for Sherman and Austin (Savage-Rumbaugh, Rumbaugh, and Boysen 1978a, 1978b; Savage-Rumbaugh 1981), there has been no systematic attempt to ascertain the extent to which productive competency equals receptive competency.

Receptive Competence: Teaching a Chimpanzee To Be a Good Listener

Mothers use language to induce their infants to participate in cooperative behaviors soon after the infant has developed the motor skills to do so. With words and gestures, mothers encourage their infants to hand them objects, to go get objects, to point to things, to go to certain locations, and to stop particular activities and begin others. Children who do not become competent at such tasks within the first two years of life are generally labeled retarded or autistic. In fact, it is usually the degree to which a child's receptive and cooperative abilities develop that determines whether he remains at home. Thus, children of normal intelligence who cannot readily demonstrate their ability to cooperate (for example, children who have cerebral palsy and who also experience severe motor difficulty) often have to be institutionalized. A moment's thought should make clear that it is not so much the absence of the ability to speak as it is the absence of the ability to effectively *comprehend* that places a child at the most severe disadvantage.

It would thus seem absolutely critical to determine the existence of listener competencies, as well as speaker competencies, in Sherman and Austin. Given the limited vocabulary available to Sherman and Austin at this point in their training (food names, tickle, out, blanket, and key), the most straightforward way of assessing receptive competence was to turn the request task around and ask them to give to their teachers the items they had requested from us.

As far as we can tell, no one else working with apes has reported giving such a simple and straightforward test of comprehension to their subjects, although it has been successfully employed with children as young as 13 months of age. This is not that surprising. Tests of receptive competence are relatively difficult to administer to apes, since they resist giving up food or items of interest to others. Although they do share food on occasion in the wild, the recipient must generally beg and plead in order to obtain some. Even then, the donor typically provides only a small morsel.

In designing our first test of lexigram comprehension we anticipated that it would not be easy for a chimpanzee to look over a tray full of foods and hand the teacher the requested food. By virtue of our past training failures, we also knew that it was not a good idea to abruptly change response-reward expectancies. Initially we asked the chimpanzees to give us only three foods—foods of low preference: sweet potato, chow, and beancake. In return for handing the experimenter one of these foods, the chimpanzee received a coin he could use to obtain any of a number of more delectable foods (M&M, banana, orange, etc.) from a nearby vending device.

The experimenter sat by the keyboard facing the chimpanzee, with the three foods placed on a table between them. The experimenter would then operate the keyboard to say "give sweet potato" and hold her hand out palm-up toward the chimpanzee and look at the foods on the table. The experimenter was careful not to gesture toward a specific food but instead to emphasize the food symbol which she had lit on the keyboard. Initially, the chimpanzee's response was to look at the teacher, repeat the request, and then wait for the teacher to hand *him* the food.

Clearly, the chimpanzees did not interpret our request as a request. The only way the teacher was able to convey to Sherman and Austin that she wanted them to hand her food was to pick up a piece of food and place it in their hands. They would then start to eat the food and the teacher would have to tell them "no" very firmly and hold her hand out in a request gesture. They would then slowly and hesitantly give her the food they were holding. Once she received the food, she would praise them, eat the food happily, and give them a coin, so that they could "buy" a more preferable food for themselves. It was necessary to repeat this procedure eight to ten times before Sherman and Austin began to understand that they were to give the food to the teacher.

Eventually Sherman and Austin began to pick up the food of their own accord and offer it to the teacher. They then quickly held their hands out to receive their pieces of money, and if the teacher were slow in giving them, or appeared to forget, they would point to the pile of coins to remind her. In fact, once Sherman and Austin caught on to the basic idea, they handed us food more rapidly than we could eat it.

Unfortunately, handing us any food in order to get paid was not exactly what we had in mind, as the foods they handed us had no clear correspondence to the food we requested at the keyboard on any given trial. Additionally, they did not seem very attentive to our requests. They handed us foods just as frequently before we asked for them and after we asked for them, as when we asked for them.

It did not help to emphasize which food we wanted by pointing to the lighted symbol. It did help to point directly to the food. Once Sherman and Austin understood that we were interested in receiving specific foods, we thought that it would be effective to substitute a lexigram such as "bread" for pointing to the bread. This, however, was not the case. While the chimpanzees occasionally responded to our keyboard requests by giving us the correct food, they showed no consistent ability to do so.

The problems Sherman and Austin experienced in giving us the foods we requested suggested that the composite task of decoding our symbol productions and searching through a group of foods was too difficult. Hence we decided to design a different receptive task, one in which they had to comprehend the teacher's message in order to engage in an action that was directly beneficial to them. Giving us a specific food benefited them only indirectly. Even though they received money when they gave the correct food, there was no intrinsically obvious reason for them to care *which* food they selected.

Comprehending Without Giving

Since they had already learned that, if the teacher entered the room with one or more foods, they could ask for the food using the appropriate lexigram we altered this situation by entering the room with food *hidden* in a container. Just before we entered the room we allowed the chimpanzees to watch us take this container to the refrigerator, where we loaded it full of food. We made this situation as salient as possible by making food barks in the process. The angle of the refrigerator door was such that the chimpanzees could not tell which food we had selected. The experimenter opened the refrigerator door just enough to get a piece of food and made certain that Sherman and Austin were watching closely the entire time (see figure 6.1). They showed keen interest in this process and positioned themselves so as to get the best view possible.

The first time the teacher entered Sherman's room with a container freshly filled with food, Sherman rushed over and tried to smell the container to determine its contents. When he could not, he gestured for the teacher to open the container. Instead of responding to Sherman's gesture, the teacher used the keyboard to say "This chow." After observing this, Sherman immediately said "Open chow" and held his hand out toward the closed container. The teacher complied and Sherman quickly consumed the chow.

The teacher then returned to the refrigerator and selected sweet potato on trial 2, followed by banana and beancake on trials 3 and 4. When the teacher stated which food she had hidden in the

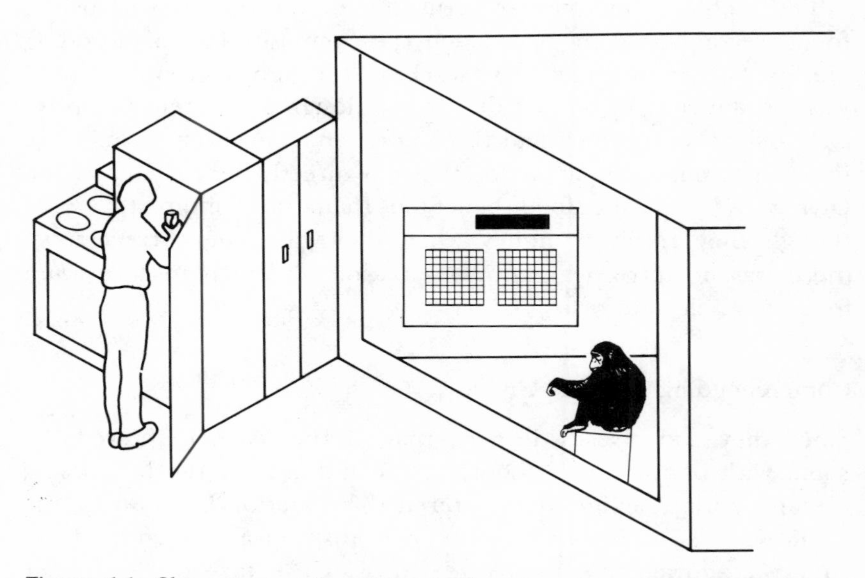

Figure 6.1. Sherman watches the experimenter go to the refrigerator and load a container with food. He cannot tell which food is being placed in the container, but he can tell that food is being put into the container for him.

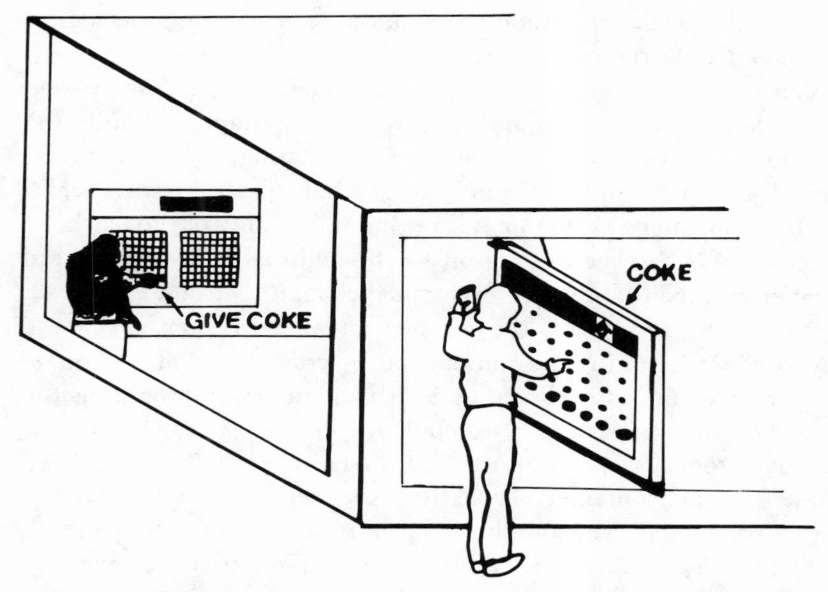

Figure 6.2. The chimpanzee looks at the symbol on his projectors which describes the type of food that has been hidden in the container.

container, Sherman immediately used the keyboard to ask for that food, holding his hand out toward the container.

On the fifth trial, the teacher loaded the container with one of Sherman's most favorite foods, M&Ms. The teacher identified the food at the keyboard by saying "This M&M." Sherman looked at the teacher and said "Give banana." The teacher said "No" and then repeated, "This M&M." Sherman observed this and said "Give bread." The teacher replied "No, this M&Ms." Sherman again observed the teacher's statement and said "Give banana." The teacher tried once more saying "No banana. This M&M." This time Sherman looked at the teacher and said "Out." Thinking that perhaps Sherman was asking to go out simply because he was failing in his attempts to request the food, she turned him back around toward the keyboard and said very emphatically "This M&M." Sherman looked directly at the teacher, looked at her keyboard statement and then said "Open sweet potato. Open." The teacher was at a loss as to how to convey to Sherman the contents of the container. She therefore replied "Yes, open" and showed him what was inside. As soon as Sherman saw the M&Ms he immediately requested "Give M&M."

Sherman's behavior on trials 1–5 revealed that he understood this situation immediately and that he was not simply imitating the teacher's behavior. Had he only been imitating, it would surely have not been so difficult simply to get him to say "Give M&M." His behavior suggests that he was testing the validity of the teacher's statements. Whatever the reason for his unusual behavior on trial 5, Sherman made only one error in the next 15 trials.

During the first trials, the teacher sat next to Sherman at the keyboard while stating the contents of the container. Later in order to eliminate the possibility that Sherman was merely lighting the same key that the teacher selected, the teacher began using her keyboard (outside the room) to tell Sherman what food had been hidden in the container. Each time she depressed a key on her keyboard, it appeared on the projectors above the chimpanzee's keyboard.[3] Surprisingly, Sherman correctly requested the food that the teacher indicated on the projectors in his room. Moreover, the same procedures worked as well with Austin as they had with Sherman.

These were unexpected accomplishments, given Sherman and Austin's previous failures on comprehension tasks in which they were required to offer food to their teacher. Just to be certain that they were not utilizing a match-to-sample strategy, we retested Sherman and Austin on a match-to-sample task in which lexigram stimuli were present on projectors above their keyboard. Both chimpanzees failed miserably.

It is noteworthy that the procedure which informed the chimpanzees of the type of food which had been hidden in a container, and the match-to-sample task, are at the surface level very similar. When identifying a hidden food, the chimpanzee must attend to a row of projectors above his keyboard. When he sees a symbol appear on those projectors, he must then select that same symbol from among a number of others located on his keyboard (see figure 6.2). In the match-to-sample task, the procedural requirements are exactly the same. The teacher displays a symbol on the chimpanzee's projectors and the chimpanzee must find that symbol on his keyboard and light it (see figure 6.3).

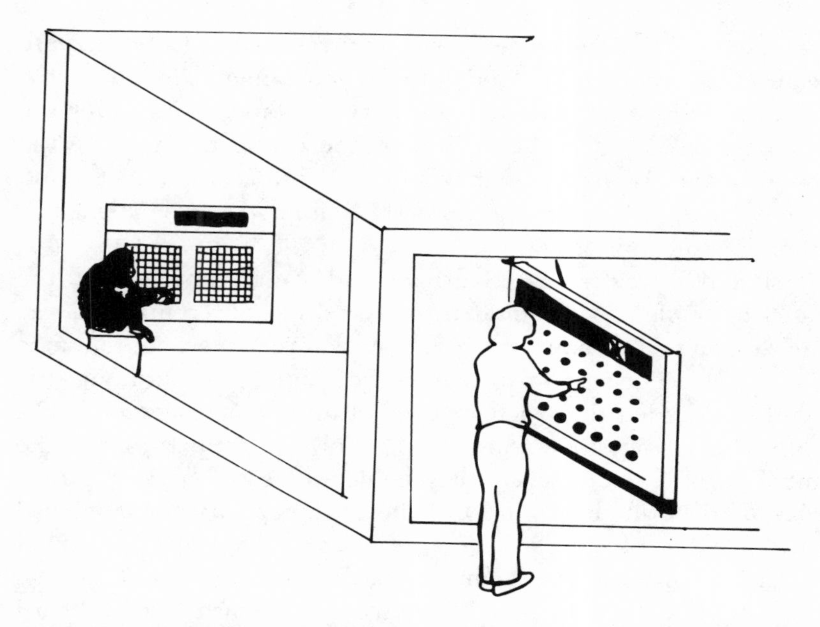

Figure 6.3. The chimpanzee looks at symbol on his projectors which he is to match by selecting the same symbol on his keyboard.

The difference between thee two situations is to be found in the communicative structure of each task. In the match-to-sample situation, there is no extra-task structure to lend meaning to the chimpanzee's symbol selection. By contrast, in the hidden-foods task there exists a large amount of extra-task information for the chimpanzee which lends communicative meaning to the situation.

Another Instance of the Limitations of a Behavioral Analysis

Sherman and Austin's inability to match lexigrams to sample, while at the same time successfully requesting hidden foods based on projected information, is but one of many counterintuitive examples which led us to look beyond the standard behavioral explanations of their performance. It became clear to us that they were not simply organisms who learned which behaviors resulted in reinforcement and which did not. In the example on hand, the teacher's use of symbols to denote the contents of a container meshed especially well with Sherman and Austin's natural desire to *know* what was in the container, and with their assumption that because the teacher had put the food into the container, she *knew* what kind of food it was. Further, there were obvious rationales for the use of symbols by the teacher and by the chimpanzees. Since the food was not visible, the symbols carried information that could not be conveyed by pointing gestures. Neither rationale applied in the match-to-sample task.

It should be clear that Sherman and Austin approached the match-to-sample and the hidden foods tasks as different problems, even though the stimulus and response requirements were the same in both tasks. Their performance on the hidden foods task suggests that they decoded the teacher's symbol into a representation of a food and that they then searched their own keyboards to find the symbol for the food which they could not see, but which they believed to be in the container. This strategy was not employed in the match-to-sample task because neither the projected symbol nor the chimpanzee's matching symbol actually referred to anything. Hence the chimpanzee had no reason to search the keyboard for a specific symbol.

If matching were employed in the hidden foods task, it would seem that the chimpanzees could employ it in an ordinary match-to-sample task, but they could not. This lends strong support to the view that they were indeed solving the hidden foods task by a representational strategy which entailed decoding and encoding from symbol to referent and back to symbol.

It may be objected that some subtle performance difference between these two tasks was overlooked and that such a difference, were it understood, could account for the disparate task performance. Procedurally, except for the hiding of the food, the paradigms were identical, as were the stimuli and responses. We saw no reason that the chimpanzees should not be able to do either task and we were, at that time, disappointed that they could not do a simple match-to-sample task with their lexigrams. We did not set out to design a study which showed that the chimpanzees had solved a communicative problem with a representational strategy and a nearly identical noncommunicative problem with a nonrepresentational strategy. We were simply trying to get the chimpanzees to look at and decode symbols. We did not, initially, view these strategies as incompatible and in fact we suspected that a match-to-sample capacity was necessary for accurate symbol decoding capabilities.

What we found however, was something quite different. Extra-task content was critical. If the extra-task structure lent meaning to our communications and the chimpanzees' communications, then these communications would be attended to and decoded accurately. Our communications transmitted information about a food that was not immediately visible, and because they did this, they were interpreted as *referring* to that food. There were indications above and beyond the obvious one of keyboard performance that such internal referral was occurring. For example, when the foods and drinks the teacher hid were highly preferred ones such as orange drink and M&Ms, the chimpanzees began to show their delight in advance of the opening of the container. For example, as soon as they saw "orange drink" on their projectors they would often grin broadly and make high-pitched food barks while searching for the lexigram on their boards. By contrast, if the lexigram was "chow" they would at times refuse to ask for it and show complete uninterest in the closed container if it were offered to them. Instead, they would look and gesture toward the refrigerator, encouraging the teacher to make another selection. These sorts of differential non-verbal reactions to various lexigrams were never seen, even in successful performance on the match-to-sample tasks which used the same stimuli.

From the Three Atomic Skills of Requesting, Naming, and Comprehending to Communication

Our digression concerning the difference between matching-to-sample and decoding a teacher's communication about a hidden food should not detract from our main focus of establishing communicative skills in Sherman and Austin. At this point in their training Austin and Sherman had mastered three simple skills which they could use with a set of symbols. They could use these symbols to request foods, they could use them to name foods (without expecting to consume them), and they could decode symbols when used by others to identify nonvisible foods. With this small range of speech act functions, could they begin to exploit the communicative power of symbols? Could they use symbols to coordinate actions between themselves and to convey to one another information that could not be conveyed by expressions, glances, and gestures? The next phase of our training program sought to provide answers to these questions.

Chapter 7

First Symbolic Communications: From Sherman to Austin, From Austin to Sherman

Symbolic Communication Between Chimpanzees

Could chimpanzees who were able to use symbols to request foods they could not see, and to comprehend information about hidden foods from the symbols of others, also use symbols simply to talk to each other? Could they ask each other to do such things as to give foods or to tickle? Although other ape-language projects had typically required apes to communicate with their teachers, there was no evidence that the apes could use their newfound abilities to enhance their communication with each other—no evidence that they communicate in ways that go beyond whatever their natural nonverbal signals might enable them to convey in contextually bound situations. Fouts (1975) and Fouts, Fouts, and Schonenfeld (1984) have reported signing between ASL-trained chimpanzees. Such chimpanzees apparently signal their desires for play, grooming, transfer of food, etc. However, these chimpanzees, as well as wild chimpanzees, are also able to convey such desires without ASL signs. Therefore, one must ask, how has their acquisition of signs broadened their ability to communicate.

Once Sherman and Austin had learned to use symbols in the different ways described in chapter 6, they did *not* spontaneously begin to communicate with one another. Sherman, seeing Austin with an orange in hand, did not, for example, stroll over to Austin and use the keyboard to say "Give orange," while extending his hand. On the contrary, if he wanted Austin's orange, he walked over and took it. If Austin wanted to keep his orange, Austin did not say "No, my orange," he ran away as fast as he could. (Which was generally fast enough to keep the orange.) Similarly, Sherman and Austin did not need the keyboard for grooming, tickling, traveling, etc. If they did not need symbols to request such activities from one another, what need might they have to communicate for which they would find symbols a necessary tool? Part of the answer is revealed by looking at studies which have focused specifically on the types of things chimpanzees without language skills can and cannot communicate.

Attempts To Determine Whether Chimpanzees Are Capable of Symbolic Communication Naturally

Menzel and Halperin (1975) conducted a series of studies on non-symbolic communication between chimpanzees. In a large open field, Menzel hid food at various locations, which were changed from trial to trial. After the food was hidden, a teacher carried one of four young chimpanzees to the place where the food had been deposited and showed the food cache to the chimpanzee. This chimpanzee was then returned to his companions who were, upon his return, all allowed in the open field enclosure along with the knowledgeable chimpanzee. Because the chimpanzees were young, none of them would wander far about the field alone. Hence, if the knowledgeable chimpanzee wanted to go and retrieve the food which he had seen hidden he had to encourage some of his buddies to go with him. The knowledgeable chimpanzee did not have much difficulty doing this since, after a few trials, all of the chimpanzees quickly understood that food was being hidden and they were ready to follow any of the group who seemed to head out in a definite direction as opposed to searching about randomly. Moreover, those chimpanzees who had not been shown the location of the food could determine it from the line of site and direction of travel of the chimpanzee who did know the location and they often set out ahead of this chimpanzee and came upon the food themselves.

The signals which the knowledgeable chimpanzee used to get the others to travel with him in a particular direction, and to allow others to overtake him and find food which they themselves had not seen, were often subtle and difficult for a human observer to discern. They were, without question, nonverbal and nonsymbolic, yet they were highly effective in establishing group coordination of travel patterns.

The study was then taken a step further. The knowledgeable chimpanzee was now allowed to spend some time with his companions once he had been shown the food cache. However, he was not allowed to accompany them into the field afterward. Hence any information that the knowledgeable chimpanzee might pass on to the others had to be transmitted in advance. As one might suspect, under these conditions, the unknowledgeable companions were unable to find the food, thereby revealing that any communication which took place about the food was of a nonsymbolic nature and required that the referent of the communication (that is, the location of the food) be present in time and space so that it could be referred to by glance or bodily orientation.

The situation which Menzel devised is intriguing because it entails many of the important components of the natural field setting which must have been faced by early hominids and which are currently faced by modern-day apes. In any primate group whose members do not always travel together (as is the case for *Pan troglodytes*) there will inevitably be individuals who have come to possess information regarding the whereabouts of valuable resources that other members of the groups do not have. For any species that retrieves and shares food, the transfer of information about a source of food would be of particular importance. Group members who did not know of a food source could be directed there and could return with food to be shared. For species without a strong tradition of sharing or carrying food, it may be sufficient to interest others in following and then to lead them to the proper location so that all may eat once they arrive.

The value of symbolic communication is thus properly seen as closely linked to social patterns (food sharing), travel patterns (home base), and general cognitive capacities (ability to work out exchange agreements in advance that are mutually beneficial to both parties). In this view, symbolic communication does not stand alone as a simple skill in conveying one's ideas and feelings. It becomes necessary to link it to a whole network of species-typical social and ecological behavior patterns, many of which become possible only because of the existence of a capacity to use symbols. The paradox is that it is the existence of such social and ecological patterns which makes symbol usage a viable activity to begin with.

The situation Menzel designed clarifies the distinction between what can be accomplished by symbolic and nonsymbolic communication. Moreover, only the simplest of symbolic skills—single words—are required to solve Menzel's second task. Syntax, metaphor, creative combinations, and all the other elaborate qualities so often mentioned in discussions of the "essence" of language would not be necessary. A few simple words would suffice, and these words, in a natural setting, could clearly have enormous survival value for the individuals who could use them.

Yet even the situation designed by Menzel requires a complex joint understanding of the world. It is necessary that the individual who has seen the food hidden know that only he knows its whereabouts. It is also necessary that he understand that others would like to have such information. Additionally, it is important that he have some awareness of the potential outcome (both for himself and for other group members) of his sharing the information about the food's location. And finally, it is necessary that the group have an agreed upon set of symbols which would be useful to transfer this information.

Moreover, any symbolic communication between two individuals is itself a complicated inferential affair which presumes intentionality (Grice 1968). For example, let us say that A has used a symbol to inform B of a certain state of affairs—or to make it possible for B to act in a way that A desires. This very act presupposes that A has information B does not, that A is aware of this, and also that A and B share a common symbol system from which A is using an element to stand for some object or set of actions B should act upon. Furthermore, it presupposes that some value *(Va)* exists for A in transmitting this information to B and that some value *(Vb)* exists for B in decoding the symbolic information and responding appropriately. Finally, it presupposes that both participants have an awareness of and intentionality regarding the nature of the symbolic communication process itself (see Savage-Rumbaugh 1984a, for a detailed operational definition of intentionality).

These are not simple matters. Many two- to three-year-old children who are fluent language users still have considerable difficulty deducing what others know and what they do not know. Young children have a tendency to presume that others see the world as they do or at least that others have access to the same external information that they have (Piaget 1954). Furthermore, children younger than two years of age, (who are already producing two- and three-word combinations and showing elementary syntax) cannot yet affirm or deny the accuracy of simple statements by others such as "It's a truck" (Pea 1982).

Shared Consequences Are Necessary for Communication

Recognizing the extensive cognitive complexities and assumptions about interindividual knowledge required to inform one's companions of the location of hidden food, we began our efforts to establish interchimpanzee communication with a rather simple task. We attempted to develop situations in which it would be obvious to Sherman and Austin that to communicate with each other would be mutually beneficial, and equally obvious that not to do so would be mutually detrimental. It did not take long to realize that "mutually beneficial" translated rather directly into "shared consequences" and that "shared consequences" translated just as directly into shared resources, the most significant of which was food.

One can distinguish between two basic types of shared consequences, those in which the participants potentially can become aware that they are sharing the consequences, and those in which they cannot. Consider, for example, a recent study by Epstein, Lanza, and Skinner (1980) involving two pigeons. One is trained to select one of three keys marked with the letters R, Y, or G. That selection serves as the stimulus for a second pigeon to select a particular color: red, yellow, or green. Both pigeons are rewarded after the second pigeon selects the correct color. Thus, the shared consequences follow the joint behaviors of these pigeons.

It is important to recognize that the procedure used by Epstein et al. does not require that either pigeon see the other at any point or that either pigeon realize that their joint behaviors produce their "shared" consequences. Indeed, their consequences are shared only in the sense that they have occurred simultaneously. The pigeons are not aware of the shared consequences of their joint actions or that reinforcement is linked to their interindividual performances.

Such "communicative interactions" are really just two individual behaviors yoked together by the experimenter. It is not possible in such a case for A to "tell" B anything (in the intentional sense). A can only act and B can only react. That both A and B must act in a certain manner for either to be correct is an external constraint. This arbitrary interlocking between the performance of the two birds is contrived by the experimenter and clearly differs from situations in which individuals know that their communications will result in shared and mutually obtainable resources, such as was the case with Menzel's chimpanzees.

Shared Consequences That Are Not Hidden from the Participants

In an intriguing study of interanimal communication with monkeys, Mason and Hollis (1962) revealed that interindividual awareness of joint consequences could be demonstrated in at least one monkey of a communicating pair under conditions designed to promote interindividual observation. Mason and Hollis constructed an apparatus in which four pairs of food carts were mounted on fixed runways on a table (see figure 7.1).

Each pair of carts was connected in such a way that movement of one cart simultaneously extended the other in the opposite direction. Handles were attached to one side of each pair of carts. The so-called "informant" monkey could see which cart was baited. The operator monkey could not. When the operator monkey pulled a cart toward himself, a mechanical coupler automatically extended the other food cart toward the informant monkey. If the informant could let the operator know which cart to choose, then both monkeys would receive food as a consequence of the operator monkey's choice, and both monkeys could clearly observe this mutually beneficial consequence.

The informant monkeys readily learned to position themselves in front of the cart filled with food and the operator monkeys readily learned to select the cart where the informant was sitting. Indeed, one of the informant monkeys began doing back flips whenever the operator monkey began to pull the correct lever (presumably in excitement over the fact that he was about to receive a portion of the food). The operator monkey observed this behavior

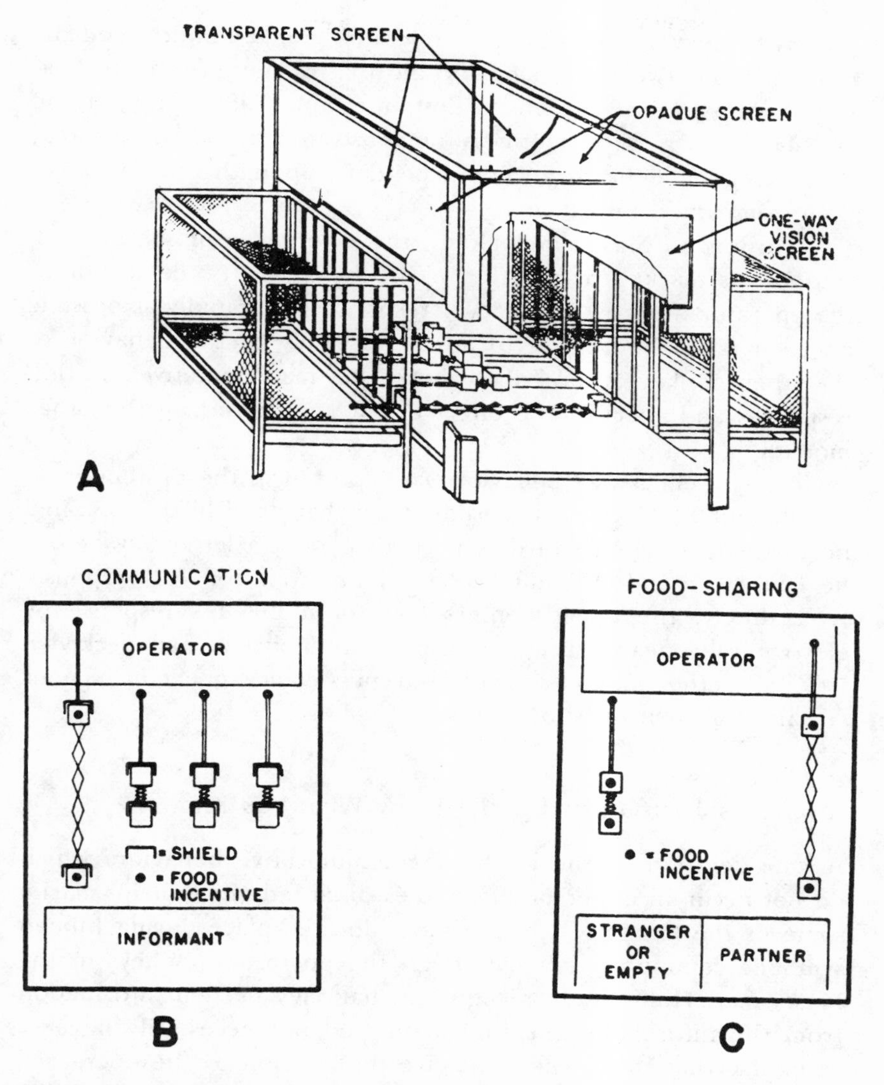

Figure 7.1. The test apparatus employed by Mason and Hollis (1962) in their study of communication of baited food containers with young rhesus monkeys.

and began to apply those observations as follows. It would touch, but not pull on, a particular lever until it observed the informant do a back flip. The operator monkey then selected that lever as the one to pull.

This strategy suggests that the operator monkey understood the relationship between his behavior and that of the informant. That is, in contrast to the pigeons in Epstein et al.'s (1980) study, Mason and Hollis's operator monkeys demonstrated an awareness that their selection behaviors were dependent upon the behaviors of their companion informants.

According to Mason and Hollis, the emergence of such a realization was marked by a rather clear change in the demeanor of the operator monkeys. Instead of responding as rapidly as possible on each trial, they began to attend carefully to the behavior of the other monkey. As described above, they made tentative selection responses and checked the effect of such responses on the other monkey.

No such "checking" behavior was reported in the Epstein et al. (1980) study. In fact, in that study, the behavior of bird B revealed no understanding that bird A needed to peck a letter key before he could respond. Initially, B would not wait for A's response. Accordingly, the experimenters had to inhibit B's response by electronically deactivating his display until bird A had pecked a letter. Whether or not signs of awareness could appear in pigeons remains an open question.

Awareness That Others Do Not Know What You Know

A consideration of the kinds of communicative behaviors which do not occur in either of the studies discussed above helps clarify some of the things which inevitably do take place during human symbolic communication. Although the operator monkeys in the Mason and Hollis study recognized that they needed information from the informants, the informants did not necessarily have to recognize that they needed to give information to the operators. Human communicators, by contrast, typically have some ideas of the types of information needed by others. We are able to make inferences regarding the state of knowledge of the listener and normally we adjust our communications accordingly.

In the Mason and Hollis study the "informants" were naïve with regard to the state of knowledge of the operator monkeys. Moreover, although the operator monkeys had learned how to utilize the behaviors of the informants, they apparently had not learned that such information was crucial to their success in the task.

Mason and Hollis cleverly deduced these facts by running several unique control conditions. In one, they placed a screen between the two monkeys. This screen could be raised by the operator monkeys in order to permit them to see the informants. They taught the operator monkeys how to raise this screen whenever they wished to do so. This training was done when the food carts were not baited. Following this training, the food carts were baited and the screen between the two monkeys was left down. If the operator monkey knew that he needed information from the informant, then he should have pulled up the screen and looked at the informant before he made his selection. None of the operator monkeys bothered to open the screen before they selected the food cart. This revealed that even though these monkeys previously had learned to attend to the informant in order to choose the correct food cart, they did not fully appreciate the importance of the informant's information. We can conclude that they had learned how to gain information from another monkey but not that they needed such information.

Subsequently, Mason and Hollis (1962) decided to teach the operator monkeys to raise the screen before they selected the food cart. Initially, they restricted access to the food carts until the operator monkey pulled the lever that raised the screen. Once they learned to do so and to look at the informant, the operator monkeys continued to raise the screen even though they were no longer prevented from selecting the carts before opening the screen. This procedure appears to have helped the operator monkeys to understand the crucial role that the informants played in their own success. After the operator monkeys had learned to raise the visual barrier between themselves and their informants, Mason and Hollis then raised the screen for the operator monkeys, but left the screen-raising levers present. Thus the operator monkeys could still pull the lever to open the screen before they made their cart choice, but they did not really need to do so. The operator monkeys did not continue to pull the lever when the informant was visible. When it was not necessary for them to do so in order to see their informant, they did not waste the effort.

Observations such as these are significant in delineating not only what a given animal can do in a particular communicative task, but also what it understands about its role and its comprehension of the causal nature of its action. Does the animal see its actions as magically successful—or does it understand how its actions affect both its inanimate environment and the behaviors of its companions? Not many studies of animal communication have asked such questions in the controlled laboratory setting though awareness and intentionality are readily attributed to animals in the field by both biologists and anthropologists.

Some Crucial Aspects of Interindividual Symbolic Communication

Before discussing the extent that Sherman and Austin learned to engage in symbolic communication with each other we wish to characterize the key features of such communication. Interindividual symbolic communication presupposes the following:

1. A attends to B's behavior and B attends to A's behavior.
2. At least one of these individuals has information the other does not.
3. Individual A is aware that she, and not B, has access to this information.
4. A's transmission of this information to individual B is of value to A.
5. Attending to the information presented by A is of value to B.
6. Both individuals share an overlapping set of experiences with regard to a common symbol system. Thus, when A uses a given symbol, it results in actions by B that A can accurately anticipate because of past experiences with that symbol and with B or with other members of the group.
7. The exchange of information between A and B is accomplished symbolically. That is, A does not demonstrate or show B the information; rather A produces a symbol which calls to mind for B particular actions or objects that are not present for demonstration.

Testing for the Requisite Skills

In order to instantiate the above features in Sherman and Austin's communicative repertoires we defined contingencies that would maximize the benefits of communicating while simultaneously minimizing the benefits of acting individually. To begin with, we changed the contingencies of the hidden foods situation (described in chapter 6) so that one of the chimpanzees, as opposed to the teacher, saw which food was hidden in the container. We then arranged to test their ability to convey symbolic information regarding this hidden food.

Before actually testing them under such conditions, however, we wanted to be sure that none of the component skills utilized in such a task would be the unfortunate result of inadvertent cueing. In such an interchimpanzee communication task, only the chimpanzees would know the contents of the container. Thus, it was essential that they be able to carry out the transmission of information entirely on their own.

Subtests to Eliminate Possible Cueing

First, we tested Sherman and Austin's ability to use information provided by the teacher. During this control test, the teacher filling the container stood outside the room and used her keyboard to describe the contents of the container. A curtain was drawn across the clear Lexan wall separating the teacher and the chimpanzee so that they could not see each other while the teacher told the chimpanzee what was in the container. Likewise, they could not see each other while the chimpanzee asked the teacher for one of the twenty different foods or drinks he had been told was in the container. The chimpanzee could see what the teacher said on the projectors above his keyboard, and the teacher could see what the chimpanzee said on the projectors above hers.

It was evident that the chimpanzees understood that the teacher was causing information about the container's contents to appear on their projectors. If the teacher delayed, the chimpanzees did not just respond haphazardly; they tapped on the wall to gain attention. When the teacher looked behind the curtain, the chimpanzees pointed toward the projectors where the symbol should have appeared. During the 20 trials of this blind test, Sherman attended to the teacher's information and asked for the correct food on all 20 occasions. Austin made only one error.

We next sought to ascertain that the chimpanzees could indeed accurately identify the contents of the container. To determine this, the teacher simply placed different foods in the container and set it down outside the chimpanzee's door, then stood behind the curtain. The chimpanzee walked over to the container, observed its contents, then returned to the keyboard to identify the food he had seen in the container. The teacher could see the selected symbol on the projectors above her keyboard, but could not see the chimpanzee as the selection was made. If the chimpanzee were correct the teacher used her keyboard to answer "yes." The chimpanzee then walked over to the container and was allowed to consume the food. During these identification tests, both chimpanzees were correct on all 20 trials.

The First Communications Between Sherman and Austin

In the chimpanzee-to-chimpanzee version of "hidden foods," one of the chimpanzees (Austin on some trials and Sherman on others) accompanied teacher A to the refrigerator and helped hide food in a container. Teacher A then closed the container very tightly and asked Sherman (for example) to carry the closed container to teacher B, who did not know which food was in the container. Teacher B then took Sherman and the food to the keyboard where Austin (who also did not know what was in the container) was waiting.

The contingencies intentionally designed into this paradigm are summarized as follows:

1. Only one chimpanzee was provided with information.
2. Both chimpanzees were required to demonstrate knowledge of the information which was given to only one.
3. Neither chimpanzee received food unless both chimpanzees produced correct requests. (Since Sherman and Austin were not predisposed to share food on their own, the teacher did the sharing, thus assuring the "joint outcome" at this point.)

In our initial test using this paradigm, Sherman and Austin alternated roles. The "informant" chimpanzee (Austin for example) was led to an adjacent room, where he observed while a container was filled with one of eleven foods or drinks: beancake, banana, chow, milk, orange drink, juice, Coke, orange, sweet potato, bread, or M&M. It took approximately one minute to walk from the room where the container was filled back to the area where Sherman waited by the keyboard. Hence, Austin had to remember what he had seen placed in the container during this interval. Even though Sherman and Austin had not received any training which required that they remember an item and respond one minute later, this delay did not prove problematic. This was unusual for chimpanzees, as they typically do very poorly on nonspatial delays of more than 15 to 30 seconds (Kohts 1923; Nissen, Riesen, and Nowlis 1938). Figure 7.2 illustrates the floor plan of the laboratory, Sherman and Austin's individual training rooms, and the path they traveled from the container filling area to the keyboard.

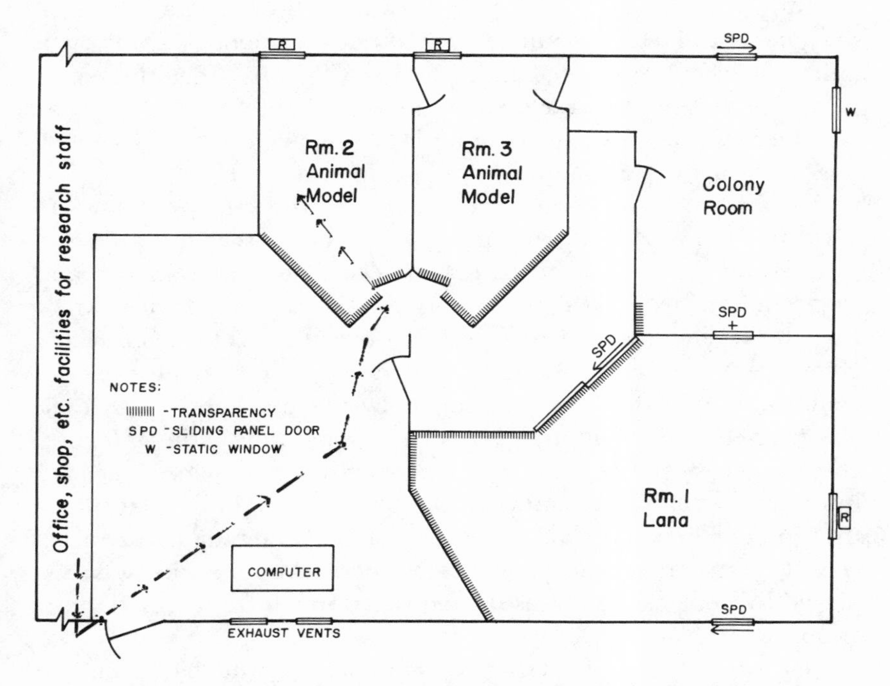

Figure 7.2. Physical layout of the Language Research Center building.

On the very first trial, Austin was the informer. Sherman seemed distracted and unsettled while Austin was out of the room, but as soon as Austin and the teacher returned with the food container Sherman became keenly attentive. He watched Austin use the keyboard and he saw Austin look expectantly at the container which the teacher held. Sherman then hurried to the keyboard and also asked for the food which he had just seen Austin request. Sherman's behavior was surprising since neither chimpanzee had imitated the other's usage of the keyboard on past occasions. It suggested that Sherman credited Austin with having some special knowledge that he needed. Since both chimpanzees requested the food correctly, the teacher divided the food between them. On trial 2, Sherman followed the teacher into the second room to watch the container being baited with food. When Sherman returned Austin also watched his keyboard response and requested the food he saw Sherman ask for. Thus both chimpanzees, from trial 1 behaved as though they understood that the symbol use of the informant served to reveal the contents of the container. (Had they been trained to imitate each other's responses, or if such imitation often occurred, we might have doubted that their performance reflected any such comprehension. However, since neither was the case, the only explanation for the behavior of the "listener" was that he was attending to the informer and patterning his response accordingly because he understood the value in doing so.)

During the first eight to ten trials, it seemed that the informer was recalling the food he had seen placed in the container and requesting it for himself. It did not appear that the informer knew that the listener needed his special information. The uninformed chimpanzee, however, was taking advantage of this information, which was coincidentally transmitted by the informer. Once each chimpanzee had been in both roles several times, the informer chimpanzee began to realize that the other chimpanzee needed information before he could use the keyboard and also that unless the other chimpanzee did use the keyboard, neither of them received any food. Comprehension of these contingencies was displayed by both chimpanzees during trials on which one or both of them erred.

For example, on one trial, the chimpanzee who had seen the container baited requested the correct food upon returning to the keyboard room. However, the uninformed chimpanzee (Austin in most cases) did not ask for the same food which Sherman had, even though he had watched Sherman closely. Apparently Austin either did not believe Sherman, or he did not want that particular food; or perhaps he was hoping that by requesting a different food, he could magically make that food appear in the container. However, once the container was opened it became clear to both chimps who had requested the correct food and who had not. Austin then tried to change his request, but to no avail since only initial requests were accepted by the teacher. In any event, both chimpanzees eagerly looked into the container, and from its contents, both chimpanzees could determine who had made the error. Following this error, Sherman began to monitor Austin's behavior most closely once he had lit the food symbol himself. In some cases, if Austin seemed hesitant or uncertain, Sherman took note and repeated his identification of the container's contents.

This experiment demonstrated that Sherman and Austin could use symbols to convey information to one another about the nature of an item that was not visually available to both of them (see table 7.1). While such exchanges did not yet have all of the fundamental characteristics of human communication noted at the beginning of the chapter, still these results were most encouraging. It was not necessary to train Austin and Sherman to use their symbol skills in this way. The important factor was the designing of a situation which permitted and encouraged interindividual communication and cooperation.

Some Additional Controls

Following this initial test, control conditions were run to eliminate the possibility that the uninformed chimp had merely learned to imitate the informed chimp. We now required the chimpanzees to use separate keyboards. Symbols were in different locations on the two keyboards, thus the chimpanzees could not respond according to position. The informer used one keyboard to identify the food he had seen placed in the container, and the uninformed chimpanzee, after watching through the window of an adjacent room, used a second keyboard to request the food. As table 7.1 reveals,

Table 7.1. Sherman and Austin's Performance on First Interspecific Communication Task

Experimental condition	*Trials (correct/total)*	*Proportion correct*
First interanimal test		
Animals use single keyboard. Experimenter knowledgeable. (Vocabulary size = 36)	33/35	.9
Control conditions		
Animals use single keyboard. Experimenters blind.	60/62	1.0
Animals use separate keyboards, observe one another through window. Experimenters blind. (Vocabulary size = 40)	36/40	.9
Animals use single keyboard, observing animal points to photograph of food following his request. Experimenters blind.	27/30	.9
Informed animal denied use of keyboard to describe contents of sealed container. Experimenters blind.	4/26	.2

this caused no difficulties for the chimpanzees, even though no training to use two boards preceded the actual test.

The most critical issue—did the uninformed chimpanzee really learn the contents of the container—was tested by asking the uninformed chimpanzee to select a photograph of the container's contents after the first chimpanzee had made his request, but before the container was opened. Again, no training was given on this task before the test. After the informed chimp had revealed the type of food in the container, three photos were placed in front of the other chimp and he was asked to point to the one which indicated the contents of the container. How did he know that this is what he was to do without any training? Since the chimpanzees had learned to use lexigram symbols to request specific foods hidden in the container, they seemed to turn naturally to the photographs as a means of doing the same thing. The experimenters facilitated the altered mode of response by gesturing toward the photos to indicate that the chimpanzee should give them one.

On the first trial of this test, Sherman (the informant) correctly identified the food in the container and three photographs were placed on the floor in front of Austin. At first Austin seemed hesitant. The teacher gestured toward the photographs, then pointed to the container and extended it toward him. Austin then picked up the photograph of juice (the drink Sherman had identified) and handed it to the teacher. On the second trial, after Austin (now the informant) had identified the food, Sherman seemed to know immediately what was expected of him. As soon as the photographs were placed in front of him, he selected the correct photo, gave it to the teacher, and pointed to the container. On all future trials, both chimpanzees quickly selected the correct photograph with no further gestural encouragement from the teacher.

In order for the uninformed chimp to correctly select a photograph of the hidden food, he would have to have received that information from the chimpanzee who had seen which food was placed in the container. If he were simply imitating symbol choice, he would not know what food was in the container. Recall that before this test neither chimpanzee had ever been taught to request food by handing the teacher photographs. Likewise, we had never asked them to pair the keyboard symbols with photographs. In fact, we had not yet used photographs with them at any point in training, though we had shown them slides and movies from time to time.

On some trials in which Sherman was the informant, he frequently had to be prevented from pointing to the correct photograph for Austin. This was because Austin tended to look over all of the photographs very slowly before making his decision. Sherman, on the other hand, always selected the correct photograph with the greatest possible speed. Sherman's eagerness to help Austin select the correct photograph interfered with our ability to test Austin's knowledge of the container's contents. Thus we told Sherman he was not to point to the photographs when they were in front of Austin. Nevertheless, Sherman's overt attempts to help Austin clearly revealed that he understood the interlocking contingencies which had been imposed on the task—i.e., that Austin had to be able to show the teacher what was in the container before she would open it. Sherman and Austin were correct on 27 of 30 trials in this test (see figures 7.3–7.8).

This last test was critical for it demonstrated that not only could the uninformed chimpanzee "do" something as a result of information provided by the informant, but also that the uninformed chimpanzee now "knew" something that he did not know before the informant told him—namely, what was in the container.

Observations of the chimpanzees as they related to one another during this test added additional evidence of their general comprehension of events. For example, on some trials when the informant misidentified the container's contents, the uninformed chimpanzee expressed surprise (eyes widened, jaw dropped slightly) to find that the food he had just requested, based on the information provided by the informant, was not in the container. The "informer" chimpanzee in such cases did not act equally surprised. Rather, he appeared to realize that he had erred in selecting a symbol or had temporarily forgotten the contents of the container.

Figure 7.3. Sherman's teacher points to the closed container, encouraging him to ask for the food which has been placed in the container. Sherman has just illuminated "give" and is searching for the food symbol.

Figure 7.4. Sherman selects the symbol for "orange drink." (Sherman did not vocalize when he saw orange drink in the container; Austin was the only one who did this.)

Additionally, Sherman remembered previous trials on which Austin had misidentified food and frequently would not ask for that food on future trials if Austin declared it was in the container. For example, if Austin said "sweet potato" when the container was filled with oranges, on the next trial that Austin said "sweet potato," Sherman was likely to ask for some other food, as though he concluded that Austin's identification of that food was not reliable. Additionally, on orange drink trials, Austin tended to make particularly high food barks as he came back to the room, even before he lit the symbol for orange drink. Sherman quickly picked up on this fact, and when Austin entered emitting such food barks, Sherman *did not wait* for Austin to identify the food with a symbol. Instead, he hurriedly said "orange drink" himself. This shows that at least one food could be communicated vocally, and once it was, Sherman had no need to attend to any keyboard information provided by Austin. However, on other trials where he did need that information, he waited until Austin identified the container's contents before he requested them.

Figure 7.5. After Sherman says "orange drink," three photographs of foods are placed in front of Austin and he is encouraged to point to the picture of the food which Sherman has indicated is in the container.

Because it was imperative to demonstrate that the uninformed chimpanzee learned the contents of the container *from the informed chimpanzee* and no other source, we ran two additional control tests. One sought to rule out the possibility (albeit an unlikely one) that the chimpanzees had somehow learned to select certain photographs when shown certain lexigrams. To test this possibility, the teacher sat in front of the keyboard and placed three photographs in front of each chimpanzee. During this test, there was no container baited with food nor was there any reason for them to presume that the teacher was using the keyboard to tell them anything about hidden food. The teacher simply lit a lexigram and held her hand out toward the photographs to indicate that she wanted the chimpanzee to give her one of the pictures. She had a dish of food by her side to use as a "reward" for correct picture selection. However, this dish did not contain foods that were in the photographs.

The chimpanzees readily understood that they were to hand the teacher a photograph, but there was no evidence that they drew a relationship between which photograph they gave and the lexi-

Figure 7.6. Austin does not point right away and so Sherman shows him the correct picture. Austin then points to this picture also (on future trials, Sherman was asked to stand back and not to point).

gram which the teacher selected. They seemed not to know why the teacher wanted them to give one photograph on some trials and another photograph on others. The chimpanzees made more and more errors in this task and finally they began to hesitate for long intervals before selecting a photo. They then started trying to solicit cues by touching photos but not picking them up, or by starting to pick up one and then rapidly shifting to another while watching the teacher closely. Finally, their inability to select the correct photograph under these conditions led to a refusal to hand the teacher any picture at all. These behaviors were reminiscent of their earlier responses to other tasks which they did not understand. They were quite different from the confident unerring picture selection we had just seen in the hidden foods test.

The discrepancy between these two tests should make clear that the chimpanzees' success on any given task depends upon their perception of the "purpose" of the task's subcomponents more than on the nature of the skills these subcomponents require. Thus, while they found it difficult to select a particular photograph in

Figure 7.7. The teacher opens the container to see if orange drink is inside. The chimpanzees are watching very closely.

Figure 7.8. Orange drink is inside, so the teacher states "give orange drink" (note projectors) and shares the orange drink between them.

response to the experimenter's arbitrary request, they found it easy to select a particular photograph when selection served to access the food in the picture. Both tasks require that the chimpanzee look at a symbol and then select a photograph of the food that symbol represents. However, in one task the reason for doing this is linked to the identity of the reward, while in the other it is not.

A final control condition was run to rule out the possibility that odor cues, instead of lexigrams, might be inadvertently revealing the contents of the container to the uninformed chimpanzee. In this case, the informant entered the room with the teacher and the baited container as usual. However, he was not permitted to use the keyboard to request the food. He was allowed to do anything else that he wanted and both chimpanzees could smell and touch the container. If, after a brief period, the uninformed chimp did not approach the keyboard and ask for food, he was encouraged to do so by the teacher. The uninformed chimpanzee inevitably did need encouragement in this, as he seemed not to want to ask for the food, presumably because he did not know what to say. It was necessary to gesture repeatedly toward the keyboard and even push him gently in that direction in order to get him to light a symbol. The uninformed chimpanzee requested the correct food on 4 of 26 trials—two of which were orange drink, which we had noted earlier could be communicated by high pitched vocalizations. The chimpanzees became so reluctant to use the keyboard under these conditions that we were forced to terminate this control task after 30 trials.

Conclusions

It is clear from these studies that Sherman and Austin were able to exploit previously learned symbol skills to communicate information to one another that could not be transmitted by typical chimpanzee nonverbal signals. These studies also produced strong evidence for the view that Sherman and Austin were aware of the nature and purpose of their communications in a way that other animals engaging in similar tasks seem not to have been. Our next goal was to determine whether Sherman and Austin could use symbols to communicate with each other when the joint consequence—obtaining food—was not controlled by the teacher, but rather by themselves.

Surely, as early humans learned the benefits of communication, they must have also become aware of the benefits of shared resources and the advantages of cooperation. Trivers (1971) has even gone so far as to suggest that it was precisely this recognition of the shared advantages of cooperation that resulted in the appearance of reciprocal altruism. Once a strategy of cooperatively sharing food resources among group members appeared, the race for bigger and bigger brains would begin. It would be fueled by the need for groups to outplan one another by cooperation. Cooperation in obtaining a goal would need to be matched by cooperation in sharing that goal. As coordination became the key to reproductive capacity, humankind would inevitably evolve toward the superbly communicative creature that walks the earth today.

Chapter 8

Giving, Taking, and Sharing: The Currency of Symbolic Exchange

A Missing Piece

The communication between Austin and Sherman described in the previous chapter was considerably more sophisticated than any previously reported between animals that use a learned symbol system. Information was clearly transferred between Sherman and Austin, but some very important questions regarding what they understood about the process itself remained. If we return to the definition of interindividual symbolic communication, we find that the sort of communication which took place between Sherman and Austin meets only some of the criteria which were set forth. To be more explicit:

1. Both individuals were attentively engaged with one another, albeit with some assist from the teachers in the initial stages.
2. One chimpanzee had access to information, the other did not.
3. It was not clear that the chimpanzee who saw the container baited knew that he, and he alone, possessed information about the container's contents. (This chimpanzee *did* know that the other chimpanzee also had to request the food, but as the study was designed, we had no way of determining what, if anything, Sherman concluded about Austin's state of knowledge or vice versa.)
4. It was valuable to the informant to transmit information, since the other chimpanzee did not act until this was done.
5. It was valuable for one chimpanzee to attend to the information presented by the other.
6. Both chimpanzees shared a common set of overlapping experiences with the symbol system being employed.

7. The exchange of information was accomplished by indirect means.

Even though the communication achieved by Sherman and Austin meets six of the seven criteria listed above, it does so only because the "value" in so acting is mediated through the actions of their teachers. It is valuable for Sherman to tell Austin what is in the container because the teachers insist that both chimpanzees request the contents by name and because the teachers share these contents between the chimpanzees. If the teachers did not ensure that the food was shared, it is not likely that transmitting information about the type of food would be of value to Sherman or Austin. To put it another way, the communication about the type of food in the container occurs not because of intrinsic values the chimpanzee possesses, but because of constraints created by the task.

Such external constraints do not make the final communication product any less real. As language began to evolve in our own species, surely there existed environmental constraints which rendered language an effective medium of exchange for both parties. It is, in fact, hard to imagine any constraints which might facilitate language unless one also presumes that an extensive repertoire of coordinated cooperative behaviors (which presumably included food sharing) was already extant in our species.

Prelinguistic Precursors of Cooperative Symbolic Communication

Not only do human beings demonstrate an ability to acquire and use symbols in a manner that far transcends the nonverbal communicative capacity of the ape, it is to be noted that, quite apart from language, they also engage in a range of *cooperative* behaviors not reported for any other mammal. For example, food sharing is a universal and elaborate behavior among all human groups. Yet it is infrequent and rudimentary in the ape (McGrew 1975; Teleki 1973; van Lawick-Goodall 1968). In general, chimpanzees "share" by allowing others to take fruit from the tree they themselves are feeding upon. Humans share by cooperatively gathering food, carrying it to a home base, preparing it, and dividing it up. Somewhere in between these two vastly different patterns there must have appeared a pattern of gathering a rather large amount of food, carrying it a few meters, and then sharing it on the spot.

Once such food sharing appeared among adults, other forms of cooperation could then follow. Obtaining food, if it were generally shared, would be advantageous not only to the individual who found the food, but to the others as well. This would clearly make it important to inform others of information about food sources. Earlier, when "every man for himself" conditions still applied, it was advantageous to inform others of food sources only when there was more food available than the finder could possibly need.

This is the typical present-day pattern shown by chimpanzees. Upon arriving at a large and plentiful food source, chimpanzees give loud "pant hoots" which serve to draw other apes to the food source. However, at small resources, the chimpanzees feed quietly.

It should be clear that the emergence of the human communication system presupposes not only a "speech and symbol producing" mechanism, but also the cognitive capacity to reason through the results of cooperating and sharing even when food supplies are limited. The environmental constraints which fostered the emergence of these skills in our species must have been rather severe, at least as contrasted with those the present-day ape faces in its natural environment. Yet, such behavioral advances were the product not only of distal environmental forces acting in concert, but also of proximal contingencies exerted by other group members. And it is the delineation of such proximal contingencies that falls within the domain of psychology.

In the case of Sherman and Austin, we were not just faced with the problem of teaching them *how* to communicate. In addition, we had to consider the broader problem of conveying to them the *value* of communicating. These values are difficult to make self-evident, in that they do not seem to be a part of a chimpanzee's natural social interactions. Because language in humans is tied inextricably to our sociocultural practices, it is difficult to create conditions in which two chimpanzees would want to communicate without simultaneously creating conditions in which they should also begin to approximate other nonlinguistic aspects of human behavior. This is not to say that we wanted to turn chimpanzees into people; we certainly did not. It is to say that it became apparent that developing language capacities as though they could exist apart from other social and cognitive capacities was not a realistic approach.

Turning Some of the Contingencies Over to the Chimpanzees

From our perspective, a remaining critical question was: could we do more than get Sherman and Austin to communicate when that was really the only appropriate behavioral avenue open to them? That is, could we get them to comprehend the value of cooperating—and consequently of communicating—in situations where they alone were in control of the contingencies? Could we extract the teacher and her regulatory role? This is, in a sense, a question every parent faces at some point, as their children internalize in increasing degrees behaviors initially structured by the parent.

It is a problem that is, for the parents of many mentally retarded persons, very severe. It is easy enough to get such children to learn how to do things for extrinsic reward; it is very difficult to move beyond that point so that such behaviors as spontaneous communications, cleanliness, etc., become self-motivated or intrinsically rewarding. The difficulty is this: while the retarded child can learn *what* to do, he often cannot understand *why* he is doing it. Similarly, Sherman and Austin had learned to communicate with one another, but they did not understand the value of doing so outside of the highly structured context which was designed to make these behaviors mutually extrinsically rewarding.

Learning To Share Food

The conclusion seemed ever increasingly inescapable that although language learning might take place without a sharing of resources, intraspecific language use would not, unless resource sharing came to be an important part of the way the chimpanzees related to one another. It was therefore determined that the next step in language acquisition should not be to increase vocabulary size, nor to encourage combinations, nor even to develop a wider range of semantic competencies with extant symbols; rather, it should be to teach Sherman and Austin how, on their own accord, to cooperatively share food resources.

It was recognized that teaching Sherman and Austin to share food would be difficult, and that any behavior acquired in such a manner might differ in many respects from the food sharing seen in our own species. Yet because the willingness to share important resources appeared so clearly tied to the motivation to communicate and coordinate actions in a symbolic mode, it became apparent that it was essential to try. Clearly, when humans began to cooperate and share resources, presumably with the aid of symbols, they did not have experimenters to teach them either of these skills. Consequently we did not view our undertaking with Sherman and Austin as a re-creation of early human competencies. Nevertheless, to the extent that Sherman and Austin did learn the cojoint benefits of sharing and communicating symbolically, they would be engaging in actions formerly considered possible only by human beings.

We began teaching Sherman and Austin to share by sitting between them with a bowl of food. As the teacher took out each piece, she simply broke it in half and gave one part to each chimp. Initially, they did not like being close together as they ate and each would take their piece and move a short distance, turn their backs, eat their food, then return for more. This eating pattern is very common for chimpanzees, both in the wild and in captivity. Indeed, one can readily tell which animals are dominant in any given group by noting those that do not move far to eat their food. It is, additionally, rather "impolite" to watch another chimpanzee eat unless one does it in a particular manner by assuming a special posture and facial expression ("peering") which says "I am watching you eat and have no intention of eating what you are enjoying" (Savage 1975).

To decrease Sherman and Austin's tendency to move away to eat their food, we gave them smaller and smaller bites until the piece of food could be consumed so quickly it was hardly worth their effort to move away to eat it. As they stopped moving away, they also began to observe that the teacher was breaking each bite into two portions. They then began to evidence anticipation that each was going to receive half of each bite. Instead of hurriedly trying to take the food from the teacher's hands, they waited patiently for her to break each piece in half. This gave the teacher time to point to the chimpanzee who was to receive the first half of each bite (this varied). They began then to attend to the pointing and to anticipate whether they were to receive their half first or second. If Sherman saw the teacher pointing to Austin, he would not hold his hand out to receive food until Austin had already received a portion. They were, in a sense, learning to anticipate turns, and thus to take turns. After several occasions of sharing food each afternoon (with both chimpanzees always getting all they wanted to eat simply by virtue of being patient), they no longer appeared nervous about eating close together. They seemed to anticipate that all food would be shared and understood that sharing did not mean giving up a chance to have food.

Once they both seemed calm and happy in this situation, we placed a piece of food in Sherman's hand and told him, by using pointing gestures, to give that piece of food to Austin. Next, we put a piece of food in Austin's hand and asked him to give that piece of food to Sherman. Both chimpanzees hesitantly complied with their teacher's gestural requests, but they did not actually hand the piece of food to the other chimpanzee. Instead, they dropped it next to him. The recipient was usually afraid to pick it up—as though he still viewed that piece of food as the possession of the other chimpanzee. It was clear that the discarder would have liked to eat the food and was "giving" it to the other chimpanzee only because the teacher was asking him to do so, not because he really wanted to.

Sherman and Austin's hesitancy to take the transferred food may reflect the absence of coordinated give-and-take games in the chimpanzee. Object transfer and food transfer, as games, occur repeatedly between human mothers and infants and become ritualized as a number of anticipatory signals evolve to mediate this transfer. As the infant grows older, objects are rolled, and then thrown back and forth. No such games are seen in chimpanzees. To take an offered object is usually viewed as an acceptance of an offer to chase. If one chimpanzee holds out a palm frond, and another starts to take it, the first chimpanzee pulls it back and disappears around a tree. Morsels of meat after a hunt may also be offered, and taking these does not result in a chase game. Such taking is, however, associated with many placating gestures and vocalizations which serve to request permission to take food. Since we were not dealing with a highly prized and rare food such as meat, and since the chimpanzees were certainly not in a state of food deprivation, begging gestures were contextually inappropriate.

We wanted the chimpanzees to learn to share as a "matter of course," but they seemed predisposed to share only under special circumstances—which entailed excitement and considerable affective display. When numerous positive affective signals were made by the giver, they seemed to mutually presume that the chimpanzee who offered food did not intend that the food be taken.

To encourage the recipient chimpanzee to take the giver's discarded food, it was necessary for the teacher to push it closer to the recipient, thereby indicating that she intended that he have it. Should the giver chimpanzee start to appear upset as the recipient reached for it, the teacher would have to ask the giver to sit quietly. Often she would have to quickly produce a morsel for the "giver" chimpanzee so that he would not feel that "his" food was being taken. After one such session, both chimpanzees appeared to feel more comfortable with sharing and they no longer displayed anger or threatening facial expressions when one chimpanzee saw the other chimpanzee begin to pick up the food he had put down.

Next we attempted to get the "giver" chimpanzee to place the food directly in the recipient's hand. (We conveyed this by picking up the dropped piece of food and pointing directly to the other chimpanzee's hand.) This worked well when Austin was the giver. All he had to do was to extend a hand full of food toward Sherman, and Sherman would eagerly reach out and take it. When Sherman was the giver, however, problems arose. Sherman would extend his hand toward Austin, but Austin was afraid to take something directly from Sherman's hand. He seemed to view that as taking the food "away" from Sherman, and since Sherman was larger, he did not want to give the slightest appearance of doing this. Sherman coped with Austin's hesitancy by adopting new strategies. He would toss the food to his smaller companion or also look away as he handed over the food. Looking away appeared to reduce the threat inherent in the situation, since if Sherman were looking at something other than the food, he was clearly not concerned that Austin was going to take the food. Sherman also adopted innovative gestures such as motioning Austin to approach, waving the food near Austin's hand, putting the food directly in Austin's mouth, etc. to assure Austin that it was okay to take food from him.

From this point on, it became standard procedure to put out two portions of food and ask either Sherman or Austin to pass out the portions, one for the other chimpanzee and one for himself. They came to anticipate readily the existence of two portions, and the sharing thereof across a wide range of contexts. In fact, sharing became the norm rather than the exception.

Food sharing was integrated in all activities from this time forward. Because a premium was placed on it in all situations, Sherman and Austin readily came to anticipate and *expect* that the other chimpanzee was going to share food with them. Consequently they, in some sense, came to view the sharing of food as a "cultural expectancy" or "right." This became evident when the chimpanzees would engage in food sharing whether the experimenters were there or not. If Austin were given food, Sherman did not just go over and take it, as he had, at times, done in the past. Instead, he persisted in attempts to request it (both by gestures and through use of the keyboard) until Austin shared. Likewise, if Sherman were given food, Austin would persist in attempts to request it and Sherman would usually share. This was true even when the teachers were not present and Sherman and Austin were unaware that they were being observed.

At times, their attempts to work out some sort of sharing of foods were unexpected and even comical. For example, on one occasion, Sherman was given a whole grapefruit and the teacher left the chimpanzees' living area and observed, through a one-way partition, what they would do with it. Austin immediately approached Sherman and gesturally indicated that he would like some of the grapefruit. Sherman looked at the grapefruit and did not seem to know what to do since it was one large round ball and not divided into two portions, as was usually the case. He hesitated for a moment and then handed Austin the whole grapefruit. Austin looked surprised that Sherman had decided to give him the whole thing, but amidst excited food barks he hurriedly sank his teeth into the grapefruit. Seeing this, Sherman threw a temper tantrum, screaming and slapping himself, apparently upset that he had given Austin the whole grapefruit and had gotten nothing in return. He shortly regained his composure however, and then walked over to Austin and gesturally requested some grapefruit. Upon seeing Sherman's request, Austin calmly tore off half of the remaining fruit with his teeth and handed it to Sherman.

On another occasion Austin had been trying to get Sherman to play chase by picking up various toys and running past him. Sherman was interested in building a nest and really did not want to chase. Austin kept trying things which would get Sherman to chase him. He put a mask over his face and shoved Sherman. He put a sock on his foot and shook his foot in front of Sherman's face. He stole one of Sherman's favorite stuffed toys off the shelf near where Sherman was sitting and flaunted it in Sherman's face. He threw a stick at Sherman. All of this was to no avail. Sherman ignored Austin and continued to busy himself with nest making. The teacher then gave Austin a coconut—to help him out a bit, since nothing else seemed to be getting Sherman's attention. Austin immediately ran over to Sherman, showed him the coconut, and ran away. That worked. Sherman chased Austin. They ran around and around the lab for 15 minutes. Finally, tired of chasing, Sherman sat down and looked at Austin and whimpered. Austin stopped running, hugged Sherman and handed him the coconut. Sherman banged it on the concrete floor till it broke in two and then handed the larger piece to Austin before Austin even gestured.

Ristau and Robbins (1982) have suggested that the food sharing by Sherman and Austin is atypical of chimpanzees and consequently it must have occurred under conditions of threat or the withholding of food. That was not the case. Indeed, withholding food would serve only to make sharing *less* likely. It is more accurate to say that we have shaped and rewarded the behaviors of food sharing, and that it has generalized to contexts other than ones in which we initially encouraged the behavior.

Does such trained sharing nevertheless still differ from sharing behavior as seen in human children? I can answer that question only by offering a personal opinion—one based on thousands of personal interchanges with these apes and a resultant sensitivity to their facial expressions and the feelings mirrored therein. To me, Sherman and Austin appear to feel very much as we do when we share. Sharing for them seems preferable to hoarding, for it indicates a state of plenty. If there is a difference to be found between human beings and chimpanzees, it is that human beings continue to share even in a state of great want. Chimpanzees, even Sherman and Austin, do not.

Linking the "Willingness To Give" to Symbolic Requests

Once Sherman and Austin had learned to share food, we returned to what had been a major stumbling block for them—the ability to decode requests of others and to give specific items in response to such requests. As noted previously, when chimpanzees use symbols to get others to do things which they request, their symbolic skills remain egocentric and unilateral. Although Sherman and Austin had learned to use symbols reliably to request foods from others—even foods they could not see—they had yet to show any ability to reverse these roles and give specifically selected foods in response to symbolically encoded requests. However, now that they were willing to share foods readily with one another, we reasoned that they might also find sharing with their teachers less objectionable and thus they might be able to attend to our keyboard-encoded requests for specific foods. Additionally, they had, since the introduction of the hidden-foods task, begun to attend more closely to the teacher's keyboard statements and apparently to decode them in some way—even if such decoding was for their own immediate direct benefit (i.e., it made it possible for them to request the correct food.)

We again tried placing a tray of various foods on a table in front of the chimpanzee and the teacher. The teacher then requested particular items, one at a time. If the chimpanzee gave the teacher the food she asked for, the teacher shared a bite of that food with the chimpanzee. In a sense, the teacher's request established a request contingency for the chimpanzee. For example, when the teacher asked for "sweet potato," the chimpanzee could anticipate being handed back a portion of the sweet potato after he selected and gave this food to the teacher. If the chimpanzee handed the teacher a food she had not requested, the teacher simply said "No" and politely returned the food to the tray. Thus no one got to eat an incorrectly selected food—neither the teacher nor the chimpanzee.

Returning the food to the tray served to emphasize to the chimpanzee that the teacher did not want *that* food, but desired another food instead. The original request was repeated until the chimpanzee did give the teacher the correct food. When this occurred, the teacher acted very happy and eagerly consumed the food while sharing a portion with the chimpanzee.

Although we observed that such sharing, in fact, made our requests for a particular food more relevant to the chimpanzee—and also provided a specific sort of reinforcement for his selecting and giving a particular food item—we did not initially share for these reasons. We shared because that had become our way of treating foods. That is, as the teachers had encouraged the chimpanzees to share all food, we had ourselves begun to share nearly all our own food with the chimpanzees, constantly modeling the sharing behaviors which we were attempting to encourage in the chimpanzees. Thus sharing in this new context was not so much a paradigmatic decision on our part as it was an extension of our effort to encourage Sherman and Austin to behave in a cooperative sharing manner with one another.

It seemed eminently reasonable that we should, at the very least, behave toward them as we were asking that they behave toward one another. Thus, sharing of one's food, no matter who you were, became the norm. (This was often problematic for new individuals who did not relish such intimacy with apes, but it seemed perfectly natural to those teachers who had been there since the chimpanzees' infancy.) In any case, as food sharing became the norm among chimpanzees and teachers, a far greater group camaraderie seemed to emerge.

From food sharing emerged joint efforts at removing food from storage areas. It became customary for the chimpanzees to go to the refrigerator with the teacher and help get out the food. Previously, they would simply have shoved food into their mouths as fast as possible if they had been allowed to go to the refrigerator. They could not have waited patiently for the teacher to point to the food that she wanted, nor hand her that food.

Out of the joint obtainment of food, emerged the joint preparation of food. Food that is to be obtained and shared must at the very least be cut into pieces, and the chimpanzees willingly began to help in this process. Sharing also prompted, on the part of the teachers, a greater interest in exactly what was eaten and how it was prepared. Ill prepared foods were not appealing. Well prepared ones were, and thus emerged an interest in cooking and a great expansion of the chimpanzees' diet. The chimpanzees themselves became quite interested in the cooking process and began to anticipate the jointly prepared meal. Attempts to take or steal food during such preparation were diminished markedly because of the anticipation of sharing once the dish was prepared.

It was then in this new quasi-cultural context that we again asked Sherman and Austin to hand us a specific food in response to a symbolic request. We were far more successful than when we had previously attempted the same task. From the start, when we used only medium- to low-preference foods (sweet potato, beancake, chow) and put out no more than three or four foods at a time, the chimpanzees attended to the teacher's request, and handed her the food she asked for. However, if we complicated the situation by using a larger number of foods, and also included highly preferred foods, the chimpanzees experienced difficulty handing the experimenter the exact food she had requested. Under such circumstances, when the chimpanzees appeared to attend to the teacher's requests, it was hard for them to hand the teacher a low-preference food (such as carrots) when, directly in front of them, they were confronted with a dish of M&Ms. It was extremely difficult for them to reach across the M&Ms to pick up a piece of sweet potato or chow. Like magnets, their hands traveled toward the M&Ms. Also, if there were more than four foods in front of them at once, they seemed rather bewildered by the visual array. If they did not quickly spot the food that was requested, they would simply hand the teacher a different one.

Even with all of these difficulties, Sherman and Austin's behavior revealed a basic comprehension of the task and a willingness to cooperate. They attended to our requests, they decoded them, and they furnished the foods we requested so long as competing responses did not interfere.

We practiced this task with them each day, slowly expanding the number of foods we placed in front of them. When they did not see a requested food, we returned the wrong food, pointed to our displayed request, and then to the correct food. With such practice, Sherman and Austin learned that, even though the teacher did not always ask for M&Ms, or bananas, or orange drink initially, she eventually did so. Accordingly, their self-restraint grew to the point where they could regularly reach across M&Ms to pick up a piece of chow without fingering the M&Ms on the way. Their search skills also increased as they learned that the teacher always asked for foods which were on the tray and that if they did not spot a food immediately, they should still continue searching. It was not long before receptive skills began to equal productive ones and soon the teacher no longer needed to share each bite with the chimpanzees: she could simply take turns, with the chimpanzees requesting the various foods on some trials while she requested specific foods on others. Thus, sharing moved from the actual splitting of a given piece of food, to sharing across time, by taking turns.

The Importance of Teaching Receptive Skills

The need to train receptive skills, even when productive skills are already well developed, suggests that these tasks may be processed independently at the neurological level, even though the same symbol is involved in both tasks (such as naming items and giving items). Although receptive and productive skills are often comparable in normal children (that is, words which are produced accurately are usually comprehended accurately), this is not always the case (Rice 1980). Additionally, studies of brain-damaged adults have suggested that the brain does, in fact, store different language capacities in different locations. In Broca's aphasia, damage is typically restricted to the posterior inferior portion of the left hemisphere. Although productive naming is poor, comprehension is generally quite good. By contrast, there are instances of pure word deafness or pure word blindness in which the patient lacks receptive competence although he can use words normally while speaking or writing (Benson 1979).

Goodglass (1980) has proposed a view of human word processing which suggests that a number of different word skills and word context associations become linked together at the level of semantic representation. According to Goodglass, the ability to name a given object depends "on a convergence of various associations, which constitute its semantic field, and which, along with the phonological form of the word, represent the verbal concept. Variability in the successful evocation of words may be due to the adequacy with which the concept is activated, [and] to the number and character of the associations which constitute the concept" (p. 652).

That is, regardless of whether a word is heard, spoken, or seen, it becomes linked across these forms by virtue of an emerging core semantic referent. This core referent is composed of groups of associations, selectively combined from past experiences. Such core semantic referents are viewed by Goodglass as the "meaning" of the word and when the avenue of access to this semantic core is destroyed, the individual will retain the ability to produce or read the word, but these processes will no longer be linked to comprehension.

If Goodglass (1980) is correct in his general thesis that production and reception are separate skills, which become linked together because of common semantic components, then the need to train both receptive and productive capacities in Sherman and Austin seems reasonable. Chimpanzees do not spontaneously produce language—and they also do not spontaneously comprehend language. Furthermore, language training of the ape apparently requires that the teacher devise training techniques which foster the development of both capacities.

Once chimpanzees have acquired both receptive and productive capacities, do they link these skills together in terms of a common referent and common associations as Goodglass posits human beings do? At this point in Sherman and Austin's training we had no way of determining whether or not they were linking these capacities together through a core semantic referent. The very fact that they had required specific training paradigms to establish naming, requesting, receptive decoding, and receptive giving skills, with the same lexigrams, clearly suggested that it was important to determine whether these various tasks were becoming linked internally, or whether they were remaining separate skills.

As we worked with Sherman and Austin in contexts which required both receptive cooperation and food sharing, it became apparent the advantages of anticipated food sharing were considerable. The chimpanzees now anticipated that food, when it was obtained, would be shared, and consequently they began paying far closer attention to each other's symbols. Such spontaneous mutual attention was extremely important in the development of their ability to use their symbol system reliably and spontaneously without the teachers' presence.

Spontaneous and appropriate attention to the behaviors of another individual is a difficult skill to train. While it is possible to shape what a chimpanzee looks at in a complex and rapidly changing visual environment, it is not possible to shape what he actually sees. For example, a teacher can point to a lexigram Austin has lit, and try to draw Sherman's attention to that symbol, but she cannot ensure that Sherman actually looks at that symbol (as opposed to the color background, the key next to it, the end of her pointed finger, etc.). Likewise, she cannot, simply by orienting Sherman's attention to the symbol, ensure that Sherman will conclude that Austin has lit the symbol and that Austin is requesting that Sherman act on a particular object.

Linking Food Sharing to Symbolic Communication: Our Initial Failure and What it Meant

These difficulties and others were illustrated clearly for us the first time we attempted to determine whether or not Sherman and Austin could use symbols to ask one another for food and decode these symbols in order to comply with one another's requests. We filmed a short segment of this initial attempt and this film clearly illustrates the problems we faced with regard to orienting inter-animal attention sufficiently to produce coordinated communication between two chimpanzees.

In this initial attempt, which took place on December 30, 1977, we filled two trays with bits of food. Instead of the teacher asking Sherman or Austin to give food, we encouraged them to ask each other for food. We intended that one tray be Sherman's (to give to Austin) and that the other be Austin's (to give to Sherman). A teacher accompanied each chimpanzee and showed him what to do. Austin was encouraged (through gestures) to attend to and comply with Sherman's requests, and likewise Sherman was encouraged to attend to and comply with Austin's requests.

The overwhelming impression which arises from watching the film of this early attempt is that the chimpanzees have very little comprehension, if any, that they are supposed to be using symbols to communicate with one another. When Austin asked for a food, Sherman was usually in the midst of running rapidly and playfully around the room and did not appear to see Austin's request. The teacher had to slow Sherman down so that she could point to the symbol Austin lit and to the tray of food. Once Sherman picked up the correct food, the teacher then had to point to Austin. Austin did not do much better on his turn. When, after a good deal of pointing and gesturing on the part of the teachers, one chimpanzee finally did give the other chimpanzee a piece of food, tickling and chasing ensued until the experimenters once again redirected the chimpanzees' attention to the food and to the keyboard. After watching this film, it was difficult to avoid the conclusion that if Sherman and Austin were left to themselves, they would most likely never use symbols to ask each other for anything.

This failure to induce spontaneous requesting and complying between Sherman and Austin made us realize how many gaps we, their teachers, had been filling in. When we requested food, we coordinated the timing of those requests with other behaviors and inserted our requests at appropriate junctures, making certain we had gained visual regard. We also noted whether each chimpanzee was attending to us and to the food. When the chimpanzee wanted to play, we did so, but we clearly kept this activity separate from requesting and eating food. Additionally, we constantly monitored all of their behaviors very carefully and responded quickly to as many requests as possible. Accordingly, the chimpanzees had to make no effort to get our attention or to encourage us to see what they said and to respond to it. We, as teachers, structurally framed our behavior, communicative and otherwise, around theirs. This made communication with us easy to achieve.

A similar sort of attention monitoring has been documented in studies of human mothers and infants (Fogel 1977). Turn-taking is evident in the communication of human mothers with their offspring even before the onset of speech (Schaffer, Collis, and Parsons, 1977). According to Schaffer et al.,

> Interpersonal synchrony in the behavior of mother and infant can be found from birth onward. . . . With increasing age, however, the nature of interpersonal synchrony changes drastically, becoming increasingly diverse and progressively complex. This applies particularly to the first two years, during which time the various perceptual-motor and cognitive abilities that become newly available at different stages of development get implicated in the manner in which mother and infant regulate their behaviour towards one another. Thus in due course multiple channels

come to function—often simultaneously—to promote the infant's communication with others, involving such means as touch and movement, visual contact and vocalizing. (p. 292)

Schaffer et al. (1977) studied, in human mothers and infants, the two features of dialogue which seemed to pose the greatest difficulty for Sherman and Austin when they were communicating with each other: the regulation of turn-taking (with its consequent avoidance of overlapping behavior) and the manner in which visual regard between participants is integrated into the verbal exchange. When communicating with each other, neither chimpanzee knew quite how to regulate turn-taking with the other, nor how to ensure the visual engagement of the other before using the keyboard. They did not act to clearly delineate speaker and listener roles as do human beings (Kendon 1973). By contrast, in their study of these phenomena in human mother and infant pairs Schaffer et al. (1977) found that:

In two-year-olds a definite relationship had emerged between looking and vocalizing—a relationship based on a marked tendency for looks to be initiated during the child's floor holding episodes rather than in response to the mother's activity. In the younger group, however, such a relationship was not apparent: whatever precise function looks play within the vocal exchange emerges only *after* the onset of language. It seems highly likely that by the age of 12 months the child has become aware of the illocutionary force of some of his behavior. . . . it is also likely, however, that intentional signalling is not an all-or-none ability that develops at the same rate in all behavioral systems. Certainly our data indicate that the one-year-old's preverbal vocalizations are not necessarily accompanied by looking at the other person; at this age, it seems, he is not yet concerned to monitor the effects of his vocal activity and search for feedback. It is only in the course of the second year that visual and vocal activities become co-ordinated, assuming the kind of integration that gives speech its obviously other-directed behavior. (p. 322)

Perhaps the difficulties Sherman and Austin experienced in coordinating their communicative behavior appeared because they were still at the level of the human one-year-old, whose intentional signaling ability is less than optimal. However, it was also true that whatever coordinating skills these chimpanzees might have acquired, they were now faced with the task of translating what they had learned from their human teachers to some form of communication that was within the competence of a chimpanzee. Additionally, the needed communicative coordination could not be linked to a "natural modality," such as vocal speech, but had to be accomplished with an artificial visual modality. No such "species transfer" is required of human children. That is, as human children begin to communicate with individuals other than close relatives, they do not also have to deal with communicating with a member of a species whose patterns of glance, timing, etc. are all very different from those of their parents. Even though Sherman and Austin had had extensive contact with each other (and with other apes), the coordination of their natural chimpanzee gestures and glance patterns with their acquired symbol system was a large undertaking.

A Successful Linkage of Food Sharing with Symbolic Communication

To begin to remedy the attentional deficits discussed above, we put Sherman and Austin in separate rooms with a window between them (see figure 7.2). The window was equipped with a small lockable porthole that permitted the passage of objects back and forth. Next, we placed seats and a table by the common window and encouraged the chimpanzees to attend to one another through it. A tray of food, clearly visible to the onlooking chimpanzee, was placed on the table in front of one chimpanzee, who was then asked not to eat the food on the tray in front of him. If he did attempt to do so, we told him "no," took the food out of his hand, and returned it to the tray. Since previous experiences led the chimpanzees to believe that they would always receive enough to eat, they were not unduly upset by this and began to comply with our requests. At times they would sneak a piece of food off the tray, but in general they cooperated remarkably well.

The other chimpanzee, who had no food in his room, needed no encouragement to look at the events taking place next door. The chimpanzee with the food did not necessarily attend to the other chimpanzee. Since he was asked not to eat the food, he seemed most interested in soliciting play from the teacher who was in his room.[1]

The "recipient" was encouraged to request food from the "giver." Additionally, his keyboard was turned on while the "giver's" was turned off. This helped to clarify roles for the chimpanzees by providing only one chimpanzee at a time with a symbolic medium. The chimpanzee with the active keyboard was clearly the speaker.

The sequence of events once the teacher entered the "giver's" room with a tray full of food is outlined below:

A. The tray of food is placed on the table next to the window and both chimpanzees are simultaneously encouraged to come and sit on their cubes by the window. Generally, they take one look at the food, follow the teacher to the window and sit down.
B. The teachers gesture to the recipient, to indicate that he is the one who gets to ask for the food. They may also gesture to the food tray or lift it up a bit to make certain that the recipient can see which foods are available to be requested.
C. They may need to gesture to the keyboard to encourage the recipient to use it to request the food he would like to have.
D. As the recipient goes to the keyboard, they encourage the giver to watch. After the recipient selects a symbol, they may point to that lighted key to draw the giver's attention to it.
E. If the giver starts to hand the wrong food to the recipient, the teacher takes the food and returns it to the tray. If the giver does not correct himself the teacher may again point to the lighted symbol.
F. When the giver selects the food, the teacher may gesture toward the recipient to encourage him to pass the food through the window.

Separating Sherman and Austin in this task accomplished the following objectives:

1. It precluded their natural tendency to get one another's attention by simply approaching and pulling on or playing with another individual and thereby forced greater emphasis on visual signals.
2. Roles were delineated as each chimpanzee was required to use his own keyboard when he was the recipient.
3. By using only one food tray (instead of two as we had initially done) the source of the food to be given was made obvious.
4. By switching roles when the food tray was empty, a clear marker event was provided for this exchange of roles.

It took no training to get the recipient to ask the giver for a specific food which he had seen on the other side of the window.

At first, the chimpanzee with the food needed frequent encouragement to attend to the other chimpanzee's requests and to comply. However, after two exchanges of roles it became clear that the more rapidly one chimpanzee complied with the other's request, the more rapidly their roles would alternate.

One of the earliest exchanges in the setting is shown in figures 8.1 to 8.7. The exchange was filmed on January 3, 1978, only four days after the previous exchange with both chimpanzees in one room—where communication between them had seemed so improbable. No training had occurred between these two sessions, as most of the teachers were away for the holidays.

The photographs show the chimpanzees attending to each other far more readily than in the single-room, single-keyboard setting described previously. They also reveal how much easier it was for the teachers to reorient the chimpanzees' attention back to the food and to the communicative behaviors of the other chimpanzee when their attention did begin to wander. So long as the interaction flowed between the two chimpanzees, the teachers did not intervene. However, if one chimpanzee made a food request and the other chimpanzee did not see it, the teachers in both rooms intervened to draw the giver's attention to the fact that the recipient had asked for a specific food. Without such intervention, the requester initially tended to give up and walk away—doing nothing to try to get the other chimpanzee to attend to and comply with his request.

The differences between the one- and two-room settings were dramatic. Not only did the chimpanzees rapidly begin to attend to one another and to comply with one another's requests in the two-room setting, but their accuracy rates (both for the type of food requested and the type of food given) were very high. Though the teachers intervened when one chimpanzee requested a food that was not present (by pointing to the foods present on the tray) and when an incorrect food was selected in response to a request (by asking the chimpanzee to return that food to the tray and look again at the projected symbol), all such trials were scored as incorrect. Trials were scored correct only when the chimpanzees selected and gave the correct food without aid from the teachers other than encouraging the chimpanzees to watch one another.

Figure 8.1. Sherman watches as Austin requests a food at the keyboard.

Figure 8.2. Austin approaches as Sherman reaches for the requested food.

Figure 8.3. Austin watches closely as Sherman lifts the food from the tray.

Figure 8.4. Sherman hands the food to Austin through the port in the window.

Figure 8.5. Austin (right) watches as Sherman (left) requests a food. Sherman is looking back toward Austin to see if Austin is attending.

Figure 8.6. Sherman watches Austin's hand as Austin selects the requested food.

Figure 8.7. Austin passes the requested food through to Sherman. Sherman is already looking at the tray to decide what food to select next.

Table 8.1 shows all instances of the chimpanzees' performance on the food-sharing task from January 3, 1978 through July 1978. Accuracy of mutual requests and responses was quite high, ranging between 80 and 100 percent. Improvement was reflected not in increased overall accuracy, but rather in the decreased need for the teachers to play a role in directing the chimpanzees' attention to one another.

As will be described in later chapters, Sherman and Austin continued to communicate and respond to one another's requests on their own cognizance even after the teachers removed themselves completely. Finally, Sherman and Austin had achieved a significant degree of communication between themselves and they controlled the contingencies through their own recognition of the value of turn-taking, joint orientation of attention, and sharing.

Once Sherman and Austin had achieved these skills, keyboard food sharing became their second-favorite activity (going outdoors remained first). They would bring us a food tray and gesture to it while making food barks as a way to ask us to cut up the food and fill the compartments. If we responded "Yes," they would run

Table 8.1.

Month	Total number of food exchanges[1]	Correct food requested		Correct food selected & given	
		Sherman	*Austin*	*Sherman*	*Austin*
January	192	90%	82%	87%	83%
February	68	84%	93%	81%	92%
March	128	98%	88%	90%	93%
April	30	100%	82%	100%	80%
May	52	94%	88%	100%	89%
June	22	93%	90%	92%	100%
July	31	85%	88%	92%	88%

[1] All food exchanges between rooms from initiation of this task in Janaury 1978 to August 1978.

around the room and pound on the walls, while giving excited pant hoots and hugs to each other and us.

The environmental and social constraints which facilitated these exchanges were, to be sure, very different from what our early ancestors—embarking on the path of symbolic expression—would have experienced, yet perhaps these needed laboratory constraints can still provide some clues. They strongly suggest that spatial separation functions to facilitate symbolic communication.

Chapter 9

Tools: The Key To Learning That Objects Also Have Names

Sherman and Austin Assign Their Own Lexigram Symbols for New Foods

In contrast to their slow start, the introduction of new food or drink names to Sherman and Austin was now quite simple. In fact, we could leave the designation of new lexigrams up to them. When a new food was brought into their room, they searched the keyboard and spontaneously used a new lexigram to request it. If Sherman were the first to select the new symbol, Austin took note of his selection and subsequently used that symbol to ask for the food or drink. If Austin were first, Sherman would soon do the same. The lexigrams *strawberry drink, lemonade, pineapple,* and *pudding* were all assigned by the chimpanzees, not the teachers. Each of these foods could be requested, named, or given by the chimpanzees without special training. Thus, once having learned the basic productive and receptive skills, Sherman and Austin were easily able to transfer those skills to new food names.

We had not anticipated the appearance of this tendency to assign new lexigrams to new foods. On the first occasion that this behavior appeared, the teacher entered the room with the new food, pudding, intending to show the chimpanzees the new symbol that she had assigned. However, Sherman and Austin looked at the new food, approached the keyboard, and began searching for a symbol. Wondering what symbol they would select, and thinking that their selection might reveal something about the foods or symbols which they viewed as similar, the teacher waited and let them search the keyboard. Instead of selecting a symbol that he had been taught,

Sherman lighted a completely new key, one that had never been used before. After lighting this new symbol Sherman extended his hand expectantly for the pudding.

Following this initial episode three more items without names were introduced, strawberry drink, pineapple, and lemonade, in an attempt to replicate this phenomenon. Before each of these new foods or drinks were shown to the chimpanzees, between four and seven new unassigned lexigrams were placed on the keyboard and allowed to remain active for one to two weeks. During that period, no spontaneous use of these symbols occurred, thus indicating that the chimpanzees were not selecting these keys randomly nor lighting them because they were novel. The new foods were then introduced in the context of a typical food request task, with the teacher holding up foods one at a time. When the new food was held up, no hint of its name was given. Sherman and Austin hesitated a bit on these occasions, and they twice called the new food by another name.[1]

When the chimpanzees used an assigned lexigram, the teacher responded by saying, for example, "No, this melon," and giving them a taste of the melon. Then she asked "What's this" and again held up the new food. Sherman or Austin then lighted one of the new lexigrams and gestured toward the new food. Sherman selected the lexigrams for strawberry drink and lemonade while Austin selected the lexigram for pineapple. Both chimpanzees spontaneously observed the lexigram(s) selected by the other, and used them accordingly when it came time for them to ask for that food. In addition, when these new foods were presented on future occasions, both chimpanzees continued to use the same spontaneously assigned lexigrams.[2]

With the appearance of this lexigram-assigning capacity, it seemed reasonable to conclude that Sherman and Austin understood that a type of unique correspondence existed between each food and a given symbol. We also began to note instances in which Sherman and Austin pointed back and forth between a food and a symbol as a means of indicating which name went with which food. Austin's ability to observe and to adopt Sherman's choice of a new lexigram (and vice versa) implies, moreover, that they had learned that the designation of new foods required not only a new symbol, but also a coordinated usage of that symbol by all parties.

The Competence To Name Foods

The rapid understanding that foods have names did not generalize to other objects. Unlike Helen Keller who, in one inspired moment, grasped the concept that all things have names, Austin and Sherman certainly did not. In fact, when we attempted to encourage them to use symbols for things other than food, they experienced a great deal of difficulty. This was not because they were unwilling to do so, or because their motivation to employ symbols was strictly tied to pragmatic, reward-based situations. Rather, it seemed that they simply did not realize that symbols could be used to stand for things other than food. It is important to clarify that Sherman and Austin could appropriately *use* a number of non-food lexigrams. However, appropriate usage and referential usage are not necessarily synonymous.

Throughout much of the previous training of food names, a number of other symbols had been available to Sherman and Austin, and they had learned to use many of these symbols reliably during social exchanges with the teacher. These additional symbols included verbs (open, give, tickle, groom, scare, etc.), proper names (Sherman, Austin, Sue, Janet, Sally, etc.), object names (key, blanket, money, etc.), states (open, out), attributes (big, little, etc.), and locations (outdoors, sink, colony room, etc.).

The teachers modeled the use of these symbols during social interactions to describe the chimpanzees' activities and to suggest actions. Likewise, the chimpanzees were encouraged to use these symbols whenever they demonstrated an interest in these activities and objects.

The chimpanzees readily learned to make such requests as "Open door" when they wanted to go out of their training rooms, "Give key" when they wanted to unlock the door between their training rooms, "Give wrench" when they wanted to help remove the front of the keyboard panel to relocate lexigrams, or "Give blanket" when they wanted to build a nest or be chased by a teacher with a blanket over her head. All such symbols were used spontaneously and appropriately in context at least once each day in ways that appeared to be the same as the reported usage of these signs by Washoe, Lucy, Koko, and other apes (Fouts 1974a, b, Gardner and Gardner 1978a, b, Patterson 1978).

In fact, if we employed the Gardners' criterion for symbol acquisition (that is, usage each day without an immediately preceding prompt for 15 consecutive days), Sherman and Austin could be said, at this point in their training, to have acquired the symbols, blanket, wrench, money, door, tickle, groom, chase, out, and outdoors. However, when we tested their competence with object names in a context-independent labeling task, they proved unable to label the objects blanket, wrench, money, and door accurately, even though they could easily label specific foods. Thus, in the domain of nonedible objects we encountered the same sort of difficulty we had originally seen with food names. That is, things could be asked for, but not "named" in contexts in which that object was not currently desired. However, at this point in training, we knew that they already had the capacity to "name" foods apart from eating them and we had expected a ready transfer of this ability to objects. We presumed that this difficulty in object labeling was due to lack of practice and felt that, with continued training, the deficit should disappear. After all, Lana, Nim, Washoe, and other apes had demonstrated abilities to label objects.

Some Unsuccessful Attempts To Teach Object Names

To provide practice, we began concentrated naming drills with objects which were already a part of Sherman and Austin's active vocabulary *(key* and *blanket).* We also added *bowl, box,* and *lock.* As long as not more than three of these objects were used at a time, the chimpanzees seemed to understand that they were being asked to select the lexigram which went with each object, and they did well, particularly when we changed exemplars no more often than every third or fourth trial.

However, as additional objects were added to this labeling task and as we began changing the exemplars each trial, performance dropped markedly. In addition, if the animals started out doing well on most items in a given training session, their performance often decreased during the session as if some sort of interference were occurring. Their motivation and attention were generally poor and they rarely gave more than a cursory glance at the displayed object, as if uninterested in the extreme.

Table 9.1 shows the number of trials given on each item before this task was discontinued because of lack of progress. We had been attempting to reach a criterion of nine consecutive correct responses on a given item, with that item being presented randomly interspersed with others. Although the chimpanzees' performance occasionally exceeded 80 percent accuracy, they did not reach our criterion. This criterion may sound high, yet Sherman and Austin had easily satisfied it with food names.[3]

Any symbol acquisition task which did not begin with functional request-based communications and which did not require the use of symbols in a way that moved beyond contextually self-evident information posed great difficulty for Austin and Sherman. This observation has been repeated with attribute names, action names, location names, and person names.

Austin and Sherman's difficulties in learning to label objects were not due either to lack of motivation or to comprehension of the task. Nor, as Ristau and Robbins (1982) have speculated, was the problem that the chimpanzees "had not been guided to press the appropriate lexigram key when a tool (or object) was held aloft" (p. 205). The teachers frequently showed the chimpanzees the appropriate key and performed interesting activities with the ob-

Table 9.1. Naming and Tool Use

	Number of trials[a]				
Naming paradigm	*Box*	*Bowl*	*Key*	*Blanket*	*Lock*
Sherman	215	215	966	704	77
Austin	—[c]	—[c]	220	390	50

	Number of trials[b]					
Tool-use paradigm	*Wrench*	*Straw*	*Stick*	*Sponge*	*Money*	*Key*
Sherman	59	9	88	21	70	49
Austin	54	55	41	34	21	20

[a] Until name-training on item was discontinued, due either to C's lack of progress or his lack of willingness to cooperate. Neither animal reached criterion with any object.

[b] From introduction of tool name until the tool was correctly requested on nine consecutive correct presentations with all tools randomly sequenced.

[c] Not attempted.

jects. Rather, the crux of the problem was that, as the teacher pointed to an object and lit a key, Sherman and Austin did not understand that the teacher was referencing the object symbolically, or that this lexigram was to be associated with the displayed object. They did understand such relationships when the object was edible. Presumably, Sherman and Austin saw little merit in referencing things that were inedible.

This difficulty should not be interpreted to mean that Austin and Sherman showed *no evidence* of understanding that objects could be specified symbolically. On the contrary, if we asked for a key *as* we were trying to open a locked door, or a blanket *as* we were making a nest, they would quickly search for and retrieve the requested item. These behaviors implied that much of the information inherent in these symbolic requests was being carried by, and inferred from, the social context. In a similar sense, their use of these symbols was also context-dependent.

At this stage in their symbol acquisition, though, "blanket" meant, e.g., making nests, playing hide and seek, or playing toss the blanket. "Key" meant access to boxes, other rooms, or other chimpanzees. Saying "key" simply because someone showed them a key was not part of the sphere of "meaning" they associated with the symbol *key*.

Normal human children seem to engage in an active process of sorting out word meaning that goes beyond the capacity of apes. As children hear similar words in different contexts, they are able to recall and analyze the same sound across a number of different contexts in order to determine its referent (Braunwald 1978). Likewise, they are aided by the awareness that a point plus a vocalization is a means of singling out a specific reference. Thus, a child might hear "ball" in the context of "Give the ball," "It's your ball," "Don't touch the ball," etc. From such instances, the referent of the sound "ball" is sorted out. However, if someone holds up a ball, points to it, and says "ball," the child will not have to sort through the meanings; he will know that the speaker is referring to the round object. The simple fact is that Sherman and Austin were unable to benefit from such instruction.

In human children, difficulties in word acquisition arise mainly when the referent is not available or clearly point-at-able. Braunwald (1978) documented such a difficulty as her daughter (Laura) learned the word "Bow-wow." The mother used this word to label the barking of the neighbor's dog, which was visible in the neighbor's yard, along with birds and the neighbor's car. For a brief period, Laura learned to label dogs, monkeys, cats, lions, cars, and airplanes as "bow-wow." Presumably the generalization of "bow-wow" to cats and monkeys reflects their similarity of appearance to dogs, while her use of "bow-wow" for cars and airplanes reflects the fact that both airplanes and cars, in addition to the neighbor's dog, were making noise as the mother said "Bow-wow." As Laura began to evidence comprehension of the word "car," her usage of "Bow-wow" for cars dropped out *before* the word "car" appeared in her own vocabulary.

When a symbol such as blanket is used with Sherman and Austin, the referent is for them less clear than it is for the human child. At this stage we had no evidence that they either spontaneously sorted out the proper referent of a symbol as they encountered it in a wider number of contexts, or that they understood, apart from foods, that when others pointed to an object and selected its lexigram, that they were, in effect, attempting to "reference" an object.

The Tool-Use Paradigm and Its Unique Advantages

In order to encourage Sherman and Austin to compare their symbol usages across contexts, as children do, and to emphasize the referential properties of symbol usage, we devised the tool-use paradigm, in which there are a number of boxes baited with food. Once food is placed inside such a box, it can be extracted only by a tool (a key to unlock a padlock, a straw to be inserted in a small hole, etc.). This paradigm has the advantage of making the physical characteristics of the objects highly salient.

In order to solve a given tool problem, the chimpanzee had to determine which of several tools he needed to extract food from a particular site. He also had to learn to use a symbol to ask the teacher for the tool he needed. The food sites were located around the circumference of the room and were introduced one at a time. The initial six tools and their uses are listed in table 9.2. Eventually nine tools were employed.

Table 9.2. Tools and Functions

Tool	Function
Key	To unlock a variety of padlocks from doors, boxes, etc.
Money	To operate a vending device for food.
Straw	To obtain liquids by threading through small holes in the tops of containers and the wall.
Stick	To dip pudding, yogurt, etc. from containers out of reach, and to push food out of a long hollow tube mounted horizontally on the wall.
Sponge	To dip into a vertically mounted, narrow hollow tube, to soak up liquids from flat surfaces.
Wrench	To unscrew bolts from the keyboard and from small bolted doors mounted in various locations in the rooms.

With the introduction of each tool, the teacher baited the appropriate food site and allowed the chimpanzee to investigate. After he had determined that he could not extract the food with his hands, the teacher drew his attention to the tool kit and showed him how to use the new tool to extract the food. The chimpanzees always observed this with rapt attention. The box was then baited again and the chimpanzee was allowed to investigate once more.

This time attempts to open the site by hand were abandoned quickly, and the chimpanzees turned to the tools. As the teacher observed the chimpanzee's gaze fix upon the correct tool, she picked it up and named it at the keyboard. The chimpanzees watched this and then immediately used that name to request the appropriate tool by saying "Give key," "Give money," etc. Upon being handed a tool they took it to the site to try to open it. Thus, introducing new tools and their symbols was a straightforward and rapid business.

The tool-use paradigm has a number of advantages for working with organisms who do not seem to understand the referential function of pointing.

1. It creates a need for a specific tool, thereby orienting the chimpanzee's attention to that object and preparing an opportunity for the teacher to reference the item while the chimpanzee is interested in it.
2. After the tools have been introduced the chimpanzee may err on future trials because he is unable to recall the specific symbol for each tool. When this occurs, the teacher never has to tell the chimpanzee that he is wrong: she simply searches through the tool kit and gives the requested tool. If the tool is not the needed one, the chimpanzee will conclude this for himself and alter his request spontaneously. Because the teacher's behavior is ultimately predictable (each symbolic request always results in the teacher's giving the tool appropriate to the selected lexigram) the chimpanzee comes to alter his own symbol use so as to discover which symbols will have the intended effect upon the behavior of his teacher.
3. This procedure makes clear that the lexigram refers to an object (the tool used to obtain the food) rather than the food itself. When the chimpanzee says "Give M&M," because he has seen M&Ms placed in the tool site, the teacher can respond by focusing her attention on the M&Ms while trying to extract them by hand. This illustrates that the lexigram "M&M" is directing the teacher's attention to the M&Ms. The chimpanzee sees that the teacher responds appropriately to his request, but is unable to provide the food because he lacks a tool. Thus, the chimpanzee turns his attention to the problem of how to refer to the tools. With attention appropriately and intrinsically directed, reference soon follows.

4. The chimpanzee's behavior in this paradigm reveals clearly the types of conceptual confusions he is experiencing. This knowledge greatly aids the teacher in taking appropriate corrective action. For example, if the chimpanzee requests a socket wrench when he needs a key, and then proceeds happily to take the tool and rush over to the key site and try to open the padlock with the wrench, it is clear that the chimpanzee believes that he has the right tool for the job. If, on the other hand, the chimpanzee discards the wrench as soon as the teacher hands it to him, and asks for another tool, it is obvious that he knows what he needs, but hasn't used the correct lexigram to request it. Overall, both chimpanzees learned to request and employ the correct tools with surprising rapidity using this paradigm.

Learning Tool Names: Functional Errors Predominate

Intermittent confusions regarding the proper lexigram to be used to request a tool typically continued until the chimpanzee became proficient in using the actual tool itself and could open the site without assistance from the teacher. The simpler the tool, the more quickly this occurred. Very simple tools, such as a straw, were learned quite rapidly and with very few errors while more difficult tools, like the magnet attached to the string, took longer to learn.

The physical similarity of the lexigrams themselves provided far fewer difficulties between the tools. For example, the lexigram stick 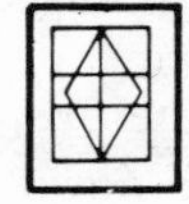 and straw  looked very similar but were rarely confused. By contrast (socket) wrench 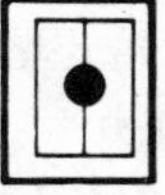 and key  did not look similar but were still repeatedly confused. Their commonality lay in their function—insert and turn against another larger fixed object. Such errors reveal that the chimpanzees understood only that these tools had to be twisted. They did not know why the lock came open when they twisted the key, or why the bolt came open when they twisted the wrench. Thus even though the actual implements looked very different, and their symbols were very different, errors between these two tools were common.

The strong prevalence of functional errors over perceptual and encoding errors, suggests that the chimpanzee is organizing information in his world from an operational perspective. This interpretation is also supported by instances in which a tool was used successfully for other than the experimenter's intended purpose. For example, during one tool training session with Sherman, a wrench was requested after the stick site had been baited with food. At this point in training, Sherman knew both of these tools and was quite familiar with their uses. Mistakes between them had been rare and thus the teacher thought this erroneous request was due probably to inattention since all the symbols had recently been relocated and Sherman had not searched the keyboard carefully before making his request. Nevertheless, in accordance with the general procedure on this task, the teacher replied "Yes, give Sherman wrench," then looked through the tools and found it.

Sherman, however, did not reject the wrench, as the teacher had anticipated. Instead, he took it and went straight away to the stick site. He inserted the wrench into the tube, just as he would have used the stick, and tried to push the food out. However, the wrench was too short (by design) and would push the food only halfway out—not far enough for Sherman to retrieve it from the other end. Sherman still refused to discard the wrench and request the stick. He carefully pulled the wrench back just to the end of the tube, then popped it hard with the heel of his hand . . . and sent it flying through the food-filled tube. He then proceeded to eat the food, which had sailed out the other end.

Sherman, of course, had not used the wrench in this way before, nor had any of his teachers. To our knowledge, he had never observed any tool being employed as a missile inside a tube. However, from this point on he began to use the lexigram "wrench" when he needed a stick and the lexigram "stick" when he needed a wrench. In order to eliminate the confusion between the functions of wrenches and sticks, we were forced to redesign this tool site making it too narrow for the wrench to be inserted inside it.

The preeminence of functional errors was again encountered when we devised new functions for a learned tool. For example, a stick site was devised in which it was necessary to insert the end of the stick through a hole in the wall and then dip it into a sticky food such as peanut butter or yogurt. The chimpanzees did not, on trial 1, use the lexigram "stick" to request the needed tool, even though they could point to the stick as the tool they wanted, and they could use the lexigram appropriately to request "stick" if food were placed in the typical stick location. It was necessary to show the chimpanzees which symbol to use to request a "stick" for this new site.

Errors also arose from time to time which appeared to result from certain foods having become inadvertently associated with particular tool sites. For example, the teacher might often place M&Ms in the stick-tube site because they were the most convenient food to load and extract from this site. We soon found that when one food tended to become inadvertently associated with a single tool site, the chimpanzee would then request that tool even when that food was placed in another site. However, even though they would say, for example, "stick" when M&Ms were loaded into the lever site, they appeared quite bewildered if the teacher actually handed them the stick. Upon seeing the teacher offer them a stick, they generally returned to the keyboard and requested the correct tool. This sort of error revealed the strong predominance of food as the item of reference in their lexigram-associative learning process. In order to keep "stick" from simply becoming another name for M&M, it was important to require the use of other tools to retrieve M&Ms. Errors such as these help reveal why teaching chimpanzees to request things by using longer strings (such as "me Nim food") does not result in greater linguistic productivity. All symbols regress to representing the food in situations where there is no way to make clear the differences between the various items of reference.

The Emergence of Labeling and Receptive Capacities

When Sherman and Austin were able to request six different tools, each to be used in two or more locations, we began to concentrate on their naming and receptive capacities with object symbols. Although transfer of naming and receptive capacities from food symbols to tool symbols did not occur immediately in either chimpanzee, it did occur in both chimpanzees within one or two training sessions. This suggests that the chimpanzees were able to use with other symbols the speech-act competencies they had acquired with food names.

Initial probes of naming and receptive competencies were done differently for each chimpanzee. Austin's teacher began by administering 38 training trials, in two days. She baited a food site and then asked Austin to give her the tool *she* needed to open the tool site. If Austin gave her the correct tool, she then opened the site and shared the food with him. Essentially, the teacher was asking Austin to act on her behalf by selecting the tool she said she needed. During the first two trials, Austin was allowed to see which tool site was baited. On the remaining trials, he did not know where the food was and he could tell which tool the teacher needed only by observing her symbolically encoded request.

Austin's teacher thereby required that he receptively decode a symbol, look through a set of objects, and give her the symbolically requested object. If Austin gave his teacher the incorrect tool, she said "no" and took Austin to the site to show him that the tool he had given her would not work. Once Austin saw where the food was, he readily ascertained which tool the teacher needed. They would return to the keyboard and the teacher would direct Austin's attention to the lighted symbol she had used to request the tool that she needed. Austin would then hand her the correct tool.

During the first session (24 trials) only three tools were used and Austin was correct on 16 trials (0.66—a little better than chance). During the next session (14 trials and four tools), Austin made only two errors.

Austin did not immediately understand that he was to look over the tools and give the teacher a specific tool in response to her request. At first, he would simply hand the teacher whichever tool was closest to him. However, after following the teacher to the food site and then returning to the keyboard, where he saw the symbol for the correct tool reiterated, Austin began to behave as though he realized that the teacher's request was specific and purposeful. He began looking carefully back and forth between the teacher's selected symbol and the group of tools before making his selection.

Austin then received 18 additional trials in which all six tools were used. Austin selected the correct tool on 16 of these trials, thereby revealing that he could use this receptive capacity with all of his tools' names.

Equally impressive was Austin's performance when naming skills were investigated after the use of the receptive tool paradigm described above. During the first 36 trials, with all six tools being randomly presented, Austin made only five errors, for a total of 86 percent correct.

Sherman did not receive training on the receptive tool task described above. Instead, his naming capacity was tested directly. During his first test, Sherman initially appeared completely unable to name his tools. On trial 89, he evidenced a sudden shift in behavior and named all the tools almost without error—even though no procedural change had been made by the teacher. This change is best described in the transcript of the teacher's comments made immediately following this session.

. . . I then began to work on naming with Sherman. I would ask him "?What is this" and encourage him to reply, prompting him initially by pointing to the stick and then to the lexigram for stick on the keyboard. At first, I attempted to keep the task very simple by working only with two tools, wrench and stick. Sherman did not seem to catch on to the idea. Instead, he simply began to alternate back and forth between the symbols, picking either symbol first and if it was not correct, he then picked the other symbol. Because of his tendency to alternate, I added a third tool. This seemed to confuse him even more and he quickly developed a preference for a single tool, wrench, and labeled everything "wrench," regardless of the tool which was shown to him. Thus he was always correct on ⅓ of the trials—those on which wrench was the tool I held up. If wrench was not correct, Sherman then would answer straw; and if straw were incorrect, he then would select stick. He did not seem to draw any sort of correspondence between the correct tool and the tool I was holding up. This was most evident when I would show Sherman the same tool several times and he would still make errors. For example, if I held up the stick and he said "wrench," I would correct him by saying, "No, this stick." He then would say "stick" and be rewarded with food. However, if I held stick up again on the very next trial, he would not call it a stick. He would continue to make errors even when I held up the same tool for four consecutive trials. Sherman's overall tally for the morning session was 34 correct trials out of

78 tool presentations. Since only three tools were used, obviously this is not very accurate performance. Also, there was no increase in Sherman's performance throughout the session. Out of the last 30 trials, he was correct on only 11 trials. My general impression was that he did not understand that the key he was selecting should correspond in any way whatsoever with the object which I was holding in front of him.

In the P.M. session we began again with the same task, starting out with the same three tools we had used earlier: straw, stick, and wrench. During the first 11 trials he behaved as he had in the morning. However, on the next trial there was a sudden and marked change in his behavior. He attended closely to the tool which I held up and began to respond rapidly and correctly. Out of the next 106 trials, he made only four errors. Not only did I use those three tools, but I was able to introduce all of his other tools, so that by the end of the session, he was readily naming magnet, sponge, key, straw, money, wrench, and stick. He had no problem with any of the other tools when I inserted them into the naming task. Whenever I introduced a new tool I would hold it up and light the correct symbol, and then ask Sherman what the name of that tool was. We would then practice that tool two or three times. (Both the introduction trials and the practice trials are eliminated from the overall tally of 106 trials.) After the introduction and practice, this tool then would be interspersed randomly with all other tools which Sherman was naming. Contrary to my expectation, Sherman did not have trouble with "money," even though money was given to him as a reward on each trial. For example, if he named straw correctly, he received a piece of money and was able to buy food with it. All tool naming was done with me sitting behind Sherman as he worked at the keyboard. He did not look toward me for prompts; however, before he began to be correct, he dawdled between trials and pasted his food on the wall. I had to encourage him on every trial. However, once he grasped the task his attitude changed completely. He quickly named each item and even attempted to get me to hurry by pulling my hands toward the keys "?What's this."

This remarkable change in Sherman's ability clearly illustrates the performance difference which can ensue as a result of the chimpanzee's somehow coming to understand what is being required by the task. Such a radical alteration in performance implies that Sherman was trying a variety of strategies to determine what to do, and once he selected the correct one, he quickly utilized it. For a chimpanzee to jump from 43% correct (with three items) to 96% correct (with six items) in the middle of training session where no procedural change has occurred is remarkable. Such rapid and complete performance changes are strong evidence of hypothesis testing. Equally puzzling as Sherman's shift on trial 89, however, is his seeming inability to name his tools up to that point.

Additional Contrasts Between Sherman and Austin

When Sherman and Austin were asked to give the teacher specific tools, *with no food site baited,* a discrepancy again showed up in their performance. This receptive task differed from the previous one used with Austin in that the teacher now had no use for a tool. She was simply requesting that the chimpanzee look through the tools and hand her the one specified symbolically. Correct responses were rewarded with food.

Austin was correct on 81 percent of the first 69 trials of this task (all given in one session). Sherman, however, evidenced a complete lack of comprehension for the first 73 trials of this task, then—just as he had done with the naming task—he suddenly seemed to understand what the teacher was requesting of him and began to perform accurately.

Two teachers had worked with Sherman on this receptive giving task and both had become equally discouraged with what appeared to be cursory attention to their requests. Sherman's total inability to give the requested tool prompted the teacher to return to the naming task to see if Sherman still knew how to name the tools. Although Sherman had been able to name his tools accurately throughout the preceding week, he could no longer do this after failing the receptive task. His teacher then returned to the functional task, in which a food site was baited and Sherman was to request the correct tool. Sherman made no errors at all on this task. The teacher again returned to the naming task and Sherman again failed. Then on one trial he unexpectedly named straw correctly (unexpectedly, because he had missed it on all preceding trials). The teacher was looking away, distracted by a noise, just as Sherman named straw. In contrast to the previous trials, during which he had seemed not to attend at all to the symbol he had lighted, on this trial *he tapped the teacher on the shoulder and as she turned around, he pointed to the symbol* (straw) *which he had just lighted.* The teacher was so surprised at this unexpected indicative behavior that she laughed, hugged Sherman, and praised him quite profusely. Sherman made *no* naming errors on the remaining 61 trials. The teacher then attempted the receptive task and Sherman performed well on that task for the first time. He was correct on 16 of 20 trials during this session and on 116 of 146 trials in the following session, and his performance increased steadily thereafter.

The above description of Sherman's performance on the naming and receptive tasks has been described in detail to show how difficult it often proves to assess an ape's competencies adequately, and how an extant skill (in this case, tool naming) can be affected both by requesting the chimpanzee to engage in a different task with the same symbols and by the affective rapport between teacher and subject. As we only trained two subjects, it is not possible to assess the importance of another factor. The brief period of receptive training to retrieve a tool needed by the teacher may have facilitated Austin's transfer of naming and request skills relative to Sherman.

Blind Tests of Tool Requesting, Labeling, and Receptive Skills

Following acquisition of tool symbols, "blind" tests were administered to ensure that the chimpanzees' abilities to request, to name, and to give tools were not cued by the experimenter. These tests were conducted in a slightly different manner for each task. In the functional task, the teacher baited the tool site with the keyboard turned off, then stepped outside the chimpanzee's room and turned on the keyboard. The teacher observed the chimpanzee's request on projectors outside the room, and then handed the chimpanzee the requested tool (see figure 9.1). In the naming task, the teacher again stood outside the room and held up a tool so that it was visible to the chimpanzee through a Lexan wall, but neither the chimpanzee nor the experimenter could see one another. The chimpanzee's responses again were monitored on projectors located outside the room (see figure 9.2).

During the blind test of receptive skills, one teacher stood outside the room and used her keyboard to indicate, by displaying symbols on the chimpanzee's projectors, which tool was to be selected. The chimpanzee observed the symbol as it appeared on his projectors,

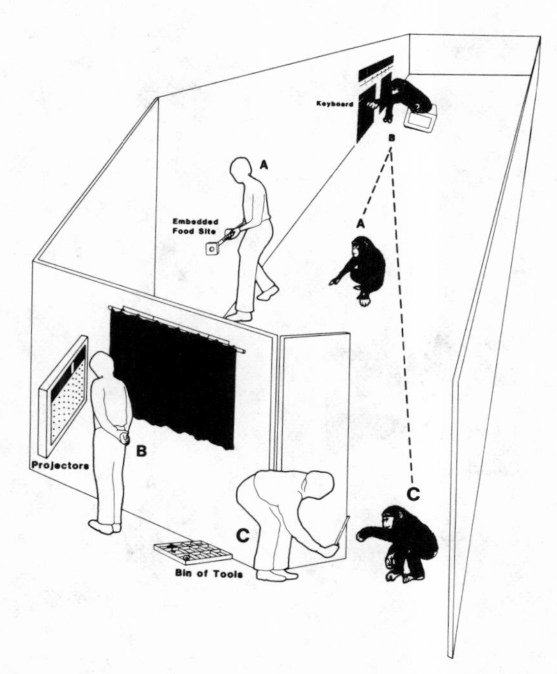

Figure 9.1. Blind test procedure used for the functional tool setting. The teacher places the food in a tool site (in this case the stick tube) while Austin watches (A), then leaves the room (B) and activates Austin's keyboard. (The keyboard is turned off to prevent Austin from requesting the tool while the teacher is present). Austin requests a tool and the teacher (noting the request on the projectors located outside the room) hands the tool to Austin (C). The purpose of this test is to eliminate the possibility of cueing Austin as he selects the symbol used to request the tool. Note that Austin cannot see the tools as he makes his request. He must recall the correct implement and select the correct symbol. Initially, this task was difficult for the chimpanzees when the tool kit was not visible as they tended to always approach the tool kit and look it over carefully before making their requests. They were never instructed to do this; however, it seemed that visually finding the tool in the tool kit aided the request process. With practice, they were able to request tools even without the tool kit present, though even then they would often hesitate and look back at the tool site as though to remind themselves of the needed implement before selecting a symbol.

chose a tool from the tool kit, and gave it to the second teacher, who did not know which tool the chimpanzee was to select. Both animals passed these tasks easily, thus demonstrating that their

Figure 9.2. Blind test procedure used in the "naming" setting. In this test, no tool site was baited. The teacher simply held up a tool as Austin came to the door of the room. Austin looked at the tool, then returned to the keyboard and "named" it. The teacher observed Austin's response on the projectors located outside the room. Initially, in this test procedure, the chimpanzees would often look quickly at the tool, then start confidently toward the keyboard, only to stop halfway and begin to stare at the keyboard as though they had forgotten what they were going to say and were trying to look over all the symbols to help remind them. We made it clear that it was permissible to come back and take another look at the tool if they needed to and they rapidly made use of this leeway. Thus instead of staring at the board when they forgot, they came back, took a second look, and "named" the tool. With practice, second looks became largely unnecessary.

abilities were not dependent upon cues from the teachers. (Table 9.3 shows each chimpanzee's total scores on each of the blind tests.)

Communicating the Need for a Tool to One Another

When the chimpanzees had acquired the individual skills of requesting and labeling, we hoped to determine whether they would

Table 9.3. Blind Tests

	Proportion and Percent Correct	
	Sherman	*Austin*
Functional task[a]	29/30 = 97%	29/30 = 97%
Naming task[b]	30/30 = 100%	30/30 = 100%
Receptive task[c]	27/30 = 90%	27/30 = 90%

[a] Chimpanzee requests tool with experimenter out of the room. Experimenter reads request on projectors outside the room.

[b] Question posed through projectors. Object to be named held by experimenter, who is outside the room and cannot see the chimpanzee.

[c] First experimenter informs the chimpanzee of requested tools through symbols on projectors. Tool is selected by the chimpanzee from tool kit and is given to the second experimenter, who does not know which tool was requested.

be able to use them to communicate to one another their needs for a particular tool. We asked: What if only one chimp had access to tools, but none of the tool sites in his room were baited with food, and what if the other chimpanzee saw the sites in his room filled with food but had no tools? In such circumstances, would they perceive the necessity of requesting tools from one another? Would the animal who had access to the tools watch the actions of the other at the keyboard? Would they willingly hand objects back and forth? Would they understand that they could make symbolic requests for objects of one another?

If Sherman and Austin could (1) attend to one another, (2) coordinate their communications, (3) exchange roles of tool-requester and tool-provider, (4) comprehend the function and intentionality of their communications, and (5) share their access to tools and the food obtained through tool use—then, by all functionally based definitions of human communication, they would have taken a significant step in using symbols very much as we do.

To answer these questions, we again placed them in two rooms separated by a large window. Both rooms were equipped with keyboards. By looking through the window of one room it was possible to observe anything that was "said" on the keyboard in the next room. While this window was covered, the teacher baited one of the tool sites in Sherman's room. The cover over the window between the two rooms was then removed, and all of the tools were located on a table in Austin's room. (On other trials a tool site in Austin's room would be baited and the tools were located in Sherman's room.) Keyboards were available to the chimpanzees in both rooms. On the first trial, after seeing the food placed in one of the tool sites in his room, Sherman went to his keyboard and asked for a tool. However, he directed this request, by gaze, toward the teacher, not Austin. The teacher responded to Sherman by demonstrating that she had no tools, then pointed to the tools in Austin's room. Sherman then redirected his request to Austin.

The roles of requester and provider were then systematically alternated from trial to trial. Both chimpanzees quickly began to orient toward each other and soon demonstrated that they understood the nature of the task and their respective roles in it. This understanding became evident when the tool provider began to hurry to the window to observe the request as soon as the window was opened. Moreover, if the teacher did not open the window so that the chimpanzees could communicate, they opened it themselves. (Recall that Mason's monkeys had failed to remove a barrier that was preventing them from communicating.)

Evidence of comprehension also appeared in the behavior of the tool requester as he began watching the other chimpanzee to see whether or not he was paying attention to the symbols being lighted at the keyboard. At times, if Austin (as recipient) appeared inattentive, Sherman (as requester) would get Austin's attention by repeating the request or by pointing to the projectors where the symbol for the tool was brightly displayed. Initially, the teacher gave some of the food obtained with the tool to the tool provider; however, the chimps soon came to do this on their own, following encouragement to share by the teachers.

The behaviors of the chimpanzees during the exchange of these various tools is illustrated in figures 9.3 through 9.22. These figures depict the entire sequence of events with a few of the tools used during the test. They reveal, in a way that words cannot, the joint concentration on the tasks of determining which tool is needed, attending to one another, and sharing with one another. Although only a single word was uttered each time, its utterance reflected a remarkable cojoint interpretation of the context. Moreover, the tight linking of symbol with contextual demands shows how symbols became a part of doing things for Sherman and Austin. This task required knowledge of how the tools functioned, skill in using them, knowledge of where they were located, skill in getting another to help one obtain them, and a willingness to reward others, all of which are reflected in the photographs. Such a complex intertwining of symbols, cooperation, and use of objects is usually seen only in man.

The experimental results, presented in terms of groups of trials per day, are shown in table 9.4. *All* trials during all instances of tool communication between chimpanzees are shown in this table. That is, table 9.4 is not presenting a series of successful performances following extensive training to ask one another for tools. Rather, it presents the whole of learning and testing that was accomplished in the interindividual setting.[4] Even on the first day that joint communication was attempted, the performance of the chimpanzees was far above chance, and as table 9.4 illustrates, their accuracy increased steadily across days. The rapid success in this complex task revealed that Austin and Sherman understood the underlying dynamics of the situation. Had it been necessary to extensively shape and teach *interindividual* communication (as had been the case with the initial learning of the symbols themselves), then we would have had good reason to doubt whether the chimpanzees even knew that they were communicating. Such doubts are inevitably present in the communication studies with pigeons (Epstein, Lanza, and Skinner, 1980).

Observations of the chimpanzees during this task suggest that their steady improvements were due to spontaneous advances in their abilities to attend to the communications of one another, and to coordinate the sending of their messages with the readiness of the recipient to receive the messages.

Noting how they dealt with occasions on which one or the other made an error often revealed more about their understanding than did correct trials in which everything went smoothly. For example, on one trial Sherman erroneously requested a key when he needed a wrench and he watched as Austin searched the tool kit. When Austin started to pick up a key, Sherman looked over his shoulder toward his keyboard. When he saw that he had selected the symbol for "key," (which was still displayed on the projectors) he rushed back to his keyboard, lighted "wrench" instead, and tapped on the projected wrench symbol to draw Austin's attention to the fact

Figure 9.3. Sherman says "Give Key."

Figure 9.4. Austin, observing Sherman's request through the window, reaches for the key.

Figure 9.5. Sherman approaches the window and Austin gives him the key.

Figure 9.6. Sherman takes the key to the padlocked food site and, holding the small padlock between the index and middle fingers of his left hand, starts to insert the key.

Figure 9.7. Austin watches while Sherman attempts to open the key site.

Figure 9.8. Turning himself, as well as his arm, Sherman manages to unlock the padlock.

Figure 9.9. Sherman brings a container of the pudding (which he found in the food site) back to Austin, sampling Austin's share along the way.

Figure 9.10. Sherman hands the container of pudding to Austin.

Figure 9.11. Austin watches as the straw site is filled with juice.

Figure 9.12. Austin (left), after asking for straw at the keyboard, goes over to Sherman. Sherman is still looking at the keyboard to see which tool Austin needs.

Figure 9.13. After giving Austin the straw, Sherman goes into Austin's room where they take turns sipping through the straw. Sherman is drinking and Austin is gesturing for Sherman to give him a sip.

Figure 9.14. Sherman complies, giving Austin a turn at the juice.

Figure 9.15. Austin then gives Sherman another sip.

Figure 9.16. Austin watches as food is placed in the wrench site.

Figure 9.17. Austin asks for the wrench, then goes to the window where Sherman has already selected the needed tool and is preparing to pass it through the window to Austin.

Figure 9.18. Sherman slips the wrench to Austin.

Figure 9.19. Austin studiously attempts to fit it over the tight bolt.

Figure 9.20. Sherman watches from his side as Austin tries to open the wrench site.

Figure 9.21. Austin carefully turns the wrench to loosen the bolt.

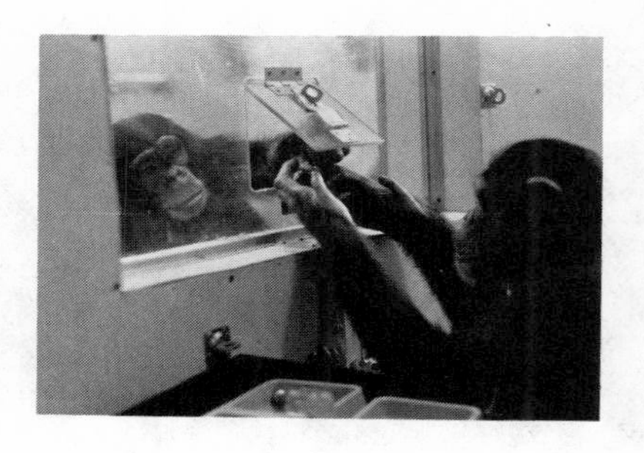

Figure 9.22. After obtaining M&Ms and bananas from the wrench site, Austin passes an M&M through the window to Sherman.

Table 9.4. Interanimal Communication Concerning Tool Transfer

	Proportion and Percent Correct											
Day	*1*[a]		*2*[a]		*3*[a]		*4*[a]		*5*[b]		*6*[c]	
Total correct	32/47	68%	38/50	76%	28/40	70%	27/30	90%	55/60	92%	3/30	10%
Sherman	36/47	77%	45/50	90%	32/40	80%	28/30	90%	55/60	92%	2/30	7%
Austin	43/47	91%	41/50	82%	34/40	85%	29/30	97%	58/60	97%	1/30	3%
Requesting errors	12		9		8		1		3		n.ap.	
Providing errors	3		5		6		2		2		27	

[a] Experimenter not blind.
[b] Experimenter blind.
[c] Controlled condition; keyboard turned off.

that he had just changed his request. Austin looked up, dropped the key, picked up the wrench, and handed it to Sherman. It seems quite unlikely that their training could possibly have conditioned such behaviors.

Controls

Following the achievement of successful communication about needed tools, we posed the same problem for the chimpanzees, but turned the keyboards off. We did so to determine that it was indeed the lexigrams which were conveying the information about the type of tool that was needed as opposed to some undetermined cue. By turning off the keyboards, and keeping all other factors constant, the importance of the ability to use these symbols could readily be ascertained.[5,6] On the first trial of this control test, Sherman, who needed a sponge, rushed to the keyboard and tried once to press sponge. When it did not light, he then went to the window and looked at Austin. Austin did nothing. The teacher sitting with Austin encouraged Austin through gestures, to go ahead and hand Sherman a tool even if he could not see a symbol. Austin handed Sherman a straw. Sherman appeared irritated and perplexed. He grabbed the straw from Austin, shook it at the teacher, and threw it down.

The performance of the chimpanzees on these trials with the keyboard turned off dropped below chance to 10 percent correct. On trials with the keyboard turned on it was 97 percent correct.[7] On trials without the keyboard, the chimpanzees, after initially refusing to try, finally adopted a strategy of either cycling through all the tools or repeatedly offering the same tool. It seemed that they simply wanted to get through these "unproductive trials," during which the teachers would not let them use the keyboard, in order to please us. But clearly they could not communicate, and they behaved as though they knew it.

Did Sherman and Austin Really Know That They Were Asking One Another for Tools?

Once Sherman and Austin had achieved the capacity to request both foods and tools of one another, and to respond cooperatively to one another's requests, critics raised the issue of whether or not they actually understood what they were doing (Bennett 1978; Epstein, Lanza, and Skinner 1980). Some offered the suggestion that the chimpanzees had simply learned "how" to get things from one another, and pointed out that such learning was different from actually "asking" one another for things. That the chimpanzees directed their requests toward one another by gaze and gesture, that the response of the giver was coordinated with the asker's request, and that requests were contingent upon the presence of a cooperative recipient, were not viewed as evidence of comprehension but rather as "trained" behaviors. These critics seemingly disregarded or did not understand that while the component skills of labeling, requesting, and comprehension were trained, the interindividual use to which Sherman and Austin put the skills was not.

Joint regard, amplification of symbols with gestures, and spontaneous correction of errors were behaviors that emerged out of the interindividual interactions between Sherman and Austin. They were not trained by the experimenter, and it was not clear that such skills could be trained in any typical sense. Sherman and Austin had to learn how to communicate with one another. In this sense, *they* may be said to have shaped or trained each other.

It is also important to recognize another issue framed by critics such as Bennett (1978): do Sherman and Austin actually "know" that they are communicating with one another or do they just go through a highly conditioned routine? This objection ignores the fact that many primate species have been shown to demonstrate a rather advanced notion of causality (Chevalier-Skolnikoff 1973; Taylor-Parker and Gibson 1979). For example, chimpanzees give clear evidence of knowing that they can use actions and gestures to cause other chimpanzees to engage in particular actions (Savage 1975; Menzel and Halperin 1975; Woodruff and Premack 1979). While appearance of the capacity to attribute causality is delayed in the chimpanzees relative to human children, who demonstrate an understanding of interindividual causality at about nine months (Mathieu et al. 1984) it is nonetheless clearly extant in the apes.

When the capacity to attribute causality appears in human children, it is manifested communicatively by the appearance of visual checking to see if they have the recipient's attention before they gesture or vocalize. Also, at about this same time, children begin to request that others act upon objects in their behalf. When such behaviors begin to appear they are not stereotyped in any sense—that is, young children (or chimpanzees in the case of Sherman and Austin) will request that the adult act on all manner of objects for them, in many different contexts, and in many different ways. Such behaviors reveal that the children or apes have now developed a concept of causality which permits them to use another individual as a tool. That is, they know that they can causally alter the behaviors of others for specific goals, just as they can causally alter the positions and uses of objects for specific goals. It is this discovery of "tool-use" in the general sense that makes linguistic requests possible.

Such linguistic requests are a function of the children's or apes' increasing comprehension of the span of their own causality. This reflects a second-order comprehension of causality; they now realize they can accomplish goals not only directly, through their own actions, but also indirectly, through their actions upon objects and upon others.

This new comprehension is different from the earlier learning that fussing produced food, changed diapers, etc., in that the individual now generalizes beyond specific behaviors and specific situations wherein act *x* produces act *y* on the part of mother. Now, they decide what they want their mother or teacher to do *in advance of her doing it* and they show her what this is. For example, when a child points to a broken toy and gestures and whines, he is requesting that the mother engage in acts which might make the toy operational. The child is able to make such requests clear to his mother even though this toy may never have been broken before and he may never have observed his mother fix it before. This sort of generalized intentional requesting appears before the one-word stage and is clearly different from the "engaging in act *x* to produce event *y*" type of causality which the child or ape evidenced at an earlier age because of the emergent cognition that generalizes rapidly to novel situations, thereby permitting the child or ape to causally alter the behaviors or ways of his own creating.

It is the emergent realization that one can causally affect the behavior of others, in a general sense, which permits true requests (as opposed to learned acts) to appear on the plane of interindividual interaction. Such intentional communication appears spontaneously not only in human beings, but in apes as well; it may appear in dolphins (Lou Herman, personal communication) and elephants, too. Its appearance is necessary to the occurrence of language as we know it, and while human tutoring may expand upon this capacity, it is not the tutoring which provides the apes with that capacity initially. Clearly they must have an innate cognitive capacity which generates an awareness that one can causally and communicatively alter the behavior of others. This ability in turn provides for the general capacity of chimpanzees to learn to make symbolically encoded requests of others.

The view that chimpanzees do know they are producing a request directed toward a recipient is strikingly confirmed by the communicative repair processes that chimpanzees bring to bear when communicative failure occurs. Such processes include: spontaneous gestural elaborations of the symbolic requests, efforts to orient the recipient's attention appropriately, reformulation of ineffective requests, and repetition of the requests following reengagement of attention. These processes will be described more fully in the chapters which follow.

It is important to point out that clarification of the issues of awareness and intentionality are not furthered by the use of terms such as "Sherman believes that use of symbol *x* will cause Austin to believe that he, Sherman, desires *x*," à la Bennett (1978). Distinctions like this are not behaviorally verifiable even with the extensive use of inference. However, an organism's emergent understanding of its own capacity for causality is behaviorally verifiable and consequently an area of investigation which can prove fruitful even with noncompetent linguistic individuals.

Chapter 10

The Intermeshing of Gesture and Symbol

Focusing on Things That Were Not Taught

More intriguing in many respects than the symbolic exchanges of information we had set about to obtain were the interindividual communicative skills which emerged between Sherman and Austin as they learned to ask one another for tools and foods. With continued practice on tasks requiring them to communicate with one another they began to attend very closely to one another's communications, to engage one another before delivering a message, to elaborate gesturally or clarify messages, and to take turns spontaneously. Consequently, their communications assumed an increasingly human countenance, not by virtue of specific training, but by virtue of use. Moreover, (with the exception of turn-taking), we were not teaching or shaping these regulatory skills in any prescribed way; rather they were apparently being developed because they were indispensable to the smooth flow of communication.

Once these metacommunicative skills became established in Sherman and Austin, we decided to study them in some detail in the food sharing setting since the chimpanzees' mutual attentiveness, awareness, intentionality, and nonverbal behavioral coordination were spontaneous and at their best in this area.

By this point, Sherman and Austin were no longer using symbols to ask for foods located in another room. Rather, they had advanced to sitting down together next to a table laden with foods. Each type of food was always divided into two portions. The chimpanzee nearest the keyboard was designated the "requester" by virtue of his proximity to the keyboard, while the chimpanzee on the far side of the table was designated the "giver." The requester could ask for any food he desired. The giver responded by selecting a piece of that food, handing it to the requester, and then taking the remaining portion for himself.

Each role had advantages. The requester could decide which food would be eaten on any given trial, while the giver could decide which piece was larger and keep that one for himself. Both chimpanzees gave all appearances of enjoying the whole thing and they cooperated readily. It was rarely necessary to remind them to ask for food (as opposed to simply taking it), nor was it continually necessary to encourage them to attend to and cooperate with one another's requests. Since two pieces of food were available, cooperation was mutually beneficial on each trial and generally spontaneous (see figures 10.1–10.10).

One might object that by requiring chimpanzees to ask one another for specific foods and to take turns, we were imposing upon them a rather meaningless set of rules that they had to follow in order to eat. To the contrary, eating was never contingent upon

Figure 10.1. Austin (requester) closely watches as Sherman eats a piece of food.

Figure 10.2. Sherman (giver) wants to eat and is directing Austin's attention to the keyboard, encouraging Austin to ask for another food to share.

Figure 10.3. After studying the keyboard, Austin asks for bananas.

Figure 10.4. Sherman selects a slice of banana for Austin.

Figure 10.5. As Austin eats, so does Sherman.

Figure 10.6. Sherman, now the requester, studies the keyboard as Austin watches.

Figure 10.7. Sherman asks for cherries as Austin observes.

Figure 10.8. Austin (giver) does not respond but looks at an empty glass. Sherman points at the cherries.

participating in food sharing bouts, yet the chimpanzees were eager to engage in food sharing. For them, sharing was an accepted and comfortable means of dealing with food, and using the keyboard to request foods was easy for them to do.

Food sharing bouts had become highly desirous occasions because so many of their favorite foods were carefully prepared and laid out for their selection. It was something like a holiday feast. The "manners" with regard to the feast were similar to those human beings use: you ask for dishes to be passed rather than just grabbing them; and you make certain everyone gets a portion. Sherman and Austin seemed to accept these "alien" prescriptions readily, just as do children, and to go on to enjoy the meals lavishly.

Food Sharing Bouts: The Perfect Blind Setting

It became *unnecessary* for a teacher to be present during food sharing bouts. This made it possible to collect data simply by setting up a video camera and leaving the room. Consequently, we could

Figure 10.9. Austin smiles and picks up the cherries while Sherman waits with outstretched hand.

generate a permanent record of everything (gestural, vocal, lexical, etc.) which occurred between the chimpanzees and simultaneously eliminate any possibility of inadvertent cueing.

Once Sherman and Austin sat down in front of a table of food, the exchanges flowed between them smoothly until all the food was gone. The duration of any given group of such exchanges was a function of the number of different types of food placed on the tray. A group of such exchanges (that is, the series of events which occurred from the time Sherman and Austin sat down in front of a table of food until the time they moved away) was termed a "food sharing bout." Within any such bout, a number of individual "food exchanges" took place. The data analysis which follows is based upon 27 such food sharing bouts which were taped over a four-day period. During these 27 bouts, a total of 255 food exchanges took place between Sherman and Austin. No teacher was present during any of these exchanges; however, all exchanges were monitored on live video from an adjacent room.

Figure 10.10. Sherman shows Austin a symbol when Austin seems hesitant to select his own.

Between seven and thirteen items (of a possible twenty different foods and drinks) were placed on the food tray during each food sharing bout. The items used were: bread, bean cake, banana, cheese, cherry, Coke, corn, juice, lemonade, M&M, milk, orange drink, orange, pineapple, pudding, peaches, strawberry drink, sweet potato, raisin, and cake. In order to discourage Sherman and Austin from simply settling into a routine of always asking for foods in the order of most to least preferred, we varied the size of the portions, the quality of the portions, their locations on the tray, and the combination of foods present during any given bout. Thus on one trial, large pieces of cake (a preferred food) might be paired with two M&Ms, two large Cokes, small cheese pieces, etc. Appendix A gives the order in which Sherman and Austin requested the various foods during each bout. It is clear that there is no "order effect."

During a given food exchange, the chimpanzees could request any food that was present on the tray in front of them. If they requested a food that was not present, they were scored in error.

The other chimpanzee could give any food that was present. If he gave the food that was requested, he was scored as correct. If he gave any other food, he was scored in error.

Of these 255 exchanges, 249 were completed while six were not (that is, a request made by one chimpanzee was responded to by the other chimpanzee in a clear quantifiable manner). During 242 of the 249 completed exchanges, keyboard symbols were used to request foods. On seven occasions, gestures alone were used to request the desired food. Thus, 92 percent of all exchanges were carried out correctly: the food that was requested was present on the tray, and it was selected and given to the requester. Sherman initiated 111 correct requests to which Austin executed 99 correct replies. Austin initiated 95 correct requests to which Sherman executed 94 correct replies. Sherman also indicated and carried out 28 correct statements. That is, he announced which food was to be selected next, then gave half of that food to Austin and half to himself. Food was shared during all but three exchanges. On these three trials, Sherman ate all of that food himself.

The overriding impression one receives from viewing the video tapes of these food sharing bouts is that although the context, structure and participants are the same from bout-to-bout, the behavioral variability from bout-to-bout is great. The chimpanzees coordinate not only the requesting and giving of foods but many other facets of their behavior as well. They often chew their food in synchrony, orient to the keyboard simultaneously, and assume reciprocal postures. Similar unconscious nonverbal coordination of behaviors, before engaging in conversation, has been found in human beings (Kendon, 1973). Presumably such behavioral coordinations bring about a synchrony of mood and attention, and serve to set the cognitive stage for the onset of communicative acts. Until these tapes of food-sharing were studied in detail, we were unaware of the significant degree of synchrony among so many of their movements. Each food sharing bout is clearly very different in tone, mood, pace, and specific regulatory and gestural behaviors from the bouts which precede it and the bouts which follow it. On each occasion, Sherman and Austin have a somewhat different rapport, different levels of interest in and attention to the communication at hand, and different methods of dealing with confusions, errors, temporary lapses of attentiveness on the part of another, etc. Three separate bouts will be detailed to illustrate this variability.

Bout I, General Description

In this first bout, Sherman is seated nearest the keyboard and thus will do most of the requesting. Austin is seated to Sherman's left and will give most of the foods. As the chimps enter they give soft barks and appear eager to start. Both are attentive. Occasionally, Austin turns to glance at the camera and back toward the door, where he entered from the adjacent room. The teachers are in the adjacent room and are watching Austin and Sherman on the video monitor. The teachers may call out if the chimpanzees approach the camera; however, this happens infrequently and any trial which was so interrupted is excluded from the data analysis.

During this bout there are a number of deviations from the basic format because Austin is not as hungry as Sherman and is often distracted from the activity at hand. Sherman must employ gestures to amplify and to clarify his symbolic requests. On one trial gestures completely replace symbols.

Bout I, Trial by Trial Account

1. Sherman requests "Pudding." There is no "pudding" on the food tray. Austin extends his hand toward the food tray but hesitates, holding his hand just above the peaches as though uncertain, and looks at Sherman. Sherman quickly points to the peaches. Austin then gives the peaches to Sherman and to himself.

(This trial shows how gestures can replace symbols, particularly when an inappropriate symbol has been selected. On the majority of trials, Austin shows no hesitation, and Sherman does not point to foods.)

2. The same sort of monitoring of the communicative situation is evidenced by Sherman again in the following trial. Austin is not looking at the keyboard as Sherman makes his request. Sherman compensates for this by pointing to the specific food (cheese) as soon as Austin turns around and then he looks directly at Austin and monitors Austin's facial expression until Austin begins to reach toward the food tray. At that moment, Sherman's eyes drop and follow Austin's hand. Austin reaches for the cheese, glancing up at the keyboard as he

selects this food, but not before. Austin appears to be checking to see if the food he is about to pick up agrees with the symbol which is displayed on the projectors.

3. On the following trial, Sherman requests cherries and Austin is watching during the request and responds immediately. (On this trial, Sherman has no need to amplify his request with a gesture and he does not do so.)

4. The following trial essentially repeats the events observed on trial 2. Again Austin is not looking as Sherman selects the symbol, and so Sherman points to the food he wants when Austin turns back around. Again Sherman monitors Austin's face, until Austin's hand begins to move toward the tray. Sherman's eyes then immediately follow Austin's hand until he receives the banana he requested.

5. Before this trial, Austin again looks away as Sherman is surveying the keyboard, but then turns back around and is watching when Sherman selects the symbol (M&Ms). Sherman does not point to the food, nor does he need to, since Austin reaches directly for the M&M's, of which there are four on the tray. Austin gives one M&M to Sherman, then leans over to eat the three remaining M&M's directly from the tray. Sherman reaches out toward Austin's face and Austin removes his mouth from the food tray. Sherman looks at the place where the M&M's were and, seeing that Austin has eaten them all, asks for another food.

6. On this trial, Austin is looking away when Sherman makes his request and because the keyboard malfunctions Austin does not see that any symbol has been selected when he turns around. Sherman seems to have noted that Austin did not attend to his request and so he points to the item he requested—lemonade. Austin glances at the keyboard and, seeing no symbol, he ignores Sherman's point, even though the pointing is very definitive and Sherman is clearly indicating the lemonade. Austin is also still chewing the M&M's from the previous trial and so Sherman stops pointing and watches Austin closely until Austin is through eating. (Occasionally, when Austin is eating, he will not respond to Sherman's requests for additional food, so Sherman simply waits.) When Austin is finished chewing, Sherman sits back up and again points to the lemonade and looks directly at Austin. Austin still does not reach toward the lemonade. Instead, he looks directly at the keyboard as though expecting Sherman to say something.

Sherman then redirects his own gaze toward the keyboard and appears to realize at this point that Austin is not going to respond to his pointing gestures and is waiting instead for him to select a symbol. Sherman looks back and forth between Austin and the keyboard, appearing to be puzzled as to why Austin will not give the lemonade.

7. Failing on lemonade, Sherman then selects "sweet potato" and quickly points to it without even looking at Austin to determine whether or not Austin is watching. Austin is, in fact, watching quite closely this time and is reaching toward the sweet potato *even* as Sherman is starting to point.

(This was the *only* occurrence of pointing during these 27 videotaped bouts which was not preceded by either hesitation or lack of response on the part of the recipient of the gesture. It was apparently elicited by Austin's lack of responsiveness on the preceding trial.)

8 & 9. On the next two trials (bean cake and orange), Austin is closely observing Sherman as the symbol is selected and he quickly responds to each request. Sherman does not gesture or look directly at Austin. He simply watches Austin's hand movements and coordinates his reaching out to receive the food with Austin's giving.

10. Sherman then requests "Coke," and Austin observes his request. However, Austin is still trying to get the last bit of his orange loose from the peel and he does not reach toward the Coke. Instead of pointing then looking directly at Austin, as Sherman did on trials 2 and 4, Sherman looks at Austin first and then points to the Coke while continuing to look at Austin. (In this case, since Austin was not looking away during the request, Sherman apparently did not presume he needed to point. However, when he looked at Austin, he saw that Austin was preoccupied with eating, and hence Sherman pointed to the Coke.) Again Sherman's gaze drops away from Austin's face as soon as Austin reaches toward the requested food.

11. When all of the items except lemonade (the item which Austin refused to give on trial 6) have been consumed, Sherman again requests lemonade. Since lemonade is a highly favored drink, it is not likely that Sherman waited until it was the only remaining item on the tray. Rather, it appears that he dealt with Austin's lack of cooperation in regard to this item in the most pragmatic fashion available: he eliminated all the other alternatives so that Austin had no choice to make. However,

when Sherman again asks for the lemonade, Austin is preoccupied with trying to pour Coke from his glass into the small cuplike container that had previously held peaches. Thus Sherman again points to the item, and this time Austin reaches directly for the lemonade as Sherman is pointing.

Bout II, General Description

In this bout, it becomes more apparent that each set of food sharing interactions between Austin and Sherman has its own unique character. Here, the roles of requester and giver are exchanged. Austin is now seated nearest the keyboard and Sherman is seated in front of the food tray to Austin's left. The change in seating is required by the teachers and is meant to signal a change in roles to the chimpanzees.

Bout II, Trial by Trial Account

1. As Sherman and Austin assume their seats, Austin is more concerned about getting something off his toe than he is about deciding which food to ask for. Sherman takes note of this and places his right hand behind Austin's neck and shoves Austin toward the keyboard in an apparent effort to get Austin to look at the keyboard and select a symbol so that they may proceed. Austin does not appear to appreciate this and he shrugs his shoulders in an attempt to get Sherman's hand off, but to no avail. Austin looks down at the food tray, then at the keyboard. As Sherman gives him a second shove, he lights "Cherries." Sherman looks at the symbol as Austin lights it, then down at the tray, but instead of selecting the food immediately, he looks back at Austin and Austin looks at him. Then Austin looks directly at the cherries. Sherman appears not to have seen the cherries when he first glanced at the tray, but after looking at Austin he follows Austin's glance and readdresses the tray of foods and drinks. He then spies the cherries and gives Austin one as he again visually checks the symbol on the keyboard and takes one for himself.

2. As soon as Sherman has the cherries in his mouth, he again puts his hand behind Austin's neck and starts pushing his head toward the keyboard. Austin glances at the food tray and starts to light "Lemonade;" however, just before Austin touches the key, he withdraws his hand as if to take the cherry pit from his mouth. As he does so, Sherman, who has been watching

Austin closely, gives him another very firm shove. Austin lights "Lemonade," then looks directly at Sherman. Sherman briefly surveys the two glasses of lemonade to determine which has the least liquid. He gives that one to Austin and drinks the other one himself.

3. Austin begins searching the keyboard for a symbol without even looking down at the food tray, apparently remembering the food which he wanted to request next from the previous trial. Austin lights "Bread." Sherman is watching as Austin lights the bread symbol and immediately puts down his glass and starts looking for the bread. Austin also looks at the tray and quickly glances from Sherman's hand to the bread and back. Sherman spots the bread and Austin watches as Sherman selects a piece for Austin, then a piece for himself.

4. Austin looks around as he eats his bread slowly. Sherman also eats the bread slowly. When Austin starts to survey the keyboard, Sherman redirects his attention from his own piece of bread to Austin and the keyboard. At this point, both hold their pieces of bread with the edges of their lips as they chew in exactly the same fashion. Austin requests "Pudding" and looks directly at it. Sherman, who was watching during the request, gives pudding to Austin and himself and both chimps swallow the last bite of bread as they prepare to eat their pudding.

5. Austin finishes his pudding first and then begins to survey the keyboard. As Austin looks it over, Sherman watches. Austin touches "Cheese." Sherman looks over the food tray but does not select the cheese immediately so Austin points to it with his left index finger. Sherman then gives a piece to Austin and to himself and their attention to each other breaks momentarily as they begin to eat.

6. Sherman again puts his hand behind Austin's head and neck and tries to orient Austin toward the keyboard. Austin ignores Sherman's push and, in fact, looks *directly away* from the keyboard. In response to Austin's "looking away," Sherman gets off of his seat, lights up "M&M," and looks directly at Austin. In keeping with this spontaneous shift of roles, Austin selects M&Ms and gives some to Sherman and some to himself. Sherman then returns to his seat and appears to wait for Austin to request the next food, thus spontaneously resuming his assigned role.

7. Austin, however, again looks *directly away* from the keyboard and when Sherman does light a symbol in response to

this turning away, Austin looks directly at Sherman. Sherman then lights "Strawberry—drink," at which time Austin selects the fuller of the two glasses and keeps that one for himself; Sherman gets the other one.

8. Austin finishes drinking first, but watches Sherman. When he notes that Sherman has drunk all of the strawberry drink, Austin asks for "Banana." As soon as "banana" is selected, Sherman puts his glass down and reaches for the banana which Austin is looking at. He gives a piece to Austin and keeps a piece for himself. Again they look away from each other as they eat.

9. Sherman is looking away from the keyboard as Austin selects the next food, "Orange." However, he hears the sound and immediately reorients to the board just as Austin looks directly at him. Sherman selects a piece of orange for Austin and for himself as Austin watches. Both again turn away from each other as they eat.

10. Just as Austin begins to look over the keyboard for the next food, Sherman glances in his direction. Austin requests "Raisin". Sherman gets down from his seat and begins eating the raisins from the tray with his mouth, instead of giving any to Austin. Austin then leans over, too, and eats his share with his mouth. As Austin is leaning over, he knocks a glass off of his seat. He stops eating, retrieves the glass, and stacks it inside another empty glass. After eating the raisins, Sherman returns to his seat and both look around the room.

11. Sherman states "Corn" and both help themselves to corn.

12. Austin moves his hand in front of the keys and as he searches for a symbol, he accidentally touches the generic term for food. (This appears to be unintentional as he does not look at Sherman and because his hand brushes the keys on an upward motion instead of moving directly to a key and remaining there briefly.) Austin then continues to look at the keyboard but waits to select a key until Sherman turns around. He then lights "Pineapple" while Sherman is watching, and Sherman selects pineapple pieces for Austin and for himself.

13. Instead of looking away while they eat, they now look briefly at each other. Sherman then looks toward the keyboard. Austin follows his gaze and selects the "Sweet potato" symbol as Sherman watches. Sherman gives this food to Austin and himself.

Bout II, Summary

Clearly, this food sharing bout is very different from the bout described previously. In this bout, Sherman and Austin have demonstrated an ability to exchange roles *during* a given bout, regardless of where they were seated, simply by the exchange of glances and by directing one another's gaze toward or away from the keyboard. Such role exchange during bouts was not taught; indeed, it was actively discouraged by the teachers. Thus, Austin and Sherman appear to have a concept of turn-taking that goes beyond their particular seating arrangements and arbitrarily imposed role assignments.

Their concept of sharing is also broader than the specific behaviors they were taught, since at times each simply takes some food for himself instead of waiting for the other chimpanzee to hand the food. This mutual taking is done in total agreement, and apparently this can be done either with the mouth or the hands and without discord, even though small pieces of food (such as raisins) are clustered tightly together in a small space. Such mutual taking was discouraged by the teachers (when they were present), thus suggesting that this is an agreement which Austin and Sherman have worked out for themselves. Their agreement here (that each can *take* a share) differs distinctly from the giving rules they were taught.

The coordination of their glances and gestures, which are essential to the flow of their exchanges, also was not taught. Sherman and Austin were encouraged to look at one another and at the symbols the other had lit; however, such encouragement on the teacher's part was often too late, since by the time we got one chimp to look at the other, the other chimp was already doing something else anyway. It is simply not possible to teach such finely coordinated interindividual behaviors in any trial-by-trial manner. It is, however, possible to set up situational contexts which foster and encourage the development of interindividual behavior (as did the food sharing situation). The coordination itself emerged from Sherman and Austin and their attempts to work together.

Bout III, General Discussion

The third and final bout to be discussed here illustrates that Sherman and Austin could freely intermingle gestural specification with symbolic specification. It also reveals that they could negotiate the sharing of food equitably in a far wider variety of ways than they had been taught.

Bout III, Trial by Trial Account

1. Austin looks around the room instead of addressing the keyboard so Sherman puts his hand on Austin's neck and shoves Austin toward the keyboard. As a result of being shoved, Austin accidentally lights a symbol, "Liquid," which is not in his vocabulary but is in Sherman's. (The keys are very sensitive and can light if a finger is held very close to them without actually touching them.) Austin then bites the keys instead of looking toward Sherman to hand the food as he would normally do after a request. This is the only request where Austin inadvertently lights a symbol which he does not know and also the only trial in which he bites the keys; apparently he is uncertain as to how to deal with this "error." Sherman, however, does know "Liquid" (as he has had separate training on this categorical term) and thus, after glancing at the food and then at Austin (who ignores him), Sherman solves Austin's dilemma by handing him orange drink—one of three liquids available at that time on the food tray.

2. They finish their orange drink almost simultaneously and Austin looks around, then looks at the keyboard. However, he does not start to light it. Instead, he looks away, then looks back toward it. Sherman, who has been watching Austin, leans forward, lights "Banana," and gives a banana to Austin and to himself.

3. As they are eating the banana slices, Austin begins looking around the keyboard as Sherman watches. Austin lights "Cheese," glances briefly at the cheese, then at Sherman. Sherman immediately selects "bread" (which is near the cheese on the tray) for Austin and for himself. Austin readily accepts and eats the bread; then he and Sherman glance at each other. Austin then reaches out and takes a piece of cheese for himself. Sherman watches and does likewise.

4. As they are eating their cheese, Sherman looks directly at Austin and Austin turns away. Sherman responds to Austin's turning away just as he did in the earlier trials of this bout. He assumes the requester role and selects "M&M." As soon as Sherman starts his request, Austin stops looking away and turns

toward Sherman to see what he will say. Sherman gives an M&M to Austin and one to himself.

5. Once Austin receives his M&M, he again turns away. Sherman selects "Peaches." As Sherman starts to pick up the peaches Austin looks back, then Austin picks up and gives the peaches to Sherman and himself.

6. When Austin finishes his peaches and puts down the container, Sherman looks at the food tray and waves his hand toward it. Austin responds to Sherman's gesture by helping himself to a piece of cheese. Sherman also takes a piece.

7. Sherman gestures toward the food tray again and Austin responds by selecting and eating a raisin. Sherman then leans forward and eats the rest of the raisins himself with his mouth.

8. As Sherman's head is down in the tray (eating the remaining raisins), Austin starts looking around the keyboard and gestures toward the keyboard with his hand as if to light a symbol. However, as Sherman raises his head up, Austin withdraws his hand and looks at Sherman. Both Sherman and Austin then look toward the keyboard. Austin again gestures with his right hand and looks toward the board, apparently encouraging Sherman to go ahead and select the food. At this point, they appear to be looking at the same key. Sherman leans back and gestures with his right hand for Austin to go ahead, which Austin does, lighting "Pineapple." Sherman immediately gives pineapple to Austin and to himself.

9. As they eat the pineapple, Sherman looks at the keyboard and Austin follows the direction of his glance and lights "Juice."

10. Sherman finishes drinking the juice first and looks at the keyboard. Austin also orients toward the keyboard and then reaches down as if to light "juice" again; however, he stops short of lighting the symbol and begins to scratch himself and look around the other keys. Sherman then leans forward, lights "Coke" and gives some to Austin and himself. (There was no juice left, only Coke. Apparently Austin was having trouble locating the correct name.)

Bout III, Summary

Throughout this bout, Austin repeatedly shifts the role of requester to Sherman by turning away from the keyboard. Sherman usurps Austin's role only on precisely those occasions where Austin turns away from the keyboard. This "turning away" is a complete head turn in which Austin rotates to face a blank wall.

The detailed analysis presented above reveals that Sherman and Austin's use of gestures is not sporadic. Rather gestures are used specifically to clarify requests. Additionally, they can be used to pass the option of requesting food back and forth, thereby functioning as indicators of role. Gestures enable Sherman and Austin to violate the *specific* exchange format they were taught while still maintaining the fundamental elements of communicating sharing.

Until we studied the taped sequences, we were unaware that Austin and Sherman had worked out their own means of signaling who was to use the keyboard, or who was to give the food. We also did not realize the highly specific and deliberate roles that gestures had come to play in the general coordination of behavioral events. Nor did we know that Austin and Sherman could work out ways of sharing food that were radically different from the specific sharing behaviors that we had taught them, yet maintained the principle of sharing.

These surprising capacities revealed that Sherman and Austin had learned considerably more than simple interindividual chains of behavior. They had learned general principles or rules regarding interindividual verbal communication, and they had acquired a general sense of role exchange and cooperative sharing.

It was this broad understanding which made it possible for the food sharing bouts to continue when one chimpanzee violated expectancies by eating both portions of a food. The other chimpanzee was able to respond by eating both portions of another food and, having shared in this way, they continued as though nothing had gone awry. Certainly, sharing of this sort (i.e., one chimp gets the pudding and the other gets the milk) is very different from the "each gets half of each food" which we taught.

Data Summary of all 27 Bouts

The insights we gained from the three bouts described above were tested by analyzing the data for all 27 food bouts—or 228 exchanges.[1] Before we did this detailed tape analysis, we were, of course, aware that the chimps often gestured during food sharing bouts. What we did not know, however, was why gestures occurred on nearly every trial during some bouts, while at other times none at all appeared. We had presumed that perhaps the chimpanzees tended to gesture more when they were hungrier and that the gestures were simply a reflection of overall increased motoric activity. We knew that Austin gestured much less often than Sherman and suspected that this was because of his subordinate rank or his younger age. When we analyzed the 27 taped food-sharing bouts, we realized our assumptions were incorrect.

Each tape was analyzed in a trial-by-trial fashion. A single observer viewed and reviewed each trial until each of the items listed below was scored and then rescored for accuracy (this scoring often required as many as 20 to 30 viewings per trial):

1. The presence or absence of vocalizations or other extraneous noises in adjacent rooms. (All such trials [27] were then eliminated from the data analysis.)
2. The presence or absence, on a given trial, of any behaviors which were a deviation from the food sharing format that had been taught Austin and Sherman.
3. The presence or absence of gestures.
4. When the gesture occurred (before the symbol was selected, before the food was selected, after the food was shared, etc.).
5. The number of gestures (if greater than one) that occurred on each trial.
6. Which chimpanzee gestured.
7. Whether or not the "speaker's" gesture was preceded by hesitation, inattention, or other overt behavioral demonstration of uncertainty on the part of the "listener" (such as hovering the hand above the food tray while looking at the other chimp).
8. Whether or not the gesture was made with the right hand, left hand, or both.
9. Whether the gesture was a pointing gesture or some other type of gesture (if other, the gesture was described).
10. Whether or not the "giver" was looking at the keyboard when the "requester" lit the symbol.

We also selected at random 45 trials for scoring by a second observer in order to establish a reliability measure. For the 45 randomly selected trials which were scored by both observers, there existed a total of 495 possible occasions for disagreement, since 11 items were scored for each trial. Overall, there were only 7 disagreements or 99 percent agreement on the items scored.

Of greatest interest in these bouts is the use of untrained gestures. Overall, 72 such gestures were observed and these gestures served several purposes. A delineation of the basic functions of such gestures and the number of gestures of each general type is given in table 10.1. A specific description of each gesture is given in appendix B at the end of this chapter.

The data unambiguously revealed that Sherman and Austin were keenly aware of each other's behavior and that their gestures were not just random occurrences or ritualistic mannerisms. Rather, the

Table 10.1. Function and Number of Gestures for Each Chimpanzee

	Number of Gestures	
Function of the Gestures	*Sherman*	*Austin*
1. To get the other chimp to light a symbol to request food.	21	3
2. To point to the specific food requested when the chimpanzee who is to give the food did not observe the request.	16	0
3. To encourage the chimpanzee who is to give the food to go ahead and do so when preoccupied with other things (stacking glasses, eating leftover peels, etc.)	9	3
4. To point to the specific food requested when the chimpanzee who is to give the food observed the request, but shows uncertainty about which food to select.	5	1
5. To request that the other chimp give a food even though no symbol has been lighted.	3	1
6. To specify which of the two portions is desired.	2	2
7. To show the other chimpanzee which symbol to light.	3	0
8. Function cannot be determined	2	0
9. To encourage the other chimpanzee to share food.	1	0

gestures fit the minute-by-minute communicative demands of the situation and showed that the chimpanzees were continuously monitoring the information present in the situation. Virtually all of the indicative pointing gestures which specified a given food occurred because the recipient of the gesture was not looking at the time the request was made. When the recipient was looking during the request, then pointing gestures occurred if and only if the recipient evidenced some further sort of uncertainty, such as waving his hand around over the food, without selecting the requested food.

On every trial, the behavior of the giver was scored as to whether he was looking at the keyboard at the time the request was made. We found that Sherman observed the keyboard during all but three of Austin's requests. Austin, however, was much more inattentive. He failed to observe Sherman's request on 20 trials. On three other trials, Austin had even gotten down from his cube and was engaged in some other activity.

Sherman was acutely aware of whether or not Austin was looking at the keyboard during his request. On 19 of the 20 occasions when Austin did not observe the request, Sherman took additional steps to ensure that Austin would have the needed information. On 16 of these trials, Sherman "helped" Austin by pointing to the specific food which he (Sherman) had just requested. On two trials, he selected the food himself, giving half to Austin and half to himself, and on one trial, he simply ate his portion of food and Austin's portion.

There was only one trial when Sherman did not compensate for Austin's inattention during his request, and on that trial "strawberry drink" was requested. Strawberry drink has a unique tonal pattern, in part, because being composed of two words, it has the longest tonal sequence. In any case, Austin appeared to know which was the correct food to select on this trial from the sound alone. Immediately after hearing strawberry drink, Austin turned around and reached for it without looking at the keyboard.

On the three occasions when Sherman was not looking during Austin's request, he turned immediately upon hearing the symbol. On these three trials, Sherman spied the symbol even before Austin had lit the period key and thus Austin's request was technically still "in progress," and Austin had no need of gesturing since

Sherman quickly selected and gave the appropriate food, thereby obviating any need for a gesture on Austin's part.

Also, Austin's latency response data (i.e. how long it took Austin to respond after Sherman made his request) supports the view that Sherman's decision to gesture or not to gesture was a function of Austin's behavior. Austin responded quite slowly (mean response time 2.8 sec.) on trials when he did not observe the request, as contrasted with his more rapid response when he had observed the request (mean response time 1.54).

Sherman did not simply gesture on trials when Austin was slow. That is, Sherman did not simply wait 1.54 seconds (or the mean time it took Austin to respond) and *then* gesture. Instead, he noted when Austin was not watching his request and he gestured *at the very moment* that Austin redirected his attention to the task. He then continued to monitor Austin to determine whether or not Austin observed and responded to the gesture.

In contrast to Austin, Sherman was no slower on the three trials during which he did not observe Austin's request than on trials when he did observe it. It is clear that Austin, who only rarely provides gestural information, does not need to do so because Sherman is so attentive to his symbolic requests that there is no need for Austin to clarify his request with gestures.

Thus we find that both chimpanzees use gestures only when their gestures serve a clear communicative end. Both chimpanzees monitor one another's behavior to determine when and about what to gesture. No gestures are superfluous or ritualistic; rather, all are well integrated into the task at hand.

Such context-dependent usage of untrained gestures clearly demonstrates that Sherman and Austin do understand the topics of their communications. That is, each chimpanzee comprehends the food–symbol relationship, the use of the symbol to specify a particular food, and the importance of coordinating their behavior so that they not only communicate their desires symbolically but also ensure that the other chimpanzee satisfies those desires.

Appendix A for Chapter 10

Trial	*Order of Requested Foods*
Bout 1	
1	Juice
2	Pudding
3	Raisins
4	Strawberry Drink
5	Liquid
6	Cheese
7	Pineapple
8	Lemonade
Bout 2	
1	Lemonade
2	Banana
3	Sweet Potato
4	Strawberry Drink
5	M&M
6	Orange Drink
7	Pineapple
8	Corn
Bout 3	
1	Lemonade
2	Liquid
3	Juice
4	Bread
5	Pudding
6	M&M
7	Cheese
8	Raisins
9	Corn
Bout 4	
1	Banana
2	Sweet Potato
3	Strawberry Drink
4	Orange Drink
5	Raisins

Trial	Order of Requested Foods
Bout 5	
1	Corn
2	Cheese
3	Corn
4	Corn
5	Pudding
Bout 6	
1	M&M
2	M&M
3	M&M
4	Peaches
5	Cheese
6	Banana
7	Orange Drink
8	Beancake
9	Beancake
10	Lemonade
11	Corn
Bout 7	
1	Juice
2	Bread
3	Peaches
4	Sweet Potato
5	Beancake
6	Corn
7	Raisin
8	Corn
9	Pineapple
Bout 8	
1	Pudding
2	Bread
3	M&M
4	Banana
5	Lemonade
6	Strawberry Drink
7	Milk
8	Cherries
9	Sweet Potato
Bout 9	
1	Pudding
2	Cheese
3	Cherries

Trial	*Order of Requested Foods*
4	Banana
5	M&M
6	Lemonade
7	Sweet Potato
8	Beancake
9	Orange Drink
10	Coke
11	Lemonade
Bout 10	
1	Pudding
2	Bread
3	Pudding
4	Juice
5	Cherries
6	Milk
7	Strawberry Drink
8	Beancake
9	Pineapple
Bout 11	
1	Coke
2	Milk
3	Sweet Potato
4	Banana
5	Orange Drink
6	Lemonade
7	Corn
Bout 12	
1	Pudding
2	Orange Drink
3	Bread
4	Lemonade
5	Sweet Potato
6	Corn
7	Banana
Bout 13	
1	M&M
2	Melon
3	Cheese
4	Bread
5	Beancake
6	Banana
7	Pineapple

Trial	*Order of Requested Foods*
8	Coke
Bout 14	
1	Pudding
2	Strawberry Drink
3	M&M
4	Coke
5	Sweet Potato
6	Orange Drink
7	Orange Drink
Bout 15	
1	Cheese
2	Corn
3	Pineapple
4	Strawberry
5	Banana
6	Milk
7	Bread
8	Orange Drink
9	Beancake
10	Cherries
11	Orange Drink
Bout 16	
1	Juice
2	Cherries
3	Peaches
4	Orange Drink
5	Sweet Potato
6	Corn
7	Banana
8	M&M
Bout 17	
1	Pudding
2	Bread
3	Banana
4	Beancake
5	Corn
6	M&M
7	Cherries
8	Sweet Potato
9	Pineapple
Bout 18	
1	Pudding

Trial	*Order of Requested Foods*
2	Peaches
3	Cheese
4	Strawberry Drink
5	Orange Drink
6	Sweet Potato
7	Juice
8	Banana
9	Orange Drink
10	Corn
11	M&M
12	Raisins
Bout 19	
1	Lemonade
2	Cheese
3	Strawberry Drink
4	Cherries
5	Beancake
6	Orange Drink
7	Corn
8	Orange Drink
9	M&M
10	Pineapple
Bout 20	
1	Lemonade
2	Pudding
3	Milk
4	Banana
5	Orange Drink
6	Sweet Potato
7	Pineapple
8	Milk
Bout 21	
1	Cherries
2	Lemonade
3	Bread
4	Cheese
5	M&M
6	Strawberry Drink
7	Banana
8	Orange Drink
9	Raisins
10	Corn
11	Pineapple

Trial	*Order of Requested Foods*
12	Sweet Potato
Bout 22	
1	Strawberry Drink
2	Lemonade
3	Peaches
4	Beancake
5	Juice
6	M&M
7	Beancake
8	Corn
9	Orange Drink
10	Pineapple
11	Banana
12	Sweet Potato
Bout 23	
1	Cheese
2	Beancake
3	Orange Drink
4	Cheese
5	Cheese
6	Milk
7	Cherries
8	M&M
9	Beancake
10	Coke
Bout 24	
1	Lemonade
2	Bread
3	Pudding
4	Orange Drink
5	Banana
6	Cherries
7	M&M
8	Orange Drink
9	Sweet Potato
10	Coke
11	Corn
Bout 25	
1	M&M
2	Peaches
3	Cheese
4	Milk

Trial	*Order of Requested Foods*
5	Cherries
6	Beancake
7	Strawberry Drink
8	Coke
9	Corn
Bout 26	
1	Cherries
2	Strawberry Drink
3	Lemonade
4	Bread
5	Pudding
6	Banana
7	Cheese
8	M&M
Bout 27	
1	M&M
2	Bread
3	M&M
4	Cheese
5	Banana
6	Peaches
7	Raisins
8	Corn
9	Orange Drink
10	Coke
11	Juice
Bout 28	
1	Liquid
2	Banana
3	Cheese
4	Cheese
5	M&M

Appendix B for Chapter 10
Gesture Description

1. Sherman requests "pudding" and points to the larger of the two portions. Austin gives that portion to Sherman.
2. Sherman lights "liquid" (which makes no sound). Austin does not know this symbol and does not respond. Sherman holds 1 hand (left) out toward Austin and gestures toward the food tray with the other. Austin still does not respond.
3. & 4. Austin is not looking during the request. Sherman gestures toward Austin with one hand (left) and toward the requested food with the open fingers of the other hand. He does not actually point at the food, but this is unnecessary since there are only two foods left and they are far apart.
5. & 6. Austin is not looking during the request. Sherman gestures toward the food with his knuckles and extends the other hand to receive the food.
7. Austin is not looking during the request. Sherman points to the projectors and points to the food.
8. Sherman takes Austin's hand and guides it toward the raisins. Austin is preoccupied with the cheese from the previous trial and is not looking at the keyboard during the request.
9. Sherman lights "corn" twice and gestures for Austin to move his hand toward the corn. Austin is not looking at the symbol as it is lighted.
10. Austin evidences indecision by waving his hand around above the food. Sherman points to the specific food.
11. Austin is not looking during the request. Sherman points to the specific food.
12. Sherman asks for "liquid" (generic term which Austin does not know). Austin does not respond. Sherman then asks for "lemonade" using its specific symbol (as opposed to the generic symbol) and points to the specific food.
13. Austin evidences indecision by waving his hand around above the food. Sherman points to the specific food.
14. Sherman gestures toward Austin's hand after Austin has taken all of the remaining M&M's instead of sharing. Sherman wants

Austin to share the M&M's which are in Austin's hand. Austin pulls his hand back and does not share with Sherman.

15. Austin is busy pouring liquid back and forth between two glasses and stacking them. He does not hand foods. Sherman points to the specific food.
16. Austin is not looking during the request. Sherman points to the specific food and repeats the request.
17. Austin evidences indecision by waving his hand over the food. Sherman points to a specific food. The food Sherman points to is peaches. The food he requested is pudding. No pudding is on the tray. Austin gives the peaches.
18. Austin is not looking during the request. Sherman points to a specific food.
19. Austin is not looking during the request. Sherman points to a specific food.
20. Austin starts to eat Sherman's share of M&M's. Sherman gestures for Austin to move his head away from the tray. Austin complies.
21. Austin is not looking during the request. Sherman points to a specific food.
22. Austin refused to respond on the previous trial. Sherman points to a specific food on this trial even though Austin is looking during the request and responds immediately following the request.
23. Austin is eating an orange peel and makes no move to respond to Sherman's request even though he is observing the request. Sherman points to the specific food.
24. Austin is not looking during Sherman's request. Sherman points to the specific food and repeats his request.
25. Austin evidences indecision by raising his right hand toward the food then withdrawing it. Sherman points to the specific food (or he starts to point but knocks over some juice as he tries to point to the pudding).
26. Austin is not looking during the request. Sherman points to a specific food.
27. Sherman points to the larger of two portions. Austin gives the smaller portion to him.
28. Austin is not looking during the request. Sherman points to a specific food.
29. Austin is watching but is eating a peel and does not respond though he observes Sherman's request. Sherman moves his hand briefly toward the tray. This is not a specific pointing

gesture but rather a whole hand motion toward the tray which appears to signal Austin to act upon the tray.

30. Sherman guides Austin's hand toward a particular key (cherry). There are only two foods left on the tray and Austin is not looking toward either symbol.
31. Now there is only one food left. Austin is looking toward the wrong side of the keyboard. Sherman guides Austin's head until it is directly in front of the correct symbol. Austin then lights that symbol.
32. Sherman pushes Austin's head toward the keyboard. The only apparent reason is that Sherman is ready to eat and Austin is not looking at the keyboard. However, Austin is looking at the food tray, apparently to decide which food to request.
33. After being pushed by Sherman (above), Austin seems to be rushed and asks for a food which is not present (peaches). Sherman looks over the tray and seems eager to give him something, but selects nothing. Sherman carefully checks the oranges to make certain that they are not peaches. Austin shakes his hand toward the tray and looks at the corner containing orange drink. Sherman hands him the orange drink.
34. Sherman pushes Austin's head toward the keyboard.
35. Sherman pushes Austin's head toward the keyboard. Afterward, Austin wipes the back of his neck where Sherman was pushing. This gesture appears initiated by Sherman's pushing.
36. Sherman pushes Austin's head toward the keyboard.
37. Sherman pushes Austin's head toward the keyboard. Austin tries to shake Sherman's hand off.
38. Sherman pushes Austin's head toward the keyboard. Prior to this, Austin looks back and forth between the keyboard and the food, but does not request a food.
39. Sherman is preoccupied with a glass in his left hand and a peel in his right hand and does not respond quickly to the request. Austin points with his knuckles to a specific food.
40. Sherman pushes Austin's head toward the keyboard. Sherman hands him the orange drink. Sherman pushes Austin's head until it is directly in front of the symbol "cherry". Austin lights this symbol.
41. Sherman pushes Austin's head toward the keyboard. Sherman pushes Austin several times, but Austin refuses to light any symbol. Finally Sherman removes his hand and Austin selects a symbol.

42. Austin extends a pronated wrist to Sherman (a typical chimpanzee gesture). Sherman puts his hand on Austin's neck and pushes his head toward the keyboard. Austin lights a key.
43. Sherman puts his hand on Austin's neck and pushes Austin toward the keyboard. Austin turns and looks at the milk on the tray. Sherman sees Austin looking at the milk and takes his hand off Austin's neck. Austin then asks for milk.
44. Sherman pushes Austin's head toward the keyboard.
45. & 46. Sherman sees Austin's request for orange but ignores it and continues eating from the previous trial. Austin gestures by extending his hand, palm down, to Sherman and then pulling his hand back as though he is requesting Sherman to move his hand toward the tray. Sherman ignores him. Austin then extends his hand, palm up, apparently requesting the food which Sherman is eating from previous trials. Sherman ignores Austin and gets down off his cube and moves away until he is finished eating. Then he returns and gives Austin the orange.
47. There is only one glass of milk left. (It is Sherman's since he did not drink it all on a previous trial.) Austin extends his hand toward the milk after requesting it at the keyboard. Sherman shoves Austin's hand away.
48. Austin is grooming his foot and does not look at the keyboard. Sherman pushes Austin's hand toward the keyboard.
49. Sherman pushes Austin's hand toward the keyboard.
50. Sherman hesitates, leans over and looks at the wrong food. He does not see the cheese which is right next to him. Austin points to it with his index finger.
51. Sherman pushes Austin's head toward the keyboard, but Austin ignores Sherman's pushing so Sherman gets off his cube and lights M&M. Austin gives him M&M's.
52. Sherman usurps Austin's turn as he states "Beancake". When Sherman begins to pick the beancake up, Austin points to a different piece and Sherman gives Austin the piece of beancake he pointed to.
53. Austin looks at the keyboard then looks at Sherman. Sherman chucks him under the chin.
54. Austin waves his hand around in front of the keys without really looking. He looks directly at Sherman. Sherman leans toward the keyboard and extends his arm, with his wrist bent, toward "cheese". Austin quickly lights "cheese." Sherman gives him cake.

55. Austin asks for cake. Sherman's head is turned and he does not see the request. Sherman turns around and chucks Austin under the chin. Austin repeats the request. Sherman gives him cake.
56. Austin starts to just take the cheese, but Sherman gestures toward the keyboard. So Austin searches the keys trying to figure out how to ask Sherman to give him cheese. Finally, he just puts his head in the tray and starts eating the cheese. Sherman pulls Austin's head out and picks up the cheese and starts stuffing cheese into his own mouth. Austin pulls Sherman's hand back and takes some of the cheese from Sherman's hand.
57. Austin and Sherman look at the keys. Austin gestures for Sherman to go ahead and select a key.
58. Sherman holds Austin's hand down to prevent Austin from selecting a key, even though it is Austin's turn. After Sherman lights the symbol, he lets Austin's hand go and gestures toward the tray for Austin to give him food.
59. Sherman lifts his hand over the milk with an iconic raising gesture for Austin to pick the milk up. Sherman gestures "up" (as the glass should go).
60. Austin is not looking during the request. Sherman points to a specific food. Austin does not see the first pointing gesture, so Sherman repeats the gesture.
61. Sherman touches the food before he requests it. Then after the request, he shakes his hand toward the food. There is no indecision on Austin's part, and Austin did observe the request though his response is a little slow.
62. The last food is eaten and Sherman appears to gesture for Austin to leave the room. Austin does and Sherman follows.
63. Austin observes the request and extends his hand to take the food but withdraws it. Sherman points to both cherries and M&M (he asked for M&M's), and vocalizes as he points. He points to M&M's once and cherries twice and he looks at Austin after pointing. Austin gives him cherries.
64. Sherman asks for M&M's and points directly to them immediately after asking. Austin is watching during the request.
65. Sherman has been having difficulty during the past few bouts finding the cheese symbol. He looks over the keyboard on this trial and then gestures toward the cheese with the back of his wrist and looks over the keyboard again. Austin then picks up

a piece of cheese and starts eating it. Sherman notices this, stops looking at the keyboard and takes his own piece.

66. Austin is not looking during the request. Sherman points to a specific food.
67. Austin is licking an empty pudding cup in the food tray but looks up as Sherman requests "Coke." Sherman looks at the coke and shakes his hand (but the hand shake is not toward the Coke). Austin sits up and gives the coke.
68. Austin is eating crumbs from the tray and is not looking during Sherman's request. Sherman asks for coke and gestures with the back of his wrist toward the coke. He accompanies this gesture with a vocalization.
69. Austin looks around the room instead of addressing the keyboard so Sherman puts his hand on Austin's neck and shoves Austin toward the keyboard.
70. Sherman waves his hand toward the food tray and Austin takes a piece of cheese. (No symbol is used.) Sherman takes the other piece of cheese.
71. Sherman waves his hand toward the tray again and Austin takes some raisins. Sherman takes the rest.
72. Austin waves his hand in front of the keyboard without looking at the keys, but he does look at Sherman; Sherman looks toward the keyboard and Austin lifts his right hand just a few inches toward the keyboard. Sherman looks directly at a symbol and shakes his right hand toward that symbol. Austin lights that key.

Chapter 11

When a Lexigram Becomes a Symbol

What Do Symbols Represent to an Ape? Designing a Strong Test

Day-to-day interactions with Sherman and Austin inevitably began to lead all those who worked closely with them to the conclusion that these chimpanzees were using their keyboards in a way that was very close to human communication. Their symbols seemed to mean things to them now and they began to use them for the purpose of representing things to others. Yet it was one thing to know this at an intuitive level and quite another to be able to demonstrate it in a scientifically justifiable manner. Critics of ape language projects (Sebeok and Umiker-Sebeok 1980; Sugarman 1983), had repeatedly suggested that many of their symbol-using behaviors were simply tricks, performed without any comprehension of the specific referent, or even of the general referential function of language (Bennett 1978).

Symbol usage by Sherman and Austin had become so frequently novel and spontaneous that it was hard to reconcile the great variety of behaviors with either "rote learning" or "imitation" explanations. Moreover, their initial acquisition of new symbols was, on occasion, so rapid as to entirely preclude either of these interpretations.

Unfortunately, their spontaneous symbol usages tended to occur in complex circumstances where other variables made it impossible

Portions of this chapter have appeared previously in the *Annals of the New York Academy of Sciences,* 1981, in an article entitled "Can Apes Use Symbols to Represent Their World?"

to entirely rule out cueing, conditioning, imitation, and simple perceptual generalization. Since the issue of representation was clearly a fundamental one for the field of language acquisition, it became increasingly important to devise a way of testing for representational capacity that was not compromised by the kinds of contextual cues that are inherent in ordinary language use.

In order to design an adequate test of the presence or absence of representation, it was necessary to rule out training and imitation entirely. That is, the chimpanzees had to produce symbol encoding behaviors that were completely novel in controlled test settings that eliminated all possibility of cueing or imitation. Only a test with these characteristics could reasonably claim to determine whether or not the behaviors in question were truly symbolic.

The test devised began with a training paradigm which required Sherman and Austin to make categorical decisions about the things they knew best: foods and tools. Up to this point in training, each food and each tool had been represented by a specific symbol. That is, they had learned no generic terms and could not ask for "food" in the general sense if they were hungry, nor could they ask simply for "tool" if they did not know which tool was needed.

We decided to train "tool" and "food" as generic terms, with the goal of eventually showing Sherman and Austin the lexigrams of specific foods and tools and asking them whether each lexigram represented an item that could be classified as a "food" or a "tool." We would, of course, somehow have to convey to them our question without ever specifically teaching them to categorize the lexigrams to be tested. That is, we would not be able to teach the tasks by showing them the answers we expected. If we were to tell them, *even once,* how to categorize a particular lexigram, then it could be argued that they were "trained" or were "imitating." In addition we would need to be certain that we were not cueing their answer in any way.

In order to satisfy all of these constraints, we devised a testing-training paradigm which emphasized the generic categories of food and tool, and which moved slowly and systematically from highly concrete instances which emphasized function, to abstract instances which emphasized representational recall and decision making. (The paradigm used is illustrated in figure 11.1, below, and will be discussed at that point in the text.)

Lana: An Abbreviated Training History

Because the question of representational function is so critical to the interpretation of symbol usage by apes, we felt it was important to attempt to measure this capacity in Lana, as well as in Sherman and Austin. Lana's training and use of the Yerkish lexigram keyboard system has been described extensively elsewhere (Rumbaugh 1977), and was quite different from that which Sherman and Austin received. Her training emphasized the use of multiple lexigram strings (Please machine give juice; Please machine make movie; Please Tim move into room; etc.).

Basically, each of these strings functioned to request something either of Lana's computer-controlled environment or of her human companions. It was not essential that Lana learn to associate different outcomes with different individual lexigrams; however, it was essential that she associate different outcomes with different *strings* of lexigrams in order to produce the particular event(s) she enjoyed (tickling, going outdoors, seeing movies, etc.), and in order to receive the foods and drinks which she liked. In many cases these strings were identical except for one lexigram (Please machine give *juice;* Please machine give *milk;* Please machine give *M&M;* etc.). In such instances, Lana did learn to associate particular lexigrams with particular events or foods.

Lana also learned to identify or "name" objects by specific lexigram—even though she was not requesting these objects (and had no desire for them)—and to identify the colors of these objects with a specific lexigram. When asked to give the color of an item Lana often volunteered its specific name lexigram as well; however, she did come to distinguish between the questions "What color of this?", "What name of this?", and give only the appropriate responses (Rumbaugh 1977).

Having learned to request foods and to name foods and objects, Lana's abilities were expanded by our requiring her to use still other strings to request that her human experimenters act upon related objects (generally containers) in such a way as to make these containers more accessible to her. In Lana's environment, this meant loading the food into a vending device which was clearly visible to her outside a Lexan wall or bringing the food into the room.

Virtually all of Lana's requests were initially taught as long strings, and emphasis was not placed upon the comprehension of individual symbols. Because of the number and variety of strings Lana learned to produce, it became possible to engage in rather lengthy exchanges with her. The topics of these exchanges were nearly always the same—that of somehow making food accessible to Lana. An example of a typical exchange with Lana follows:

6/12/75
Tim. Lana want what [to] drink (9:35)
Lana. Lana want drink milk in machine (9:36)
(Tim puts half a pitcher of milk into the dispenser and leaves the other half outside Lana's room in full view.)
Lana. Please machine give milk (9:40–9:44—repeatedly until milk dispenser is empty)
Lana. You put milk in machine? (9:44)
Tim. Milk in machine (9:44—there is a small amount left)
Lana. You put more milk in machine? (9:44)
Tim. Yes (9:45)
Lana. Please machine give milk (9:46—repeatedly)

Of the symbols used in the above sentences, only one was tested independently—milk. Lana could name or request milk even when it was paired with other foods. However, the remaining words were learned in context-dependent strings. For example, the string "Lana want drink milk in machine" was used in three basic contexts: (a) early in the morning when Tim typically gave Lana milk; (b) when the experimenter stood outside of Lana's room holding a container of milk; or (c) when requests for other drinks (e.g., juice, Coke) were not rewarded.

Lana's requests for specific foods tended to be dependent upon either the visual presence of that food or on her knowledge of a routine linked to time of day (e.g., milk typically given in the morning; chow typically given in the evening). Lana did, however, often use her food symbols quite freely, and in a novel way, to request foods when there was some reason to believe that the normal routine might be set aside, such as when visitors were in the lab.

It should be clear from this brief description of Lana's lexigram training that she was required to do things very differently from

Sherman and Austin. Her consequent usage of lexigram symbols was distinctly different from the sort of usage seen in Sherman and Austin.

Lana: Additional Training

In order to compare the symbolic representational capacity of Lana with that of Sherman and Austin, it was necessary to teach Lana additional vocabulary items so that the testing procedures for all three animals could be identical. As this study began, Lana had not learned to request or to use tools as had Sherman and Austin, though she could accurately name objects.

A brief comparison of the specific skills Lana was taught (either before or during the current study) and those Sherman and Austin were taught is given in table 11.1. As this table indicates, not only was Lana taught a number of things Austin and Sherman were not, she was also taught things which were similar at a much later point in training. At the outset of this phase of training, it was believed that Lana had learned more vocabulary and combinatorial skills than had Sherman and Austin, and that she would do as well or better than they would on any test which evaluated the representational component of her lexigram usage.

Lana had learned to label and give the color of objects such as bowl, box, shoe, ball, can, and cup without having any specific use for these objects, and, in fact, without ever having actually desired or requested these objects. Since she seemed to do this readily, with high reliability, it was anticipated that it would be relatively easy for her to acquire tool names. In fact, since Lana had learned to use symbols to label foods, objects, and colors, without the extremely concrete experiences Sherman and Austin had needed, it seemed probable that Lana was functioning at a more abstract level than they were.

In any case, Lana began learning how to request tools by using the procedure that had proved successful with Sherman and Austin. Identical tool sites were placed in her living area, they were baited with food, and Lana was encouraged to ask the teacher for the tool she needed to obtain the food. As each tool was introduced in this functional paradigm, we also included that tool in Lana's regular naming review.[1]

Table 11.1. Symbol-Use Skills Acquired by the Chimpanzees Sherman, Austin, and Lana

Sherman & Austin	*Lana*
(a) Not taught	(a) Use multiple symbols & sequence all symbol productions
(b) Request specific foods to eat	(b) Request specific foods to eat
(c) Label foods without eating them	(c) Label food without eating them
(d) Respond to information about hidden foods provided by symbolically encoded statements of others	(d) Not taught
(e) Give foods in response to symbolically encoded request of another	(e) Not taught
(f) Cooperatively divide and share food with another chimpanzee by means of symbols	(f) Not taught
(g) Request objects for a specific use—tools used to procure foods	(g) Not taught until after skills (h)–(j) were acquired
(h) Label objects without using them	(h) Label objects without using them
(i) Not taught	(i) Label colors of objects
(j) Not taught	(j) Give either color or name of object as requested
(k) Give objects in response to symbolically encoded requests of others	(k) Not taught
(l) Cooperatively request of and give tools to another chimpanzee by means of symbols	(l) Not taught

Tool "naming" had not been attempted with Sherman and Austin till later in training, since they were unable to name tools when they were first introduced. Lana, however, was able to name the tools almost immediately. She could, in fact, often accurately name tools she could not request. For example, if she needed a key to open a padlock, she would frequently look at tools, then stare blankly at her keyboard, seemingly with no idea of what symbol she wanted to select. However, if the experimenter merely held up the tool (for example, the key) and asked "What's this?", Lana would immediately answer "key," then proceed to request "give key" and to use the key appropriately.

Lana's requesting difficulty did not seem to be due to a lack of understanding of which tool she needed. She could, if given access to the experimenter's tool kit, easily select the appropriate tool and use it to obtain the food. What Lana seemed to be unable to do was to recall the symbol *when she needed to ask for the tool.* Apparently, the use of a symbol such as "wrench" in the request context was not the same for Lana as the use of "wrench" in the naming context.

Interestingly, Lana exhibited the inverse of the problems Sherman and Austin had encountered earlier. They could initially ask for their tools, but not name them. Lana, by contrast, could often name her tools, but not ask for them.

Lana's difficulty was repeated with each new tool. Even after she had been shown many times how to request a tool, she would still watch closely as a tool site was baited with food, and then do nothing. The experimenter would often have to encourage Lana to approach and use her keyboard. In response to such coaxing, Lana would hesitatingly point first to one symbol and then to another while looking at the experimenter to see if she had found the correct symbol. Yet, when the experimenter picked up a tool and asked "What's this?", Lana would proceed directly to the keyboard of her own accord and name it correctly without any hesitation.

It was difficult to believe that Lana could name things that she could not ask for, and the teachers who worked with her tended to feel that she was "making the tool task difficult" for them, and "that she was doing this on purpose because she did not like to ask for her tools or did not like them." Yet the pattern was so persistent and recurrent that it was unlikely that Lana's difficulties occurred for such reasons.

The gap between Lana's abilities to name and to request tools was finally bridged when we altered the procedure by teaching her to point to the tool which she needed. Such pointing enabled Lana to respond almost as though the request task were, in effect, a naming task. Since she typically knew the correct tool, but could not recall the symbol to use to request it, she came to bridge the naming-requesting gap by pointing to the tool which she needed. The pointing behavior helped her to recall the correct symbol, which she then used to request the tool. It seems clear, however, that even though Lana did learn to request tools successfully, she went about it quite differently than Sherman and Austin had done.

Like Sherman and Austin, Lana lacked receptive competence with tool symbols—even after she had acquired productive competence. It was necessary to go through training similar to that described in the preceding chapters for Sherman and Austin, before Lana was able to select the correct tool accurately in a receptive task.

Lana's inability to give the correct tool when requested was unexpected, since she responded appropriately to many other requests (such as "Lana make music," "Lana open window," etc.). She was also able to decode the questions "What color [is] this?", "What [is the] name [of] this?" And she reliably and appropriately answered questions such as "What [does] Lana want [to] drink?" and "Question what Lana want [to] eat?"

Lana's behavior during such "conversations" revealed that she attended closely to what was stated by the experimenter, and that she typically included in her ensuing statements one or more of the lexigrams which had been used by the experimenter. However, her inability to give specific items upon request suggested that while Lana could repeat, rearrange, and often respond appropriately to the experimenter, she did not possess receptive comprehension in the sense of being able to act appropriately on a particular item within a set of objects. Apparently, elaborate symbol manipulation and receptive comprehension (as evidenced by behavioral cooperation in acting upon verbally specified objects) are not one and the same.

The Paradigm, Part I

The training paradigm used in the present study is shown in figure 11.1. As this figure illustrates, the chimpanzees first learned to sort objects according to function, then to label these objects according to function, next to label photographs of the real objects (according to function of the object), and finally to label symbols of the object (according to the function of the object). At each point in training the chimpanzees were tested for their ability to transfer their categorizing skills to novel objects.

Because this training spanned a number of months and involved various tests of a large number of their vocabulary items, it was important to ascertain that the chimpanzees' labeling, requesting, and receptive skills indeed remained intact throughout the duration of the categorization training. Figure 11.2 depicts the performance of the chimpanzees during the regular review tasks which accom-

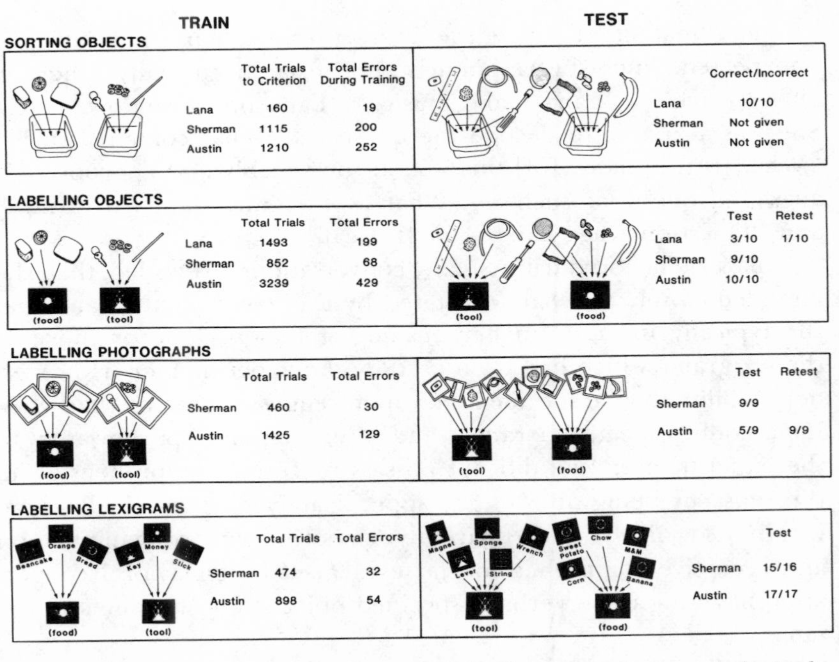

Figure 11.1. The training paradigm used in the categorization study is depicted in the flowchart. The animals learned items on the left and were tested in a blind setting with the items on the right. The numbers of total tests and total errors are given for training on the left of the chart, and the number of correct trial 1 selections are given for blind tests on the right of the chart.

panied categorization training.[2] Vocabulary review sessions during training included well known and new items, and the occasional variability of the review scores reflects the addition of new items.

Categorical sorting of foods and tools was begun by requiring the animals to sort three foods (orange, bread, and beancake) into one bin, and three tools (key, money, and stick) into another. The bins were identical and were placed directly in front of the chimpanzees. One piece of food was placed in the food bin, and one tool was placed in the tool bin to serve as a marker of the sort of items to be placed in that bin. All three training foods and all three training tools were used equally often as markers. The chimpanzee was then handed a food or a tool and asked to sort it into

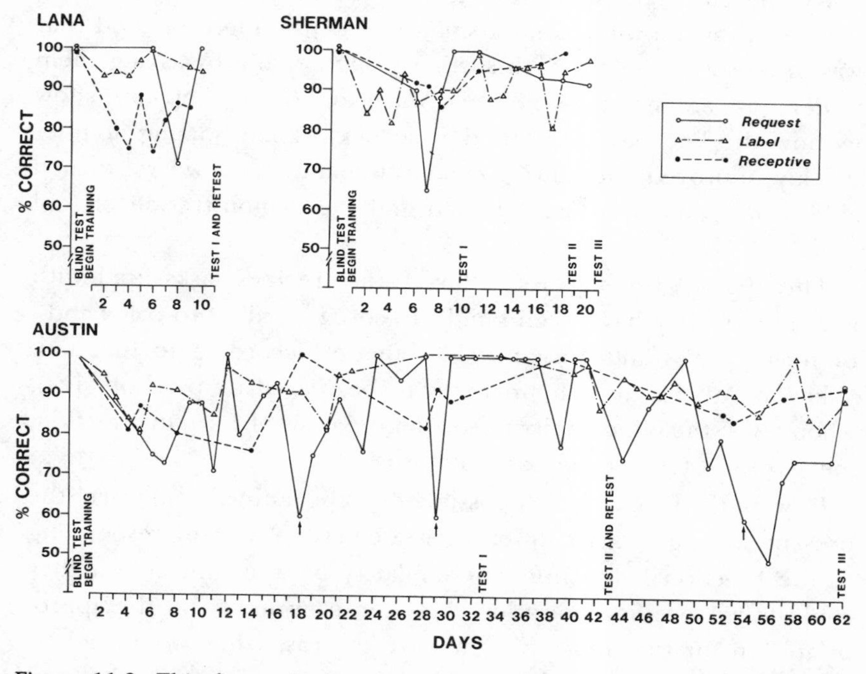

Figure 11.2. This figure depicts the performance of the chimpanzee during regular naming, requesting, and labeling tasks throughout the categorization study. The work took the longest to complete with Austin, thus his performance is shown over a broader span of time. The purpose of this figure is to illustrate that overall, the chimpanzees used these specific word names regularly, reliably in a number of different ways throughout the categorization teaching and testing.

the correct bin. None of the foods or tools resembled each other physically, thereby precluding a match-to-sample sorting response. The dimension along which items were to be sorted was strictly functional; the foods could be eaten and the tools could not.

The sorting task was difficult for both Sherman and Austin, but Lana picked it up almost immediately. Sherman and Austin extended the food and objects toward the bins in a very vague and hesitant manner, constantly checking the experimenter's eyes and face to determine if they were about to make the correct decision before they actually committed themselves by placing the object in one bin or another.

We therefore attempted to increase the saliency of the sorting dimension by allowing them a small bite of the piece of food that was to be sorted before each sorting response, and by asking them to demonstrate manually the use of each tool (for example, to show us how the key was to be used in a lock). Their sorting abilities quickly improved with this procedural change, and we were then able to drop out the bites of food and the demonstration of tool use.

This type of sorting task differs from previous tasks used with apes in that they have been taught to sort according to color and/or form (Hayes and Nissen 1971) but not according to function. Viki (one of the first chimpanzees to be reared by psychologists), when given the opportunity to sort on the basis of form or function, inevitably sorted on the basis of form.

It would, of course, be possible for the animals to learn the present sorting task by different strategies. Instead of responding to the functional dimension of similarity between the items, they could form specific associations between each item and the appropriate bin for that item. If such were the case, they would not be expected to generalize to new items. The fact that (for Sherman and Austin) training was facilitated by, and even initially dependent upon, emphasis of the *functional* distinction between food and tool implied that they were classifying these items along a functional dimension, but tests of generalization were needed to ascertain this more clearly.

The Paradigm: Part II

When both chimpanzees reached a sorting criterion of 90 percent or better across 60 trials, we introduced the lexigrams *food* and *tool.* These lexigrams were initially located on the keyboard in the same positions as the sorting bins (tools on the right, foods on the left) and the sorting bins placed directly under the appropriate lexigrams. The specific names of the training foods and tools (orange, beancake, bread, money, stick, and key) were rendered inactive and the chimpanzee was asked to sort a food or a tool into the proper bin and then to select either the "food" or "tool" lexigram.

Once the chimpanzees learned this skill, the bins were removed, and the chimpanzees were asked to label the items categorically without sorting them. Training in this phase continued until each chimpanzee met all of the following criteria:

1. Ability to label all training items correctly without eating the food or using the tool (thus the *function* of the item had to be recalled instead of being immediately demonstrated by the teacher or acted out by the chimpanzee).
2. Ability to label all training items correctly *on trial 1,* after the food and tool lexigrams were relocated on the keyboard (control for learning the position of the lexigram as opposed to its physical configuration).
3. Ability to label all training items correctly with the experimenter out of the room (control for inadvertent cueing).
4. Ability to label all training items correctly under the conditions listed in 1–3 above, for two consecutive sessions of more than 25 trials at 90 percent accuracy or better (control for reliability of performance).

Assessment with Novel Items: Phases I and II

The number of training trials needed to acquire these skills again varied widely across the three chimpanzees, as can be seen in figure 11.1. Austin's learning difficulties stemmed from a number of unpleasant experiences and illnesses during this period. He contracted chicken pox, which interrupted the training sessions for two weeks. When he began again after his illness, he seemed to have forgotten the categorization task entirely and it was necessary to start all over.

As each chimpanzee reached the training criteria outlined above we presented them with ten foods and tools *not used* during training and asked them to label each additional item as "food" or "tool." These items were familiar to the chimpanzee and all had had specific lexigram symbols previously assigned to them which the chimpanzees knew well. (That is, the chimpanzees knew the specific names for these items but they had never before been asked, or told, whether the item was a food or tool.) The items were presented once to each chimpanzee, in random order, interspersed with 10 test trials of training items. Before the presentation of the first test item, the animals had to respond correctly to the training items for 20 consecutive trials. This warmup procedure was used to ensure that as the test began the chimpanzees were attentively engaged in the task. During the test the experimenter continued to present trials on training items (orange, beancake, key, stick, etc.) to make the test situation seem as familiar as possible, and to provide a continuous baseline of performance on known items during the test.

In order to preclude inadvertent cueing, the experimenter remained outside the room during the entire test. The chimpanzees approached the door, looked at the object displayed by the experimenter, then reentered the room and labeled the item on the keyboard. The experimenter could see neither the keyboard nor the chimpanzee, once the chimpanzee had left the doorway. (The specific names of the ten novel exemplars were deactivated during this test, but all other keys remained on.)

Austin correctly identified all the novel items on their first presentation. Sherman correctly identified all the items except sponge—the tool he occasionally eats portions of as he uses it. (Austin does not eat sponges.) Lana correctly identified only three items (chance performance would be five items). Thus it appeared that Sherman and Austin had acquired a functionally based concept of "food" and "tool" that was generalizable to new items, and symbolically encoded in a categorical manner; Lana had not.

Probing the Differences between Lana and Sherman and Austin

Lana's surprisingly poor performance raised three questions. Had the differences in the training which Sherman and Austin had received versus that which Lana received enabled them to:

1. Conceptualize functional relationships that Lana did not.
2. Symbolically encode referential relationships that Lana perceived but did not encode at a symbolic level.
3. Recall specific food–food or tool–tool associations that for some reason they had mastered—but that Lana had not during previous training?

Three additional tests were run in order to answer these questions. First, Lana was simply retested to determine whether or not the first test might have been in error. On this second blind test, Lana correctly identified only one test item. The one item which Lana correctly identified was not the same as any of the three items she had correctly identified on the first test. This suggests she was guessing on the trials during both the initial test and the retest, and furthermore, that on the retest she did not even remember the answers which had been previously correct.

Lana's poor performance suggested that perhaps she had approached the task of learning "food" and "tool" with a noncategorical strategy. In order to test this possibility, we asked Lana to simply *sort* the ten foods and tools along with her training items. Thus, instead of presenting Lana with a straw (for example), and asking her *to label it* as a food or a tool, we presented her with a straw and asked her *to sort it* into one of two bins. We turned off Lana's keyboard and followed the same pretest criterion (20 correct consecutive sorts of training items). We then interspersed the *same ten test* items used in the two tests described above. (This test was also given blind.) Lana simply came to the door, took an item from the experimenter, and then went back into her room and sorted it into one of the bins.

Lana sorted *all ten* test items correctly on trial 1. Accordingly, her failure on the earlier tests cannot be attributed to an inability to conceptualize the functional relationships between the foods and the tools. Lana appeared to enjoy this test and sorted all the new items without any hesitation whatsoever. This attitude contrasted sharply with that which she had displayed during the two earlier tests, in which she was asked to categorically *label* these same items. During those tests she had been hesitant even to go toward the keyboard when the experimenter had shown her a test item, and she had to be encouraged repeatedly to try to label it.

These observations suggest that Lana was able to evaluate her own competence regarding the task she was asked to perform. Lana seemed to be aware that she did not know which symbol to select and thus did not want to try. Instead, she appeared to want to be told which symbol was correct (as she had been told with the training items).

The results of this test indicated clearly that although Lana could group items into two conceptual categories, foods and tools, she did not seem to attach symbols to these functional groupings in the same way that Sherman and Austin did. For Sherman and Austin, the food and tool symbols became more than the sum of our training procedures—they represented conceptual categories. For Lana these symbols functioned only as they had done during the training procedures.

Generalization of Food and Tool to Nonnamed Items

In order to determine the breadth of Sherman and Austin's food and tool categories, we presented them with 28 additional items (14 foods and 14 tools) which had not been used in any previous training task. These items which were, generally speaking, common household items are listed in table 11.2. Sherman categorized 24 of these 28 items correctly; Austin was correct 25 times. All but one error resulted from labeling tools which were typically used in food preparation as "foods."

The Paradigm: Part II

Since Lana's inability to encode test items categorically implied that it would be fruitless to move from real objects to photos, only Sherman and Austin were advanced to the next stage of training. In this phase, photographs of the training items were substituted for the real items. Since photographs cannot be eaten, the edible-inedible distinction became abstract. The photographs were encased in lexan for protection. As the chimpanzee was shown each picture, he had to look through a layer of lexan, recognize it as either a food or a tool, and then label it appropriately.

At first, photos of the objects were taped to the real objects themselves. The photos were removed one at a time during the first few training sessions. Sherman readily transferred the cate-

Table 11.2. Categorization of Unnamed Items

Unnamed Items Presented for Categorization Testing	
Tools	*Foods*
Scrub brush—A	Ice cube
Shovel	Peanut
Screw driver	Celery
Juice squeezer—S	Can opener
Steel ball bearing	Peanut butter
Cage locking pin	Jelly
Spoon	Raisins
Cooking pan—A	Cabbage
Hammer	Grapefruit
Sink stopper	Cucumbers
Knife—S	Chimp crackers
Scissors	Turnip
Cutting board—S	Potato (white)
Can opener—A & S	Lemon

Note: Errors made by Sherman are marked S; those made by Austin are marked A.

gorization skills acquired with real objects to photographs. By the end of the second training session, he accurately labeled all of the photographs without being shown any of the actual items (see figure 11.1). Austin experienced difficulty. As the photos of stick, key, and money were separated from the actual objects, he continued to label them as tools. However, when the photos of food were removed from the food, he also began to call these photos tools.

It was necessary to pair the actual objects and foods with their photographs for many trials in Austin's case before he responded accurately. All of the photographs were, of course, inedible, and it is possible that this caused Austin to have difficulty retaining the basic dimension of distinction—items which can be eaten versus those which cannot.

Often Austin's sessions would start out well, as he quickly and correctly named the training photographs. However, he would then appear to become bored and begin to treat all of the photographs as plastic objects and simply label them "tool." In order to curb his tendency to label everything "tool," it was necessary to get out the real foods and tools and show them to Austin along with pictures.

Phase II: Assessment

Training on these six photographs continued until both chimpanzees could label the six training photographs accurately and until they reached the same criterion with photographs that they had achieved with actual objects. Test photographs were then presented. As before, test items were randomly interspersed with training items and the experimenter was out of the room.

The test photographs were of the following items: wrench, magnet, straw, sponge, M&M, banana, sweet potato, chow, and corn. Sherman labeled all nine test photographs correctly (100%). Austin, however, labeled only five novel photographs correctly (55%).

Austin's poor performance suggested that he had not been able to conceptualize the training photographs differentially—that is, as edible or inedible—and that he had, instead learned the training items in an associative manner. This interpretation is corroborated by the relatively large number of trials which Austin needed to get through the photograph phase of training. However, another possibility was that Austin was not looking *through* the lexan cover at the new pictures but rather at it, and hence was simply guessing on these trials because he did not treat these photographs as representations of foods or of tools. (We had repeatedly encountered this problem with Austin during training.)

In order to delineate more clearly the reasons for Austin's failure to label the test photographs correctly, we retested Austin using the test objects themselves in place of the photos. Test objects were interspersed with training trials on which photographs of the training items were presented. If Austin were now responding with a noncategorical strategy, he would not be likely now to label the test items correctly even though they were not photographs. To our surprise, Austin labeled each of the novel items without error and without hesitation.

We therefore assumed that Austin had simply treated all the test photos as pieces of lexan, and guessed on these trials instead of looking closely at the actual photo and making a categorical response. The lexan casing surrounding the photographs frequently reflected back enough light to render the enclosed picture invisible from certain angles. Sherman always accommodated for this glare by moving his head to change his line of regard. Austin did so less consistently.

During a retest, we encouraged Austin to look carefully and slowly at each picture as we rotated the angle of the photo. Austin then proceeded to identify the nine test photographs correctly. Since there had been *no interim* training opportunity for Austin to have learned the correct response to these particular photos, it is reasonable to conclude that Austin, like Sherman, was able to categorize not only test items but also photographs of test items.

The Paradigm: Phase III

The final phase of this study was the most critical of all, for it alone was unequivocally a test of the representational value of Austin and Sherman's symbols. Unlike all other tests of what apes' symbols mean or represent to them, this test included the following constraints:

1. It required a completely novel response, one never before given by the chimpanzee, or even by the *experimenter*, when working with the chimpanzee.
2. It was administered under blind conditions.
3. It required that the chimpanzee use only one symbol to classify another, thereby forcing him to refer cognitively to the specific referent of one symbol. From the perceptual characteristics of the symbol, which had to be recalled from memory, since they were not present, the chimpanzee then had to assign that item to a functional categorical class when that class itself was represented only by a symbol.

Training in this phase began as did the previous ones. Trial 1 data are critical in this sort of test, and it was important to be certain that the chimpanzees treat each task similarly from the beginning, even though a different class of stimuli was being employed. Accordingly, we returned to the three training foods and the three training tools. Initially, the lexigrams for these items were taped to photos of the items—again, to provide a bridge between the levels of stimuli presented. This bridge was provided between levels of abstraction because it was possible that, without them, Sherman and Austin might perceive each different set of stimuli as a new task.

During training with the lexigrams, we began removing the photos and asked the chimpanzee to categorize the Yerkish symbol alone as either a food or a tool. Sherman rapidly transferred and within three sessions had met our training criterion. Indeed, he was able to categorize the training lexigrams themselves (without photos) from the first session at 90 percent accuracy.

Austin again had his problems. Whenever he made an error he refused to look at the stimuli and simply alternated between the keys "food" and "tool." His lengthy acquisition phase reflects the fact that every time the experimenter attempted to remove the photograph from the lexigram, Austin's performance would deteriorate rapidly. He seemed bothered by any change in the task, even those changes not directly related to the performance of his task, such as where he sat, whether the door was open or closed, etc. Thus, virtually any change in the overall situation would elicit a simple alternation response between the keys "food" and "tool," regardless of which item was shown to him.

Fortunately, it was possible to determine when Austin was not attending to the task at hand, by observing whether he would even look at the item shown to him. In fact, Austin would alternately select and light either the food or the tool key, whether or not he was even shown a lexigram to categorize. The ritualistic nature of Austin's alternation strategy was a clear and unmistakable indication to the experimenter that Austin was not attending to the task and that continued training trials would prove fruitless.

At such times it was always necessary to stop and play with Austin, to be very affectionate toward him, and to regain a rapport of positive affect. Typically, the training task could then be resumed. However, if the experimenter evidenced displeasure regarding Austin's performance, Austin would soon lose interest in the task and would either refuse to continue or else begin to alternate and ignore the items shown to him.

The performance of Sherman and Austin during the lexigram training phase suggests that only Austin needed this training as a bridge between levels of stimuli.

Phase III: Assessment

Once the chimpanzees had reached the training criterion, they were presented with the novel lexigrams for the objects presented in figure 11.3. As on previous tests, the experimenter was not in the room and novel items were interspersed randomly with training items. It should be emphasized that these chimpanzees had never before been asked to make a categorical assessment of the symbols they were shown. In some cases (starred), they had never even made a categorical assessment of the real objects these symbols represented.

On trial 1, Sherman was correct for 15 of 16 symbols, Austin on 17 of 17. Thus, both Austin and Sherman were able to make a categorical decision regarding arbitrary symbols, using *symbols* they had never experienced during categorical training. They demonstrated a capacity to respond categorically to the symbols M&M, banana, wrench, magnet, etc., just as they would have responded to the objects themselves.

Controlling for Perceptual Similarity

Is there any way Sherman and Austin could have categorized these lexigrams correctly other than by knowing what they represented? The possibility of physical similarity between the individual specific lexigrams and the categorical lexigrams must be raised. Could the tool names possibly look more like the symbol for tool and the food lexigrams look more like the symbol for food? Table 11.3 shows each lexigram used during the final test and gives the number of elements that each particular lexigram has in common with both the food and tool symbols. The table then predicts whether Sherman and Austin should select "food" or "tool" when shown each lexigram, if they were making their choice based on physical similarity between the sample and the "food" and "tool" symbols. As table 11.3 illustrates, on that basis the chimpanzees would have been wrong in six cases, correct in six instances, and would have responded at chance in five cases (the sample resembled food and tool equally), for an overall score of about 50 percent. Since Austin made no errors and Sherman made only one error, it is apparent that they did not use physical similarity of specific lexigrams to make their selections of generic lexigrams.

Categorization Test

Food or Tool?

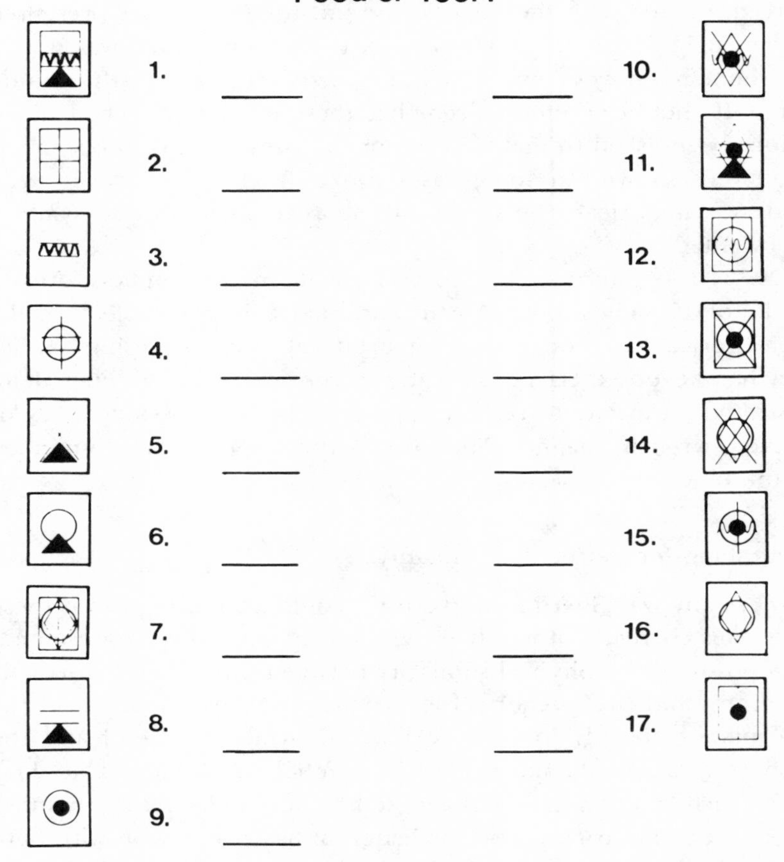

Figure 11.3. These are the actual symbols shown to Sherman and Austin during the final phase of the categorization test. This test, given with blind controls, required that Sherman and Austin look at each symbol and determine whether it was a member of the food or tool category, then use the appropriate lexigram (food or tool) to declare their answer. Thus no real items or pictorial representations were used in this phase of the test. Questions were posed using symbols, and answers were given using symbols. The reader can see the arbitrariness and difficulty of the test by attempting to fill in the blanks with the words "food" or "tool" for him or herself. The correct answers are 1) Tool, 2) Tool, 3) Tool, 4) Food, 5) Tool, 6) Food, 7) Tool, 8) Food, 9) Food, 10) Food, 11) Tool, 12) Food, 13) Food, 14) Food, 15) Food, 16) Food, 17) Tool.

Table 11.3. Physical Similarity Between Categorical Lexigrams ("Food" and "Tool") and Specific Lexigrams

Categorical Items (Tool or Food)	*Elements in Common with Food Lexigram*	*Elements in Common with Tool Lexigram*	*Number of Uncommon Elements*	*Physical Similarity Prediction*	*Correctness of Prediction Based on Physical Similarity*	*Correctness of Response*	
						Sherman	*Austin*
1. Candy (M&Ms)	2	1	1	Food	Correct	R	R
2. Banana	0	1	1	Tool	Wrong	R	R
3. Wrench	1	1	1	Either	Chance	R	R
4. Chow	0	2	1	Tool	Wrong	R	R
5. Corn	1	1	1	Either	Chance	R	R
6. Sweet potato	1	1	2	Either	Chance	R	R
7. Magnet	1	2	1	Tool	Correct	R	R
8. Carrot	2	2	0	Either	Chance	R	R
9. Pudding	1	0	1	Food	Correct	R	R
10. Pineapple	0	1	1	Tool	Wrong	R	R
11. Straw	0	2	2	Tool	Correct	R	R
12. Lemonade	0	1	1	Tool	Wrong	R	R
13. Sponge	0	2	0	Tool	Correct	W	R
14. Strawberry drink	0	1	1	Tool	Wrong	R	R
15. String	0	1	2	Tool	Correct	R	R
16. Austin's room	1	0	1	Food	Wrong	+	R
17. Lever	1	1	2	Either	Chance	R	R
					Total Correct	15/16	17/17

NOTE: This table has appeared previously in "Can Apes Use Symbols to Represent Their World?", in the *Annals of the New York Academy of Sciences* (1981).

It should also be noted that the one error which Sherman did make (sponge) would not have occurred if he had selected the closest physical match. "Sponge" shares two elements with tool and none with "food." Yet Sherman categorized the lexigram sponge as a food, just as he categorized the real sponge as a food. Thus, both Sherman and Austin's correct responses and errors demonstrate that they were not solving the task based on any perceptual similarity between the sample lexigrams and the food-tool lexigrams.

A Four-Way Categorization Task

In an effort to determine whether it would be possible to expand the bidimensional categorization system of "food" and "tool," we broke the edible class into two divisions, "food" and "drink" and the inedible class into "location" and "tool." Only Sherman was given training in this four-way task, as Austin had not acquired location symbols at this point. The additional categories of locations and drinks were chosen because they reflected the only other word classes Sherman had learned to label accurately at this point in training.

Training of these additional categorical labels followed the procedures used for the training of foods and tools, except that now Sherman was required to make a four-way judgment (food, tool, drink, or location) and both real objects and photographs were intermixed as exemplars from the start.

Following acquisition of the four-way categorical distinction, Sherman was presented with photos and/or actual instances of three novel foods (yogurt, frozen lemonade, and a lemon), five novel drinks (liquid lemonade, oil, juice, milk, and strawberry Kool-Aid), two novel locations (sink, and evening housing areas), and one novel tool (wheel). The different number of exemplars of each category reflects the fact that these were the only available well-learned novel items left in Sherman's vocabulary at that time. Sherman categorized ten of these items correctly. His only error was milk, which was termed a food. Following this test we attempted to teach Sherman to categorize milk as a liquid. This additional training notwithstanding, Sherman insisted on classifying milk as a food.

Categorizing: A General Skill

Categorization appears to have been performed quite readily by the chimpanzees. It must be a rather general and typical way of organizing information about their environment, for Sherman and Austin could not have learned all they needed to know to categorize any food or tool on the basis of training with only three items in each group. The training did not teach them to categorize; rather it simply conveyed, in the absence of words, what was expected of them in the task.

Following this training we found that Sherman and Austin would readily sort all manner of objects and could, in fact, sort even on the basis of a single attribute, such as size, shape, color, texture, or form. Ability to sort on these perceptual dimensions required no specific training. Sherman and Austin were also able to spontaneously sort photographs into groups of objects such as cars, animals, or people, with no additional training. They were able to sort up to eight items simultaneously, thus going far beyond the sorting capacities reported of other apes (Ettlinger 1982; Hayes and Hayes 1951).

Austin and Sherman's training, in apparent contrast to Lana's, served to produce a representational mode of symbol usage and acquisition. Although many of Lana's symbols were surely representational, she did not approach the acquisition of the food-tool task with a representational strategy. Lana's performance, as contrasted with that of Sherman and Austin, revealed clearly that different chimpanzees who learn symbols may not have learned the same thing. Training variables are exceedingly important. It is necessary to question the degree to which vocabulary items really stand for things as used by language-trained apes, particularly in projects where no explicit effort has been made either to teach or test comprehension. In any event, the paradigm described in this chapter should prove useful in determining whether other language-trained apes can respond to symbols as true representations of objects.

Chapter 12

Anecdotes: Chance Occurrences, or Important Indicators

This chapter was written in collaboration with Rose A. Sevcik

Novel Combinations and Extensions of Meaning

Readers who have been accustomed to finding anecdotal reports regarding novel usages and interesting novel combinations in the ape-language literature may be puzzled by the absence of such accounts in this book. Indeed, they may have concluded erroneously that Sherman and Austin use their keyboards only to ask for a very limited number of foods and tools, in a very limited number of contexts. This is hardly the case.

The fact of the matter is that Sherman and Austin began to use their keyboards freely and spontaneously, just as chimpanzees taught American Sign Language (ASL) use their signs. Likewise, Sherman and Austin began to spontaneously form many novel and contextually appropriate combinations. While we have avoided describing these in our published reports, we don't consider such behavior to be unimportant. Indeed, spontaneous occurrences are arguably the most important sort of data we have. However, they can be interpreted only within a framework that first documents the reliability, extent, and usage of individual symbols.

Anecdotes of chimpanzee combinations are inherently appealing and scientifically inadequate. When chimpanzees spontaneously say something completely unexpected, novel, and contextually appropriate, one cannot help but feel that they have revealed far more about what they know and how they view their world than is the case during strictly controlled tests which must be administered so cautiously and carefully that they tend to stifle creative symbol usage.

The difficult aspect of anecdotes, however, is that spontaneous creative and unexpected usages are impossible to predict in advance and to re-create at will. It is impossible to rule out inadvertent cueing in such situations. The complex cognitions that are often hinted at in anecdotes seem to be at the very edge of the ape's ability. When the chimpanzee is attentive, alert, and totally interested, and when the circumstances make previously learned behaviors inadequate, conditions may suddenly gel and result in symbol usage that goes far beyond anything the ape has ever been taught.

This chapter is about some of those instances wherein we have seen the limits of the apes' cognitive capacity pushed in subtle ways. These instances serve as important reminders to all who are working with chimpanzees that we are dealing with very intelligent beings.

However, we present our anecdotes against a backdrop of solid data. Appendix 12.A contains a complete corpus of all combinatorial utterances from which the novel observations that we report have been drawn. These contextual data are also buttressed by the data presented in the previous chapters concerning Sherman and Austin's competence with individual symbols. We know, in the case of the accounts to be presented, that neither the teacher's comments immediately preceding symbol usage, nor the chimpanzees' training history can satisfactorily explain these spontaneous combinations.

At the time that the utterances to be described occurred, Sherman and Austin were using the keyboard vocabulary described in chapter 3. This vocabulary was fairly broad and applicable to a wide range of activities. All of the symbols were always illuminated on the keyboard. Some symbols were used in rather limited ways (for example, "glass," which was generally used only when we asked them to name a drinking container), but others were used in a very broad fashion (such as a video monitor, which was used to ask for movies, TV, slides, and also to request that the channels be changed or that other slides be shown).

Whenever they are helpful, we preface our descriptions of what actually occurred in the instances reported below with a somewhat extended account of what the chimpanzee knew at that particular point in training, what his previous experience with regard to these particular symbols had been, and what kinds of things were going on at that time in the lab.

Objections To Reporting Unusual Utterances

Seidenberg and Petitto (1979) have objected to the reporting of unusual utterances on the grounds that unwarranted concentration is given to isolated instances which are overinterpreted, and which, on the whole, exaggerate the apes' abilities. Seidenberg and Petitto's citation, in the quote below, refers to an instance in which Washoe, observing several swans swimming in a pond, signed "water-bird." They argue that:

> It is the absence of a substantial corpus of ape utterances which is the most serious omission. In contrast to the child language literature, there is no listing of all ape utterances which occurred during a single period of time. By failing to provide a corpus, the Gardners and others obscure significant aspects of their subjects' behaviors, and make it difficult to independently verify their claims.
>
> This problem is seen most clearly with respect to the anecdotes that are frequently cited (e.g., water bird). In the absence of a corpus, one cannot determine whether such sequences were synthesized through the application of linguistic rules, or merely the result of the ape acting as a random sign generator which happened to emit some interesting-looking strings. The water bird example loses much of its force if the ape combined each of these signs with a large number of other signs (e.g., water shoe, water banana, cookie bird, etc.). (Seidenberg and Petitto 1979:182)

Although we agree that a corpus of utterances would be helpful in determining whether or not a combination such as "water-bird" reflected more than a random combination of signs, it also is true that—even if a corpus included some other combinations (such as water-shoe, water-banana, cookie-bird)—it would still be impossible to judge, from a corpus alone, whether any of those combinations were true attempts to express a meaningful symbolic utterance.

The addition of contextual information to the corpus of utterances would help the reader to make a better judgment regarding Washoe's intended meaning, yet a "judgment" would still be required and would itself be open to question. For example, when Washoe signed water-bird, water and birds (swans) were both present, and the swans were in the water. However, as Terrace et al. (1979) suggest, even though the context would lend itself to the interpretation that Washoe's choice of the "water-bird" combination was correct, she could simply have been signing "water" because she had learned to produce that sign when near water and "bird" because she learned to produce that sign near birds.

Fouts (1974a), in fact, generally did query Washoe as they walked along by the pond and often also queried her when birds were spotted. Perhaps, as Washoe encountered both together, she simply produced two signs in response to two independent stimuli. Neither a corpus alone, nor a corpus which provides contextual information, could clearly delineate exactly why Washoe signed water-bird when she did.

Still, Fouts' interpretation of the signs would be significantly strengthened if carefully collected data were presented to show that Washoe could, unequivocally, do the following:

1. Use the symbol "water" to request water when she wanted drink (i.e., request skill).
2. Label water when she saw water, in a glass, in a pond, etc. (i.e., naming skill).
3. Use the symbol "bird" when she saw birds in a variety of different contexts (i.e., context-free naming skill).
4. Point out or show a bird to others via signs (statement or indication skill).
5. Pick out, in response to symbolic requests posed by others, real birds or photographs of birds from among other animals, (receptive skill).
6. And most importantly, that she *did not* spontaneously sign "water-bird" when, for example, she was shown photographs of birds and water at the same time, or when she saw nonaquatic birds near, but not *in* water.

Without such data, one can only speculate as to why Washoe produced this combination when she did.

While a corpus of utterances is often presented when language is studied in young children, it must be recognized that the acquisition processes in normal human children are distinctly different from those of chimpanzees. Therefore, any corpus of utterances presented for trained chimpanzees cannot be readily equated with a similar corpus presented for children, who acquire language spontaneously.

Signs must be molded and rehearsed regularly with chimpanzees, and any complete body of utterances they emit will contain a significant number of symbols which are in the process of being intentionally molded and trained, as well as utterances being tested. A complete corpus will also contain a large number of symbols which have occurred because the experimenters have deliberately altered either their own behavior, or the setting, to elicit them.

Hence any corpus of chimpanzee utterances confounds utterances which have occurred under such "tutorial circumstances," and those which were truly *spontaneous* on the part of the chimpanzee. We need to find some means for identifying both spontaneous utterances and those which have been encouraged, tutored, or otherwise purposefully trained by the experimenter (and/or parent) before we can compare a corpus of children's utterances with chimpanzees' utterances. Terrace et al. (1979) utilized a discourse analysis with Nim, the purpose of which was to rule out utterances that could be attributed to the teachers' prior behaviors. According to Terrace et al., this analysis revealed that Nim was not capable of meaningful syntactical constructions on his own. Terrace, however, did not deal with the appropriateness of single signs or simple combinations of signs. Had he done so, he might have found that Nim's individual signs were not being used with representational competence.

Sherman and Austin's Utterances: The Corpus

As a backdrop for the unusual utterances cited in this chapter, we have compiled, in appendix 12.A, a corpus of utterances which includes all lexigram combinations produced over a three-month period by Sherman and Austin between October 29, 1980 and January 31, 1981. Each utterance in the corpus was evaluated, immediately after it occurred, as either correct or incorrect (given the context) and that evaluation, coded along with the utterance, became a part of the permanent computer record.

Since combinations are our main focus of interest here, we have deleted all single word utterances from the corpus in appendix 12.A.[1] Sherman and Austin, incidentally, were not required to produce combinations. Hence all of these combinations were events that went beyond the training requirements. Although the teacher often structured the situation to elicit symbol use—for example, by showing the chimpanzee a food—nothing was done to elicit simple combinations. They often combined simple gestures, such as "give" or "go," with lexigrams.

One thing that the corpus of all combinations in appendix 12.A cannot reflect is any sense of what a day of communication with Sherman and Austin is like. For example, how often and in what ways do they normally use the symbols they know to make their wishes and needs known? The only thorough way in which an accurate picture of this sort can be obtained is, of course, to work with the chimpanzees. However, since only a small number of people can have this experience, a description of a typical morning or afternoon session with Sherman and Austin will have to suffice.

In the account which follows, an afternoon was selected during the time when Austin and Sherman were six and seven years of age, respectively. This afternoon was neither exceptional nor mediocre. It reflects Sherman and Austin's attempts to get the teachers to do what they would like, as well as the teachers' attempts to provide linguistic models. It also reflects the teachers' attempts to get them to do well on whatever task had been planned for that day.

Description of an Afternoon

13:35 Sherman and Austin are brought out of their housing area into the kitchen area by the teachers (Royce, Steve, Sue and Liz). They immediately rush to the door to see if we remembered to lock it—we did. Then they start running everywhere.

Liz asks "Tickle?" (13:36:31) as she looks at Sherman. Sherman ignores this and says "TV" (13:36:33) with a questioning expression on his face. (Sherman likes to watch TV and movies and slides of other chimpanzees, and often spontaneously uses the lexigram "TV" to request this.) Liz replies "No TV, Tickle" and tickles Sherman.

The teachers have planned today to engage the chimpanzees in a play session and to encourage the chimpanzees to use a photograph to specify which teacher they want to chase and tickle them. This session is planned for today since four teachers will be present, thus offering the chimpanzees a variety of play partners to choose from.

Photographs of the teachers are placed on the table in front of the keyboard and the chimps can choose the experimenter they want to play with by selecting her or his photograph. They show the photograph to a teacher positioned by the keyboard who they wish to tickle, for example "Tickle Royce" (13:36:57), "Tickle Sue" (13:46:37), "Tickle Liz" (13:46:49), "Tickle Steve" (13:55:21), "Tickle Liz" (13:55:45).

Sherman tires of tickling and decides that it is time to have something to eat. He walks over to the keyboard and says "Cold" (13:56:05). Liz nods her head and starts toward the refrigerator. As she opens the door, Sherman looks over at her and says "Carrot" (13:56:16). Liz takes a carrot out of the refrigerator and hands it to Steve, who is standing next to Sherman by the keyboard. Steve shows the carrot to Sherman, who says, "Go carrot" (13:56:16). Go is scored as an error, and Steve answers saying, "Yes, give Sherman carrot" (13:57:13). Sherman starts eating the carrot, but while he is still eating, looks over at Liz, who is still by the refrigerator, and asks again "Carrot" (13:56:05). Liz shakes her head "no" since Sherman has a very large carrot already. The teachers then begin to play with Austin as Sherman is busy eating the carrot. While their backs are turned, Sherman asks "Go TV" (14:02:02). Sue looks at him and asks "What?" (14:02:06), as she did not see his request. He repeats, "TV" (14:02:09). She responds "Yes TV" (14:02:16), turns to Steve and then states "Steve go bathroom get TV" (14:02:32). (The TV and movie projector are kept locked up in the bathroom storage area when not in use.) Steve brings the TV out, and some movies of other chimpanzees are shown to Sherman and Austin. They display at the movie images.

When the movies are over, Sherman rolls into Steve's lap and begins playbiting him. Steve responds in kind and Sue comments "Tickle Sherman, Steve" (14:19:08). Sherman then goes to the keyboard and again asks "TV" (14:19:48). Since the chimpanzees have become quite rambunctious during the movies, the teacher decides that slides of other chimps might be more appropriate at this point and so answers "Yes TV" (14:19:52), this time turning on the slide projector.

The symbol TV was introduced with the intent of using it for television only. However, Sherman spontaneously used this symbol to ask for movies and slides also. He said "TV" and then pointed to the device which he wished to have turned on. He also insisted on using the symbol "TV" as a request for the teacher to change the subject matter on the TV, the movie projector, or the slide projector. If he enjoyed what he was seeing, he watched it intently. If not, he paced around the room in an agitated manner and repeatedly said "TV" until we put on another movie, a different videotape, or another slide carousel.

As Sherman and Austin are watching slides, Liz goes to the refrigerator and opens the freezer to take out some food in preparation for the afternoon activities. Sue comments on this activity, saying "Liz get cold food" (14:20:00). Austin looks at Liz and asks "Austin give cold" (14:22:55) which is interpreted as a request for the food that Liz has just taken out of the freezer.

Liz goes to the keyboard and shows Austin the food saying "Cold peaches" (14:23:00). Sherman hurriedly approaches when he notices the food. Austin asks "Give peach" (14:23:04), and Sherman follows immediately with "Peaches cold" (14:23:09), and holds his hand out for some. Liz replies "Give cold peach" (14:23:17), and gives them both some. They both request seconds, Sherman again saying "Peaches cold" (14:23:32), and Austin saying "Give peaches cold" (14:23:46). When the peaches are gone, Sherman and Austin both spontaneously request carrots saying simply "Carrot" (14:23:59), and Liz answers "Yes" (14:24:22), and gives them both carrots.

The afternoon training session then begins. During this session, one experimenter goes to one of five randomly assigned locations. The chimps must label the experimenter's location and are rewarded with food for doing so (9 trials for each chimp are given).

When the food which the experimenters have been using as a reward for correct responses is consumed, Sue tells Sherman and Austin that "Liz [will] get hot food" (14:45:50). As Liz starts to leave the room Sherman asks "Go food" (14:45:55) because he wants to go with Liz to prepare more food. Austin also hurries to the keyboard and says "Give Austin food hot" (14:46:04), and looks toward Liz. Presumably he also wants to go with Liz. Sue says "sink" to remind them that this is the location to which Liz is planning to go. Sherman pushes Austin aside and asks "out sink" (14:46:31). Sue says "Yes" (14:46:35) and Sherman is allowed to accompany Liz to the kitchen.

At the sink, Sherman goes to the keyboard there and requests "Peanut butter" (14:50:45). Liz nods and gets out some peanut butter to heat up with some raisins. While this is heating Steve takes some frozen bread (which will have peanut butter and raisins spread on top of it) and shows it to Sherman saying "This cold" (14:51:14). Sherman repeatedly asks for bites of the frozen bread by holding his hand toward the bread and saying "Cold bread" (14:51:27), "Cold" (14:51:42), and "Bread cold" (14:52:01). Steve gives Sherman bread in response to each request.

Liz, Steve and Sherman go back into the training room and resume the location task. After another nine trials Sherman decides he would like another food and spontaneously requests "Go jelly" (15:10:32). Sue replies "Yes, get jelly" (15:10:43) and goes out and returns with jelly for Sherman. She then says "TV" (16:04:28), and she turns on the slide projector for Sherman.

Sherman and Austin watch slides for a couple of minutes, then Sherman requests "Food" (16:06:58) and gestures toward the food sharing table. Sue nods and the foods are gotten out and the food table is prepared. (Just one portion of each food is placed on the table this time, and Sherman and Austin take turns. Sherman takes more turns.) They proceed as follows, readily complying with each other's requests.

Austin: "Give Jelly" (16:11:56) [Sherman gives Austin jelly]

Sherman: "M&M" (16:12:07) [Austin gives Sherman M&M]

Austin: "Juice" (16:12:17) [Sherman gives Austin juice]

Sherman: "Peanut Butter" (16:12:29) [Austin gives Sherman peanut butter]

Sherman: "Cake" (16:12:48) [Austin gives Sherman cake]

Sherman: "Liquid vitamin" (16:13:18) [Austin gives Sherman vitamin]

Austin: "Milk" (16:13:56) [Sherman gives Austin milk]

Sherman: "Corn chips" (16:14:39) [Austin gives Sherman corn chips]

Austin: "Orange" (16:15:27) [Sherman gives Austin orange]

Sherman: "Raisin" (16:15:44) [Austin gives Sherman raisin]

Austin: "Banana" (16:16:06) [Sherman ignores this request since there are no bananas on the food table and he asks Austin for another food]

Sherman: "Coke" (16:16:39) [Austin gives Sherman coke]

Austin: "Melon" (16:17:17) [Sherman gives Austin melon]

Sherman: "Sweet Potato" (16:17:25) [Austin gives Sherman sweet potato]

Sherman: "Give beancake" (16:27:42) [Austin gives Sherman beancake]

Sherman and Austin are, by this time, quite full, happy, and playful. The teachers take them back to their housing area where they remain and play with the chimps for another 15 or 20 minutes before giving them many large blankets with which to prepare their evening nest. At 17:00, the teachers leave and the chimpanzees settle down for the evening.

The unusual utterances produced by Sherman and Austin which constitute the remaining portion of this chapter all occurred on afternoons similar to the one described above. The general activities may have differed, but the mixture of training tasks with spontaneous requests, and the negotiations about what would be done, were typical of every day.

The corpus in appendix 12.A lists all utterances made outside of training tasks. It also reveals whether each utterance seemed correct given the context and whether or not it was a partial imitation of the teacher's preceding utterance. It is therefore possible to determine which sorts of utterances were likely to be spontaneous and which were not. Overall, during this three-month period, Sherman produced 107 compound utterances that were complete imitations of the teacher, 167 compound utterances that were partial imitations of the teacher, and 104 compound utterances that were novel and nonimitative. Of Sherman's total utterances, 95.5 percent were judged appropriate to the context. Austin produced 97 compound utterances that were complete imitations of the teacher, 233 compound utterances that were partial imitations of the teacher, and 140 compound utterances that were novel and nonimitative. Of Austin's total utterances, 99 percent were judged appropriate to the context. (To determine "imitation," we looked at the teacher's first prior utterance and also at her second prior utterance: Terrace et al. 1979.)

Spontaneous Novel Utterances

The spontaneous utterances which are described below are not the only such utterances produced during this period. They were selected because, in each example, we have reason to believe that the chimpanzee clearly understood the message, since nonsymbolic behaviors accompanied and clarified the message. Moreover, each message was spontaneous and not imitative. Yet it must not be concluded that the chimpanzee could easily *replicate* these utterances in controlled test settings or that he was certain enough of his symbol usage to continue to negotiate the meaning or rephrase it if pressed to do so.

Example 1: "Glass Strawberry—Drink"

New clear Lucite plastic glasses were purchased for the chimpanzees to drink from, but no lexigram had ever been assigned to drinking containers of any sort. During the morning session Sherman had been allowed to use one of these new glasses. He carried it all around the lab with him and treated it like a prized possession. In the afternoon session, the teacher entered the room carrying one of these new glasses, an old glass, and a pitcher of strawberry drink. Sherman saw this and approached the keyboard and lit *"New-symbol" (i.e., an unassigned lexigram on the board), "strawberry—drink",* then pointed to the new glass and looked at the strawberry—drink.

Since Sherman knew the lexigrams "give" and "strawberry—drink" and used them regularly without fail to ask for strawberry—drink, the experimenter concluded that he had used the new lexigram to refer to the new glass, which he wanted to use, and had combined it with "strawberry—drink." The teacher then glossed that lexigram as "glass." In this case, not only did Sherman form a novel combination, he also assigned a new symbol as he formed that combination, and he clarified his meaning by a nonverbal indicative gesture.

Example 2: "Go Sink Milk, Glass Get, Give Glass Milk, Get Go Give Sink Pudding, Go Coke, Glass Coke"

This example illustrates Sherman's spontaneous and correct use of a novel sequence of lexigrams to request a desired set of actions or objects from the teacher. The combinations, which are presented below, all occurred during a single afternoon, and they were not unlike the combinations observed on many other days. This day was selected because it included the continued use of the "glass" lexigram, which had been assigned by Sherman on the previous day.

Sherman had just finished using the string and the magnet to obtain food, and he requested "Go sink milk." The teacher replied, "Yes, go sink; get milk," and took Sherman to the sink where they obtained milk and a glass. Upon returning to the keyboard Sherman said "Glass get," as he reached out toward the teacher's hand that was holding the milk bottle. The teacher replied, "Yes, glass milk" and poured some milk into Sherman's glass. As soon as Sherman finished drinking the milk, he held out his glass and asked, "Give glass milk." When he finished again, he put the glass on the floor and asked, "Get go give sink pudding."

The teacher and Sherman went to the sink, got some pudding and placed the pudding in a tool site. Sherman then requested, "Give lever." The teacher gave him the lever which he used to obtain the pudding. Once the pudding had been obtained and consumed, the teacher asked "Question go get Coke," and Sherman replied, "Go Coke." Sherman did not specify "sink" on this occasion as he had done previously, so the teacher said "Coke sink?" but Sherman did not respond at the keyboard, though he gesturally indicated that he wished to accompany the teacher. They brought a can of Coke and a glass back to the tool site, and Sherman said "Coke" and pointed to the can. The teacher replied, "Yes, this Coke," but did not give Sherman any, waiting for him to specify "glass," as he had done previously with milk. Sherman then gestured to the Coke can, clearly desirous that the teacher pour some Coke into the glass. The teacher affirmed the portion of Sherman's request which corresponded to the context correctly, saying "Yes, this Coke." Sherman then replied, "Glass Coke" and the teacher replied, "Yes give Coke glass," and gave Sherman some Coke in the glass.

This series of combinatorial requests demonstrates that the occurrence of appropriate lexigram combinations and the use of gestures is combined to clarify communicative intent.

Example 3: "Go Food"

Sherman had just been introduced to the lexigrams "food" and "tool" in the study described in chapter 11. At the time of example 2, he had learned to call three foods "food" and three tools "tool." He had not yet had *any* experience using the lexigrams "food" or "tool" in any other context.

On this occasion the teacher had stopped working with Sherman and had gone into an adjacent housing area to give grapefruit to two new chimpanzees in the lab. Sherman had not learned a name for grapefruit, as we rarely had them in the lab. Sherman watched the teacher take the grapefruit to the other chimpanzees, and when she returned, he spontaneously asked "Go food" and gestured toward the adjacent room where the teacher had just taken the armload of grapefruit. The teacher took Sherman to the adjacent housing area and let him visit with the other chimpanzees and have some grapefruit.

In this example Sherman demonstrated an ability to generalize the meaning of a new lexigram very rapidly and appropriately. In addition, he combined it with another lexigram in a way that conveyed more than the use of either lexigram alone could have conveyed.

Example 4: "Austin Room-3 Scare"

Sherman and Austin were being tutored in adjacent rooms and Austin's teacher (in room 3) was putting a mask over her face and pretending to scare him. This was a game which both Austin and Sherman enjoyed and would repeatedly request by lighting "Give scare" at the keyboard. Sherman could hear this game being played and could see some of what was going on by peeking under the door. His teacher noted his interest in what was happening in the adjacent room and commented "Scare Austin." Sherman then asked "Austin Room-3 Scare" and gestured toward Austin's room. This combination, although it included two symbols used by the teacher, was clearly an expansion of the teacher's statement which added the specific information that Sherman wanted to go to the *room* where this was happening. This combination itself was completely novel, never having previously been used by Sherman, Austin, or any teachers.

The teacher replied, "Yes, go scare Austin," and she and Sherman went to the adjacent room to participate in the game. Sherman did not want to stop playing and return to his room. Consequently, when he was led back to his room, he immediately used the keyboard to request "Go scare."

Again, this was a novel combination for Sherman. It also was a request to reengage in an activity he had just enjoyed and his means of requesting a reinstatement of the activity differed from the way in which he previously requested the same activity ("Austin Room-3 scare").

Since Sherman's initial request had been granted, one would expect him to simply repeat it if symbol combinations were guided by simple reinforcement parameters. Instead, Sherman formed a second completely novel and correct combination, illustrating that he could use his limited vocabulary quite flexibly to emphasize, in the first instance, that he wanted to go to a different location, and in the second that he wanted to reengage in the game he had just been playing.

When the teacher replied "No" to his second request, Sherman refused to eat the food which was offered or to return to the sorting task which he had been doing happily before going into the next room to play "scare."

Example 5: "Give Wrench"

Austin and the teacher were doing a "statement" task in which the teacher placed twelve or more objects on the table in front of Austin. Austin looked at the items, then lit any symbol he chose, such as "beancake," and then selected that object (or picture of that object) and gave it to the teacher who placed it under the table.

On this occasion, one of the objects on the table was a tool (magnet) which Austin had been having difficulty identifying correctly over the past week. Austin indicated the items one by one, until finally, only one item was left, the "magnet." He looked at the keyboard a long time, then finally said "give wrench," but instead of picking up the magnet he leaned under the table and pointed to the wrench, which had already been correctly identified and placed under the table. The teacher looked under the table, found the wrench and gave it to Austin. Austin took the wrench with a look of satisfaction and placed it back on the table. He then said "wrench" once again, but this time he picked up the wrench and handed it back to the teacher.

This example illustrates that Austin can spontaneously use his symbols in novel ways to alter the "rules of the game" in such a way as to maximize his own correct performance. That is, by ignoring difficult items like "magnet," and requesting that easier items be placed back on the table, he increased the probability of being able to name the items correctly.

Example 6: "Austin Go Give Hit, Austin Go Hit"

Austin and the teacher were practicing the actions which Austin knew (clap, stand, tie, tickle, chase, hug, hit, groom, and cover). During this session, the teacher would light an action key, such as "stand," and Austin would respond by standing. Correct execution of some actions like "stand" and "clap" were rewarded with food; others, like "tickle" and "chase," were reinforced with the named activities themselves.

Sherman and his teacher were in the adjacent room working on the identical task. The teacher in Sherman's room said "hit-chase" and started chasing Sherman around the room while playfully pummeling him with a stick. Sherman thought this was great fun and whenever the teacher would stop, he would say "chase," and then pull the stick toward him as he drew away. The sounds of banging, chasing, and laughing carried easily to Austin's room.

Austin kept looking over toward Sherman's room obviously distracted from his task. He finally turned toward his keyboard and said "Austin go give hit" and gestured toward Sherman's room. The teacher said "Yes" and opened the door. Austin immediately raced into Sherman's room and joined in the game.

Austin was allowed to play for about five minutes and was reluctant to return, but the teacher held him by the hand and led him back. As soon as he was back in his room, he said "Austin go hit." The experimenter said "No" and resumed working with Austin.

These spontaneous combinations revealed Austin's ability to use "hit" in appropriate combination with other symbols, even though he had never before combined "hit" with other symbols.

Example 7: "Collar" + National Geographic *Photo of Koko*

Two teachers were working with Sherman and Austin on locations. In the course of this training, one teacher would go to a specific location (sink, outdoors, playroom, etc.) while the second teacher remained with the chimpanzees by the keyboard. The chimpanzees were asked to label the location of the teacher who was moving about.

During one of these trials, Sherman refused to label the teacher's location, saying "Collar" instead. He then walked out of the room and directly to the area where the collars the chimps had to wear outdoors are usually kept hanging on the wall. "Collar," was thus often used as a way of asking to go outdoors, and after saying "collar," the chimpanzees would often go search for them and put them around their necks while pant-hooting in anticipation.

On this particular occasion, Sherman saw no collar hanging in the usual spot, and so, after looking carefully, he then walked to the toy shelf (where collars are sometimes inadvertently left) and searched through the toy bins. After about a one-minute search, he noticed the *National Geographic* magazine which he had been looking at with the teacher earlier in the day.

Koko, the signing gorilla, was the main feature in this issue. When Sherman found this issue on the shelf, he tapped the pictures of Koko with his hair standing on end (pilo-erection) and swaying back and forth bipedally. He then picked up the magazine, spied his collar on the floor and returned to the teacher.

He held out the collar to the teacher with one hand and pointed to a picture of Koko with the other. When the teacher said "Yes, go outdoors," Sherman repeatedly uttered high-pitched barks of excitement and started running around and around the lab with the magazine.

Going out-of-doors was, of course, the most favored activity of Sherman and Austin and it was not unusual for Sherman to be excited about going outside. What was unusual in this case was that even in his great excitement he still carried the photographs of Koko and pointed to them repeatedly as he looked toward the teacher, who nodded and said "yes" to Sherman.

The teacher knew that Sherman was combining the presentation of collar and Koko's picture to communicate something rather specific, but was uncertain as to what. The teacher also knew that when she answered Sherman's request in the affirmative he appeared to be extraordinarily pleased. As the teacher put the collar on Sherman he continued to point at the picture and bark very loudly.

Outdoors, he continued to carry the picture while walking bipedally, and he headed directly for the gorilla quarters which were located near the adult-ape housing on a hill behind the language lab.

On a few previous late evening occasions Sherman had been allowed to go to this area and view the gorillas, an event which both scared and thrilled him. However, in general, he was not allowed to go there since his presence excited all of the large apes housed there and interfered with any research that might have been going on. Generally, if Sherman tried to lead his teacher toward this area, she would refuse to allow him to go.

The closer Sherman got to the gorilla housing area, the faster he ran, the louder he barked, and the tighter he held the pictures of Koko. Thus Sherman's intent and expectancy became clear to his teachers. From his point of view he had asked to go to see the gorillas and he had been told "yes." However, the teacher had not initially understood the request completely and the acquiescence with regard to going outdoors had led Sherman to believe that he could also go see the gorillas.

When it became apparent to the teacher that Sherman was intent on proceeding to the gorilla area, it became necessary to say "no." As Sherman saw the teacher stop and point in the other direction, he dropped to a quadrupedal stance, put down the *National Geographic,* stopped vocalizing completely, lost his pilo-erection, and appeared completely dejected. He showed no further interest in the magazine or in being outside.

Although Sherman had not used the keyboard, he had certainly made combinatorial request. It had involved the use of a symbol and a photograph to produce a novel communicative request to go to a very specific location, and one that he could not request at his keyboard alone. The marked change in Sherman's demeanor when he realized that his request was not going to be granted was dramatic. His immediate discarding of the magazine at this point further supports the view that he had indeed been intentionally using the magazine to communicate his desire to go and see the live gorillas.

It is important to point out that Sherman and Austin had never been taught to use photographs as a means of requesting to go to locations. Moreover, this particular photograph had been brought to the lab only that day, and no one had ever coupled it with any sort of training.[2]

Example 8: "Food" + Food sharing table"

Food sharing, which Sherman and Austin engaged in at least once each week after they had reached the level of cooperation described in previous chapters, was their most favored activity. They appeared to enjoy the preparation ritual associated with food sharing almost as much as consuming the food itself.

This preparation involved getting out all the food and dividing it into equal portions, then retrieving the special small compartmentalized table which was always used in this task. Each of the compartments in the table then had to be filled with two portions of food. The chimps always helped with every step, tasting food the whole time. The food sharing table, when fully prepared, appeared festive and once it was finally properly set, the chimpanzees were often so excited that they would run around the lab for two or three minutes pant-hooting loudly and displaying before they were able to settle down and eat. They would then simultaneously approach the keyboard and begin requesting and sharing food.

On this particular day, it had been some time since we had had a food-sharing bout, as we had been concentrating on new things. It was near the end of the day and both chimpanzees appeared quite uninterested in the locations-training task which was in progress.

On one trial, Sherman ignored what the experimenter was doing, went to the keyboard and said "collar" as a request to go outside. The teacher replied "No" and attempted to redirect Sherman's attention to the task at hand. Sherman ignored the teacher and began to play with some objects on the floor, so the teacher turned her attention to Austin. A few minutes later, Sherman approached the keyboard and said "food." The teacher hesitated, not understanding exactly what Sherman wanted, since there was food available but he did not seem interested in it.

Sherman then went directly to the corner of the room where the food-sharing table had been placed on the shelf for cleaning several days before. He pulled the food table off the shelf and placed it directly in front of the keyboard, pointed to it, and lit the "food" symbol again. Realizing that Sherman was requesting the activity of *food + sharing,* not just "food" itself, the teacher answered "Yes, food," and gestured to the table, whereupon Sherman and Austin immediately hugged one another and began to run around the lab loudly pant-hooting and displaying with great excitement.

The combination of the symbol "food" and the bringing of the food table was a clear request for the *activity* of food-sharing. The meaning of this request was not understood initially and so Sherman clarified it by coupling the symbol with the showing of a key object. It is important to note that it was not the food *per se* which Sherman was requesting, since he could have asked for and received any of these foods at any time. Rather it was a change in the activity of the group. He was bored with what he was doing and quite clearly wanted to change the activity of the entire group rather than just eat a different food.

Example 9: "Straw Give Scare, Outdoors"

This combination was produced by Sherman in response to seeing one of the teachers light a sparkler inside the lab. The sparklers were brought in one day as a novel item to see what Sherman and Austin would think of them. Sherman appeared quite fearful, showing pilo-erection and an open-mouthed bared-teeth grin. He rushed to the door to go out, but finding the door locked, he hurried back to the keyboard and said "Straw give scare, outdoors," then raced back to the door. When it was opened for him, he hurried out and stayed out.

Presumably, "straw" was used because the shape of the sparkler was long and thin, like a straw. Why Sherman commented upon this and upon his state of being scared is difficult to ascertain. However, he used "outdoors" as a request and was clearly relieved when outside.

This was a completely novel combination and one of the longest combinations which Sherman had ever made. We had never required that Sherman explain why he wanted to go outdoors and had he simply pointed to the door to ask that it be opened, it would have been opened for him. This was an example which was possible to repeat. The following day, the teacher again lit a sparkler. Sherman immediately said "Straw give scare, outdoors," displaying exactly the same fearful behavior as he had on the previous day.

What Type of Training Experiences Led to Such Combinations?

The examples presented above do not reflect the only occasions, or even the most interesting occasions in which Sherman and Austin have communicated novel things. Similar occurrences happen several times each week.

In each of the instances cited above, all the experimenters involved viewed these occurrences as completely novel. That is, no attempt had been made to teach the chimpanzees to use their symbols (or objects) in these ways, and no one had ever demonstrated such usage in the presence of the chimpanzees. Additionally, request skills, naming skills, and indicative skills had been demonstrated for the lexigrams used in these combinations.

Discussion

This attempt to place unusual utterances within a larger body of data is unique in the presentation of ape language data and it is hoped that, as such, it will help clarify the sorts of things which apes can do, and the settings within which these behaviors are seen. It is also hoped that these observations, the corpus, and the description of a day's activities will give the reader a clearer sense of what it is like to do research with these chimpanzees.

Appendix for Chapter 12
Total Spontaneous Utterances Across a Three-Month Period

Sherman

	No Imitations	*Partial Imitations*	*Complete Imitations*
Austin M&M	1		
Austin Room-3 Scare		2	
Austin stick	1		
Austin tickle		1	
Austin tickle give chase		2	
Banana cold	1		3
Bite chase	1		
Bite peanut	1		
Bread cold-X		3	
Butter hot raisin			1
Butter raisin			1
Carrot Steve		1	
Chase beancake wrench-X	1		
Chase Steve	1	1	
Chase Steve tickle	1		
Chase tickle Steve		1	
Chase tickle			2
Chase yes Sherman		1	
Chase TV			1
Cheese M&M	1		
Coffee this-X		1	
Cold bread		1	
Cold melon			2
Cold peanut			1
Collar go out tickle	1		
Collar outdoors yes-X		1	
Corn give			1
Corn pee-X	1		

Utterances are broken down by type; either no imitation of preceding two utterances, imitation of some of the words of one of the preceding utterances, or complete imitation of the preceding utterance. Incorrect utterances are marked by an X.

Sherman *(continued)*

	No Imitations	*Partial Imitations*	*Complete Imitations*
Corn vitamin	1		
Food cold Steve		1	
Food cold			1
Food corn M&M			1
Food glass		1	
Food go sink peanut		1	
Food hot		1	1
Food hot Steve liquid		1	
Get go give sink pudding	1		
Get scare go outdoors-*X*		1	
Get Room-1	1		
Get TV			1
Give Austin coffee hot-*X*		1	
Give Austin cold-*X*		1	
Give Austin hot food-*X*		1	
Give Austin lever-*X*	1		
Give cold			3
Give beancake	1		
Give butter		1	
Give cake	3	1	
Give coke	1		
Give corn	1		5
Give corn Sherman		1	
Give food		3	1
Give food hot		1	2
Give food liquid hot		1	
Give glass milk	1		
Give go food hot open		1	
Give go vitamin	1		
Give groom	1		
Give hot	2	3	1
Give hot lemonade			1
Give hot Steve		1	
Give jelly	1		
Give lemonade	1		1
Give lemonade hot	1		1
Give light			1
Give M&M	1		
Give milk	1		
Give milk liquid	1		
Give peach	1		
Give peanut	1		1

Sherman *(continued)*

	No Imitations	*Partial Imitations*	*Complete Imitations*
Give pineapple	1		
Give playroom	1		
Give playroom carrot		1	
Give pudding	1		
Give Sherman hot			2
Give Sherman peanut	1		
Give Sherman vitamin	1		
Give sink cake		1	
Give sink M&M		2	
Give sponge			1
Give strawberry drink hot			1
Give Steve jelly		1	
Give Sue		1	
Give tangerine	1		
Give tickle		3	
Give TV			1
Give vitamin	2	2	5
Give vitamin Sherman		1	
Glass coke		2	
Glass get	1		
Glass strawberry drink	1		
Go carrot	1	1	
Go cold peanut		1	
Go coke		3	
Go corn	1	1	2
Go corn sink M&M			1
Go food	1	1	
Go hot			2
Go give jelly		2	
Go give lemonade hot		2	
Go give sink peanut		1	
Go give vitamin	1		
Go jelly		2	1
Go lemonade hot		2	
Go liquid		1	2
Go liquid vitamin		1	
Go orange		1	
Go out open		1	
Go outdoors collar			1
Go out outdoors	2		
Go peanut		2	
Go playroom	1		

Sherman *(continued)*

	No Imitations	*Partial Imitations*	*Complete Imitations*
Go scare	1		
Go Sherman peanut		1	
Go sink	1	1	
Go sink coke	1		
Go sink food	1		
Go sink milk	1		
Go sink outdoors bathroom		1	
Go soap Room-1		1	
Go TV	3	1	
Go TV scare			1
Go tangerine soap		1	
Go tickle	2		
Go vitamin	9	10	3
Go vitamin hot		2	
Groom carrot-X	1		1
Hose this		1	
Hot Austin banana		1	
Hot banana			1
Hot coffee	1	1	
Hot corn			1
Hot give butter peanut		1	
Hot food			4
Hot lemonade Sherman		1	
Hot liquid Sherman		1	
Hot give			1
Hot liquid			2
Hot liquid Sherman			1
Hot peanut		1	1
Hot raisin			1
Jack bite cheese		1	
Jack peach	1		
Jack Room-3		1	
Jelly M&M	3		
Jelly open	1		
Jelly peanut	1		
Jack cheese			1
Key sink		1	
Lemonade yes		1	
Lemonade hot			1
Liquid hot			1
Lever open	1		
Lever Pal		1	

Sherman *(continued)*

	No Imitations	*Partial Imitations*	*Complete Imitations*
Light TV	1		
Liquid hot Steve		1	
Liquid TV	1		
Liquid vitamin	2		
M&M jelly	1		1
M&M Steve	1		
Molasses go		1	
Molasses this	1		
Money sink-X		1	
Oil peanut-X	1		
Orange drink give vitamin	1		
Orange peanut	1		
Out jelly	1		
Out scare	1		
Out sink		1	
Outdoors scare	1		
Pal Steve		2	
Peach cold			6
Peach jelly		1	
Peach carrot	1		
Peanut food		1	
Peanut give	1		
Peanut give liquid		1	
Peach pudding			1
Peanut Steve	1		
Peanut tangerine		1	
Pee TV-X		1	
Question chase	1		
Question give peanut		1	
Question peanut	1		
Question peanut question	1		
Question scare	1		
Question sink out	1		
Question strawberry drink	1		
Question tickle		1	
Room-1 chase	1		
Room-3 tickle			1
Royce Sherman tickle		1	
Sherman bread	1		
Sherman chase			1
Sherman cheese		1	
Sherman cherry-X	1		

Sherman *(continued)*

	No Imitations	*Partial Imitations*	*Complete Imitations*
Sherman coke	1		
Sherman cold		1	
Sherman corn	1		
Sherman go peanut		1	
Sherman food hot			1
Sherman give hot			1
Sherman give jelly			1
Sherman give juice			1
Sherman hot			1
Sherman hot coffee			1
Sherman jelly			2
Sherman jelly give			1
Sherman lemonade hot			1
Sherman M&M	1		
Sherman peanut		1	2
Sherman pineapple	1		
Sherman tickle Royce		1	
Sherman TV			1
Sherman vitamin			2
Straw give scare outdoors	2		
Strawberry drink lemonade			2
Sponge lemonade	1		
Straw vitamin		2	
Strawberry drink give			
Swing chase	1		
Swing open	1		
Swing playroom	1		
TV bathroom	1		
TV jelly	1		
TV light	1		
TV open	1		1
TV scare			1
This peanut	1		
This tickle		1	
Tickle Austin	1		
Tickle chase	1	3	3
Tickle playroom milk		1	
Tickle Sherman			1
Tickle Steve	1	4	4
Tickle vitamin		1	
Tickle yes tickle		1	
Vitamin corn		1	

Sherman *(continued)*

	No Imitations	*Partial Imitations*	*Complete Imitations*
Vitamin liquid			1
Vitamin open		1	
Vitamin Steve	1	1	
Vitamin TV-X		1	
Water hot			2
Wrench food	1		
Yes cherry		1	
Yes go Sherman molasses		1	
Yes Jack	1		
Yes jelly	1		
Yes liquid		1	
Yes peanut	1		
Yes peanut Steve	1		
Yes raisin	1		
Yes Steve		1	
Yes TV		2	
Yes this		1	
You collar	1		

Austin

	No Imitations	*Partial Imitations*	*Complete Imitations*
Austin banana hot		1	
Austin get lemonade		1	
Austin get shot	1		
Austin give	1	1	
Austin give Austin soap		1	
Austin give butter hot	1	1	
Austin give chase		3	
Austin give cold		2	2
Austin give gone		1	
Austin give hot	1	2	
Austin give hot coffee			1
Austin give hot lemonade		1	
Austin give juice	2	1	
Austin give liquid hot			1
Austin give M&M		1	
Austin give oil Jack		1	
Austin give out		1	
Austin give peach		1	

Austin *(continued)*

	No Imitations	*Partial Imitations*	*Complete Imitations*
Austin give peanut	1	1	1
Austin give playroom			1
Austin give sink corn		1	
Austin give Steve		2	
Austin give strawberry drink	1		
Austin give TV		2	
Austin give tickle	2	4	
Austin give tickle Steve		1	
Austin give vitamin	5	7	
Austin give vitamin food	1		
Austin go get food			1
Austin go get light	1		
Austin go give hit	1		
Austin go hit	1		
Austin lemonade hot		1	
Austin peanut		1	
Austin Steve	1		
Austin swing		1	
Austin TV		1	
Austin tickle	7	10	9
Austin tickle chase		1	
Austin tickle Steve			1
Austin tickle swing		1	
Bathroom corn		1	
Butter vitamin	1	1	
Can opener glass		1	
Cheese give strawberry drink	1		
Cherry stethoscope phone tickle	1		
Chimp tickle Steve			1
Coffee Steve			1
Collar Austin outdoors		1	
Corn give	1		
Food give vitamin		1	
Food hot			1
Food stick	1		
Give Austin		1	2
Give Austin banana peanut		1	
Give Austin butter		1	
Give Austin butter hot		1	
Give Austin Carrot		1	
Give Austin coffee hot		2	
Give Austin cold		4	

Austin *(continued)*

	No Imitations	*Partial Imitations*	*Complete Imitations*
Give Austin collar		1	
Give Austin corn			1
Give Austin food		3	
Give Austin food coffee hot		1	1
Give Austin food hot		1	
Give Austin give playroom		1	
Give Austin hot		3	
Give Austin hot juice		1	
Give Austin hot peanut		1	
Give Austin jelly	1	1	
Give Austin juice		1	
Give Austin key playroom			1
Give Austin lemonade hot		1	
Give Austin melon		1	
Give Austin peanut	2	1	
Give Austin playroom soap		3	
Give Austin playroom soap chase		1	
Give Austin scare		1	
Give Austin sink jelly peanut		1	
Give Austin Steve		4	
Give Austin strawberry drink		1	
Give Austin tickle	1	3	
Give banana cold			1
Give beancake	1		
Give bread cake		21	
Give butter		3	1
Give cake	1		
Give carrot		1	
Give chase	2	1	
Give cheese	1		
Give coffee		1	
Give coffee Austin hot			1
Give coffee hot	2	1	
Give cold	1		3
Give coke			1
Give corn			8
Give corn vitamin		1	
Give food Austin hot			1
Give food hot	2		2
Give food liquid hot		1	
Give glass		1	2
Give glass juice			3

Austin *(continued)*

	No Imitations	*Partial Imitations*	*Complete Imitations*
Give glass vitamin		1	
Give go food	1		
Give go hot banana		1	
Give go light		1	
Give go liquid hot food		1	
Give go molasses		1	
Give go outdoors collar		1	
Give go playroom Austin soap			1
Give go playroom carrot soap		1	
Give go swing		1	
Give hot coffee	1		
Give hot orange		1	
Give jelly		3	
Give juice	2		
Give key	3		
Give lemonade	1		
Give lemonade hot			1
Give light	1		
Give light vitamin		1	
Give liquid coffee hot			1
Give liquid food		1	
Give liquid food hot		3	
Give liquid hot	1	4	1
Give liquid hot coffee		1	
Give liquid vitamin		1	
Give milk		1	
Give molasses		1	1
Give open juice	1		
Give peach	5	2	
Give peach cold			2
Give peanut	4		4
Give playroom	3	3	
Give playroom soap		2	1
Give playroom Steve		1	
Give scare		1	
Give shot	1	2	
Give sink	1		
Give sink lemonade		1	
Give soap swing playroom		1	
Give sponge	1	1	
Give stethoscope	2		
Give Steve	2	5	

Austin *(continued)*

	No Imitations	*Partial Imitations*	*Complete Imitations*
Give Steve soap		1	
Give straw		2	
Give straw tickle Austin		1	
Give strawberry drink		1	
Give strawberry drink lemonade		2	
Give Sue		1	
Give sweet potato	1		
Give swing		3	
Give tickle	2	7	
Give tickle no	1		
Give tickle Steve		1	
Give tickle yes		1	
Give vitamin	26	32	17
Give vitamin Austin		1	
Give vitamin open		1	
Give vitamin peanut	1		
Give vitamin Room-3		1	
Give vitamin Steve	1	1	
Give wrench	1		
Give yes Austin Steve		1	
Go coffee hot		1	
Go corn		1	
Go give banana hot		1	
Go give coffee hot	1		
Go give hot		1	
Go give out sink		1	
Go give outdoors		1	
Go give playroom soap			1
Go give Steve	1		
Go give vitamin		2	
Go key			1
Go out		1	
Go peanut hot		1	
Go phone	1		
Go sink pee	1		
Go swing		1	
Go hot	1		
Go TV		1	
Go tickle	1		
Go tickle Room-3		1	
Go vitamin		3	
Got give butter		1	

Austin *(continued)*

	No Imitations	*Partial Imitations*	*Complete Imitations*
Got go give		1	
Got open		1	
Hot coffee			1
Hot food			1
Hot give Austin straw		1	
Hot lemonade			1
Hot orange			1
Hot peanut	1		
Hot water			1
Jelly give vitamin	1		
Jelly oil	1		
Key wrench	1		
Lemonade cold			1
Light go get light	1		
Liquid coke			1
Liquid juice			1
Liz tickle			1
Melon cold			1
Orange drink give food		1	
Outdoors no		1	
Peanut cold			1
Pineapple hot			1
Question Austin tickle swing		1	
Question give food		1	
Sherman give peanut		1	
Sherman Sue			1
Steve Austin tickle			1
Steve give food		1	
Steve this Steve	1		
Steve tickle		1	
Strawberry drink give liquid vitamin		1	
Strawberry drink peach		1	
String key	1		
Swing Austin			1
Swing sponge	1		
This give peanut	1		
Tickle Austin	1	1	
Tickle Austin chase		1	
Tickle give vitamin	1		
Tickle Steve		1	
Vitamin shot		1	
Water hot			1

Austin *(continued)*

	No Imitations	*Partial Imitations*	*Complete Imitations*
Water hot liquid			1
Wrench key	3		
Yes get Austin vitamin	1	1	
Yes molasses		2	
Yes swing		1	
Yes tickle Sue		1	
Yes vitamin	1		

Chapter 13

Video Representations of Reality

This chapter was written in collaboration with Elizabeth Rubert

Another Mode of Representation: The Video World

A number of ape facilities, including the Atlanta Zoo and the Yerkes Primate Center, have placed television sets near their ape-caging quarters in hopes of alleviating the boredom that inevitably sets in when caged apes have nothing to do. The interest shown by the apes in TV, however, appears to be rather fleeting. Television does little by way of entertaining them and it is not possible to infer from their behavior that they are seeing anything other than changing patterns and forms. They typically show little overt behavioral response to video images, though they do gaze at the screen intermittently.

Television also appears to be relatively uninteresting to apes who have been reared with a good deal of exposure to TV—such as those reared in human homes (Temerlin 1975) or in signing projects. There is no report, for example, that Washoe, Nim, or Koko watched television or signed about things which they observed on the TV screen.

Lucy, a female chimpanzee who was reared as a member of a human family and who acquired a vocabulary of 125 signs (Fouts and Mellgren 1976; Temerlin 1975), was exposed to television from birth. She was encouraged to attend to particular television scenes that members of her surrogate family knew she always responded to in real life (such as children and dogs romping together out-of-doors). Her interest in television was always brief and she did not respond to the televised images with the same interest she displayed with real figures. It was not possible to conclude from her behavior that she processed the images on the TV screen as representations of anything with which she was familiar.

Lucy's lack of interest in video images cannot be attributed to their two-dimensional nature, since she clearly enjoyed looking at and responding to still photographs. When shown a photograph of a luscious ripe apple, for example, she would attempt to take a bite of it. In fact, there were few magazines to be found in the house which, on pages depicting delicious fruits, did not bear the marks of having been "tasted" by a chimpanzee.

Lucy also responded to magazine photographs of dogs and cats (usually by thumping on them) and to photographs of people hugging, kissing, or dancing with intent gazes. As long as the photographs were novel, her interest in looking at magazines ranged from 20 to 30 minutes. She generally did not gaze at the television for more than two to three minutes.

Washoe (Gardner and Gardner 1978b), Koko (Patterson 1978), and Viki (Hayes and Hayes 1951) also enjoyed looking at photographs. Washoe could label slides of bugs, dogs, and flowers in blind tests, even when she had never seen those particular slides before (Gardner and Gardner 1978b), and Koko enjoyed looking at Viewmaster reels. It is puzzling that apes who readily react to still images with appropriate behaviors do not demonstrate a similar responsiveness to video images.

Given the apes' lack of overt behavioral response to video images, it is difficult to arrive at any firm conclusion about how they process moving video images. Do they interpret them as representations of reality? Do they think that there are Lilliputians inside the box doing things with one another, do they think that what they see is really happening—but only in a different location, or do they not even "think" about any of these things at all?

Sarah and Video Representations

Without really addressing any of these queries and simply assuming that chimpanzees "understand" that what they see on TV is a representation of reality, Premack and Woodruff (1978) have undertaken a series of studies using televised images in their work with the chimpanzee Sarah. A rather detailed discussion of Premack's work will be given below, since it is the only previous study which bears directly on the issue of the perception of video images by apes.

Premack's approach consisted of presenting Sarah with a 30-second video "problem." For example, the tape might show one of Sarah's trainers appearing to be very cold as he stands by the furnace, or trying to spray a cage with a hose that has not been connected. It is not clear why Sarah saw these situations as "problems" for the trainer or why she herself wished that the human trainer could solve these problems.

Premack provides no information to suggest that Sarah attended to video images any more than did Lucy or the apes at Yerkes. That is to say, he describes no behavior on Sarah's part which would suggest either that she understood or reacted to the images displayed on the TV.[1]

After playing the 30-second tape, Premack placed an image on hold and gave Sarah two 35mm photographs. The photographs depicted two solutions to the problem the actor on the video tape had encountered. For example, if Sarah had just seen the tape of the actor trying to clean the cage without the hose, the frozen image would show the hose and Sarah would then be given a photograph of a key in a padlock and a photograph of a hose properly attached to a water faucet. Her job was to select the correct picture, of the hose in this case, and give it to the trainer.

According to Premack, Sarah, in selecting the photograph of the hose, was making her choice to solve the actor's dilemma. This interpretation raises numerous questions: Why should Sarah want to solve the actor's problem? How are we to know that Sarah viewed the actor as possessing a problem? Why was it necessary to freeze-frame the video image and present Sarah with what appeared to be either an identity or associative match-to-sample task? Why not just have the objects available and let the tape run till Sarah handed the teacher the needed item if she really wanted the teacher to solve the problem? Even more simply, if Sarah, in fact, (a) understood the problem, (b) knew how to solve it, and (c) wanted to solve it, why not simply let her enter a room where a trainer is actually attempting to clean the room with a hose and show the trainer that the hose needs to be connected?

Unfortunately Premack provides answers to none of the above questions. Furthermore, it is readily apparent that Sarah could have chosen the correct photograph for other reasons. Given Sarah's previous training history with physical and associative match-to-sample tasks, she could readily have selected the photograph of the attached hose because it was a better match to the hose in the frozen image than the other alternative, a photograph of a key.

By still-framing the image as the sample, and presenting two photo alternatives, Premack, in fact, encouraged a match-to-sample orientation to the task. Also, it seems odd that Sarah should think that her choice of photographs would have any effect on the actor whatsoever. Why should she draw such a conclusion? She was never shown any tape of her photograph being taken to the actor, or of his using her photograph in any way to solve his problem. Indeed, was Sarah to think the actor would actually solve the problem if he received the picture she selected? Would he look at the picture she sent and realize that he had forgotten to connect his hose? Is it realistic to suppose that Sarah thinks that she can remind her trainers to connect hoses by sending them photographs? Is it realistic to think that Sarah even cares whether a trainer had his hose attached or not?

This is not to say that chimpanzees do not know that keys are required to unlock padlocks, or that hoses need to be attached to water faucets in order to operate. They do. I once observed Lucy work diligently for 20 minutes to attach a hose properly to a faucet, and then to attach a spray nozzle to the end of the hose. Why? So that she could spray me with the water. I have also had Sherman and Austin steal my keys on numerous occasions and then proceed to let themselves out of locked areas while my back was turned. I have even observed Sherman use a key to let Austin out of a locked room.

However, an ape's competence to solve real world problems is not necessarily what is being tapped by Premack's use of video tapes. It seems more likely that he is testing Sarah's ability to make complex identity and associative visual matches to still-frame video images.

We made a specific test of this possibility by devising a similar task for Sherman and Austin. This task required them to select one of two items when shown a "sample" item. No "problem" was presented along with the sample item, just the sample itself. We presented 28 different items and asked them to select which one of two additional objects was to be paired with each sample item. All testing was given in a blind situation. One experimenter sat outside the room with the sample object and another experimenter, who did not know the sample on any trial, sat inside the room with the alternatives.

The chimpanzee observed the sample, went into the room, selected an alternative, and carried it out of the room to the first experimenter. The pairs of items and the alternatives were presented once to each subject and are listed in table 13.1. No training was given before this task, and no reward was given during it. The chimpanzees made their choices quite spontaneously and remained self-motivated throughout. Sherman and Austin were both correct on 25 of 28 trials.

Sherman and Austin were able to make these selections easily (without training or food reward) even though the experimenter presented *no* "problem" to be solved, but rather simply showed the chimpanzees a "sample" object and two alternatives. Their previous experiences using these items in daily living, or observing their teachers use them, had enabled them to understand which items worked together.

Had we presented Austin and Sherman a "problem" instead of just the items themselves, no doubt their choices would have been similar. For example, had we tried to cook on the hotplate without a pan, they would still have selected a pan, when asked to choose between a pan and a hammer. However, it would not be accurate to conclude that simply because they selected the pan that they perceived the teacher had a problem. It is more likely that they would have found it interesting, and watched what the teacher was trying to do, as the food spilled and burned on the burner. The point is that knowing that pans go with, and are used on, hotplates is not the same as knowing that one has a problem if one is trying to use a hotplate without a pan.

In any case, it would seem that a more accurate way to assess Sarah's problem-solving capacities would be to first engage her in real problem-solving situations, not video ones. If, in the real world, she did go about helping others solve their problems, then her ability to deal with TV-presented problems could be broached more

Table 13.1. Single-Trial Data on Unrewarded, Spontaneous Associative Match-to-Sample Task

Sample Items	*Choice Items*[a]
foot	shoe vs. key
automatic food dispenser	dispenser tray vs. hose nozzle
vaseline jar	thermometer vs. lock
slide tray	slides vs. scrub-brush handle
wastebasket	plastic liner vs. hammer
padlock	key vs. scoop
scrub brush	scrub-brush handle vs. phone body
thermos with hole in top	drinking straw vs. paper
pounding toy	mallet vs. dispenser tray
bolt	wrench vs. shoelace
head	hat vs. stock
can (used to hold washers)	washers vs. dowel (for juicer)
string (usually used with sponge)	sponge vs. nail
telephone body	receiver vs. bottle top
pencil	paper vs. thermometer
hammer	nail vs. slide
utensil rack (S)[b]	serving spoon vs. hat
bottle	bottle top vs. wrench
dipping platform (usually used with stick)	stick vs. straw
hasp	padlock vs. towel rack
magnet (S, A)	metal vs. shoe
hose	spray nozzle vs. washers
paper towels (A)	towel rack vs. mallet
hotplate (A)	pan vs. plastic liner
can	can opener vs. sponge
shoe (S)	shoelace vs. telephone receiver
food bin	food-scoop vs. can opener

[a] The correct item is listed on the left. (Right-left positions of the items when presented to the subjects were randomized.)

[b] S and/or A indicates incorrect response by Sherman and/or Austin.

assuredly, though care would still need to be taken in interpreting how she processed and reacted to moving images.

Sherman and Austin's Initial Uninterest in Television

Like other apes, Sherman and Austin initially showed little more than passing interest in commercial television. Also, as we tried to engage their interest in live video presentations of things happening just outside the lab, or in an adjacent room (as Premack reportedly did with Sarah), we found that they did not seem to understand that they were watching a portrayal of something happening elsewhere in their world. Other chimpanzees when tested have manifested a similar lack of knowledge regarding video images (personal communication, Emil Menzel).

Speculating that perhaps the quality of the image, the size of the image, and the typically human-oriented subject matter of the material on TV might all contribute to their lack of interest, we decided to try movies and tapes of other chimpanzees. We chose some of apes with whom Sherman and Austin were acquainted and some of apes with whom they were not. The movies were large, easily visible, and in color.

At first, Sherman and Austin watched the movies with interest for only two to three minutes. However, we continued to show them movies for approximately one half hour each day for several months and encouraged them to watch by exclaiming and vocalizing when interesting social scenes appeared. Most of the movies were films of other chimpanzees taken for purposes of documenting particular behavioral characteristics of apes (Savage 1975; Savage and Wilkerson 1978), and we had available a rather large supply of different movies.

During this period of time, the chimpanzees attention span increased steadily until they were watching the movies during the entire half-hour period with constant interest. They also began to evidence *overt behavioral responses* to various events which they observed on the screen. For example, they would display when they saw other males begin to display. They attempted to inspect female swellings and attempted to bite the screen when particular chimpanzees appeared. Furthermore, they clearly discriminated between movies they had seen a number of times and new movies, ignoring familiar ones, but attentively watching new ones. They no longer needed the teachers to encourage them to look at the screen, but rather spontaneously requested "movies" and evidenced disappointment when they were over.

Responses to Live Video Images

At this point, we introduced color video equipment and found that even though our monitor was small (only eight inches measured diagonally) the chimps could identify and label, with lexigrams, moving images of foods and tools as readily on the monitor as they could identify slides of real objects. No new training was required to enable the chimpanzees to label video images just as they did real items.

Sherman and Austin now displayed fascination with live or real-time video images which depicted a teacher in an adjacent room. We began by showing them a teacher opening a package of food and beginning to eat it with obvious relish. On the very first occasion on which they saw such live video scenes of their teacher eating food in the adjacent room, both chimpanzees spontaneously used the keyboard to request the food they saw the teacher eating.

In response to these spontaneous requests, one chimpanzee at a time was allowed to go to the adjacent room. When he arrived the teacher gave him some of the food and gesturally asked him to take some back to the other chimp. The chimpanzee who remained behind observed all of this very intently on the monitor and looked toward the door of his room as he saw the other chimpanzee start to return on the TV screen. The situation was repeated a number of times, allowing both chimpanzees opportunities to request and retrieve food for themselves and their companions by observing what was happening on the TV set. Neither chimpanzee needed any encouragement to attend to the TV. Both chimpanzees watched all the activities in the adjacent room with keen interest.

Sherman and Austin's intrigue with video portrayal of the activities of one another and the teacher in the next room, coupled with the overt response of running into the adjacent room when given permission to retrieve the food, clearly revealed that they knew what it was they were seeing on the monitor. During this first session of live video presentation, Sherman spontaneously requested 19 of 20 different foods correctly and Austin spontaneously requested 11 of 14 different foods correctly.

They were able to respond symbolically to these live video images so easily it was possible to test them formally with control in the very next session. During this test, both chimps were to "label" the item they saw an experimenter (E_1) select on the TV. A second experimenter (E_2) sat with them as they watched the TV, but the TV was turned toward the chimps and away from E_2 so that E_2 did not know what items the chimpanzees saw. When the chimps recognized the item held up by E_1, they went into a third room, out of sight of both E_2 and E_1, and named that item on the keyboard. E_2 could see the symbol they selected on the projectors outside the room. Once E_2 saw the symbol that the chimpanzee had selected, she then moved into the room and looked at the TV to determine if he were correct.

Nineteen different foods and ten different tools were used during this test with Sherman and Austin. In addition, five different actions were used with Austin (but not Sherman, as Sherman did not know these action terms). The actions (bite, groom, funny face, stand, clap) were demonstrated on TV by the teacher, and Austin was to label what the teacher was doing. Sherman was correct on 29 of 31 food presentations and 26 of 27 tool presentations. Austin was correct on 25 of 30 food presentations, 26 of 28 tool presentations, and 9 of 11 action presentations.

Thus the world on the TV monitor could serve, for Sherman and Austin, as a representation of the real world. They knew that they could move into the world depicted on TV, act upon it, and then go back out of it. In further exploration of their capacity to comprehend what they saw on the screen, we found that they could observe a tool site being baited on TV, ask for the correct tool, then go into the other room and use the tool appropriately. Likewise, after observing on TV where items were hidden, they could locate them in real space.

Following the use of live video images in problem-solving settings, Sherman and Austin began for the first time to show interest in commercial TV programs. They became interested in watching scenes of zoo animals, Muppets, professional wrestling matches, young children playing, etc.; and for the first time it was possible to use television as a means of entertaining them when we were not working with them.

Self-Recognition Via a TV Image

Following the emergence of this rapt engagement with novel and interesting TV scenes, Sherman and Austin began to show evidence of recognizing *themselves* on TV. Austin was the first to do so, and since we were taping him at the time we have an excellent record of the first occasion on which Austin evidenced behavioral awareness that the chimpanzee on the TV screen was himself.

Austin was sorting colored blocks and drinking orange drink as we were making a tape to document his sorting skill. As always during taping sessions there was a video monitor near the camera operator so that he could see the image being recorded. After calmly sorting for about 30 minutes, Austin casually glanced over at the television monitor and suddenly appeared to recognize himself. He began staring intently at the screen while bobbing up and down and making funny faces. Next, he approached the TV and positioned himself just inches from the screen and began to scrutinize his lip movements as he ate a muscadine and drank ice water (see figures 13.1 to 13.4). As he drank the water, he swirled it around on his lower lip and watched it almost, but not quite, fall out in the TV image.[2] For the next twenty minutes, Austin continued to watch himself on the screen as he experimented with different body postures, facial expressions, and methods of eating. He swirled orange drink around on his lower lip, which he watched by looking down over his nose. As he glanced over at the TV monitor he suddenly realized that he could see the orange drink in his mouth better there than by looking down over his nose.

Following this display of assiduous interest on Austin's part, we tried to intrigue Sherman with his own image, but to no avail. Sherman gave no evidence of showing that he understood that the camera was depicting him, no matter what we tried. Several months later, however, he spontaneously recognized himself on the TV set and at that point he engaged in a long series of exploratory behaviors similar to those which Austin had displayed previously.

While chimpanzees have previously been reported to recognize themselves in mirrors, self-recognition on TV has not been re-

Figure 13.1. Austin making a funny face by curling his lower lip underneath his chin (a feat only chimpanzees can accomplish) while observing the resulting change in his visage on the television screen (which is showing him his live image). Photo from video tape.

ported. It is important to distinguish between video and mirror images. Video images are not normally reversed. Further, their size can vary. Typically, they are considerably smaller than a mirror image. Yet another difference is dictated by the location of the camera. The chimpanzee might see his back, his profile, etc. rather than a direct facial view. Finally, a TV image does not have to be live. Under those circumstances the image does not move in response to any of the subject's movements.

With no special training, Sherman and Austin learned to interpret all of these situations and to respond to their own images with great intrigue and pleasure. They also came to differentiate live from taped portrayals of themselves spontaneously by testing the video image when they saw themselves on the screen. That is, they made faces, stuck out their tongues, waved their hands and feet, looked at their mouths, first on the TV monitor, then in the mirror, then on the monitor, then in the mirror, etc., groomed themselves as they watched the monitor, and even pretended to

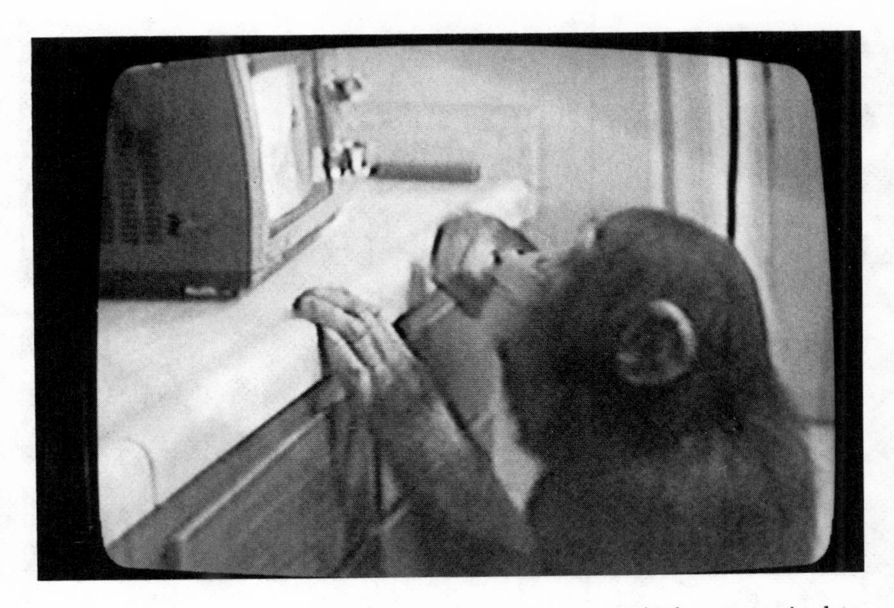

Figure 13.2. Austin observes the fine lip movements which are required to remove muscadine seeds from the fruit—using only the lips. Again, this is a feat only a chimpanzee can accomplish, and apparently even a chimpanzee appreciates a little help in the form of visual feedback from a television monitor. Photo from video tape.

feed themselves as they watched their images on the monitor. It was apparent that they could differentiate themselves from one another and from other apes, because such "image testing" occurred only when they saw their own images on the monitor, not when they saw the images of other apes.

Once they ascertained, through such testing, that they were viewing taped images of themselves, they typically showed little interest in the monitor. By contrast, when they determined that their images were live, they often continued to "play" with the image for some time. In fact, they became so preoccupied with viewing their own live images on TV that at times we were unable to make tapes of activities with the monitor present. They would stop whatever they were doing and begin to watch themselves on television. When we removed the monitor, they would try to see themselves through the eyepiece of the video camera, which of course was hopeless, since every time they moved behind the

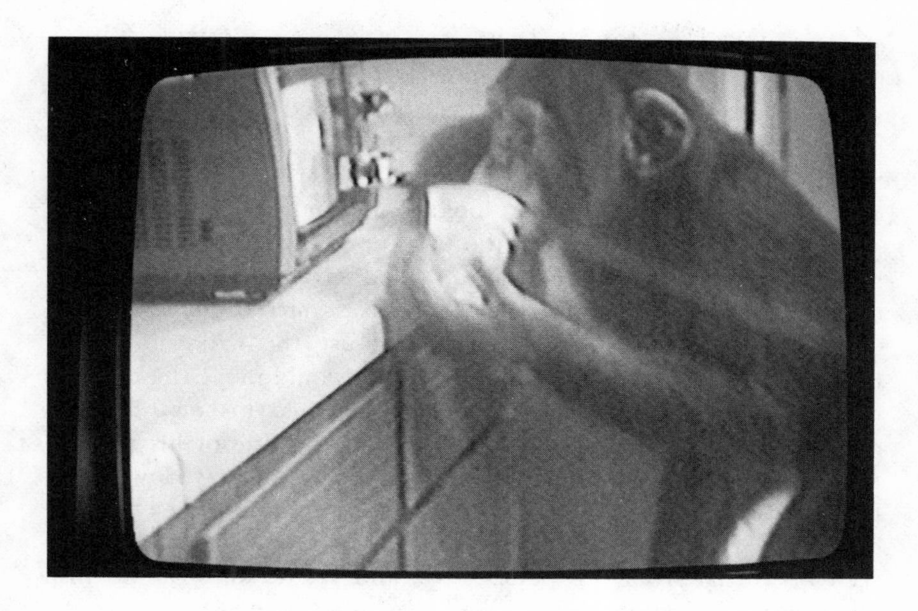

Figure 13.3–13.4. Austin tries to find the best possible angle to view himself as he swirls ice water around on his expandable lower lip.

camera, they were automatically out of the picture. They knew, though, that at times they somehow got "into the camera," since we often played back a scene we had just taped to check that the equipment was working properly.

With increased experience they began to use TV for things that could not be accomplished by looking in a mirror. For example, Austin became fond of using the TV monitor to look down into his throat. The video camera was capable of giving him a perspective that he could never hope to attain with a mirror. When it was observed that Austin was attempting to use the video image to look inside his throat, the camera operator would move the camera directly over Austin's head. Austin would hold his mouth wide open toward the camera while looking out the side of his eyes at the monitor. In this manner, he could see all the way down into his throat, something he could not accomplish by adopting a mouth-open posture in front of a mirror. Austin had never previously seen his throat from this perspective, and yet he knew it was his. His throat, at certain angles, filled the entire screen.

Austin attempted to get an even better view at this point by retrieving a flashlight and attempting to shine it down his throat while watching the monitor. This proved difficult. He could not keep the beam of light directed right at his throat while watching the TV. When he would look away from the TV to orient the flashlight he could not see his image, then when he looked back at his image, he could not keep the light shining in the proper direction. He would, nonetheless, continue working at this self-set task for 30 or 40 minutes before giving up (figures 13.5–13.7).

Sherman's experience with his live video image also led him to prefer it over looking into a mirror. On one occasion, while he was using a hand mirror to put on red crayola makeup, he gestured for his teacher to turn on the television camera for him. He then discarded his mirror and repositioned himself in front of the television set. While intently watching his live image, Sherman continued to apply makeup to his face, repeatedly painting his lips until they were big and red. He then bobbed up and down and shook his head vigorously. With his eyes still glued to the TV, he stood up and began to swagger bipedally about the room. After a bit, he sat down in front of the TV and resumed his makeup efforts. Curling his lip upwards and outwards, he applied makeup

Figure 13.5. Austin tries to shine a flashlight down his throat while watching the effect in the television monitor. Photo from video tape.

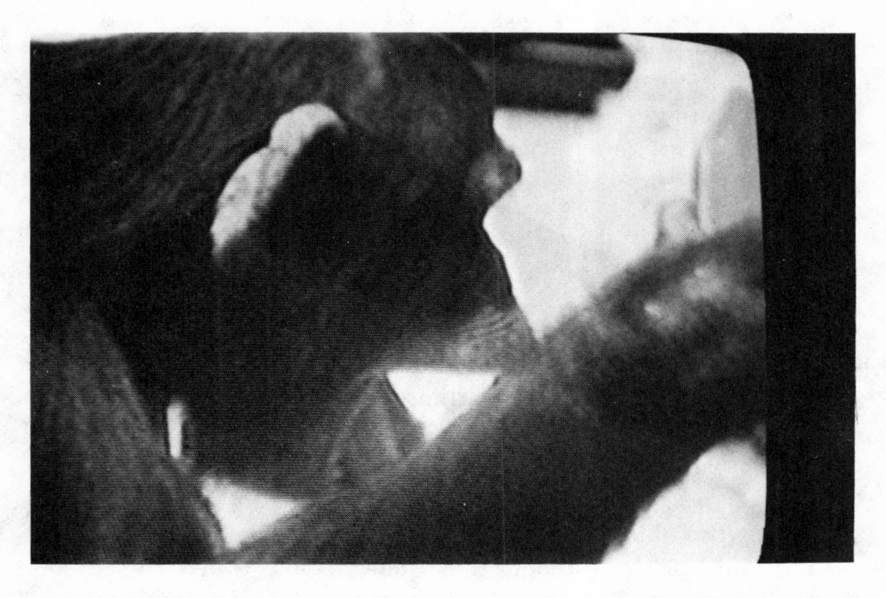

Figure 13.6. Unable to see well, he moves closer to the television for a better view. Photo from video tape.

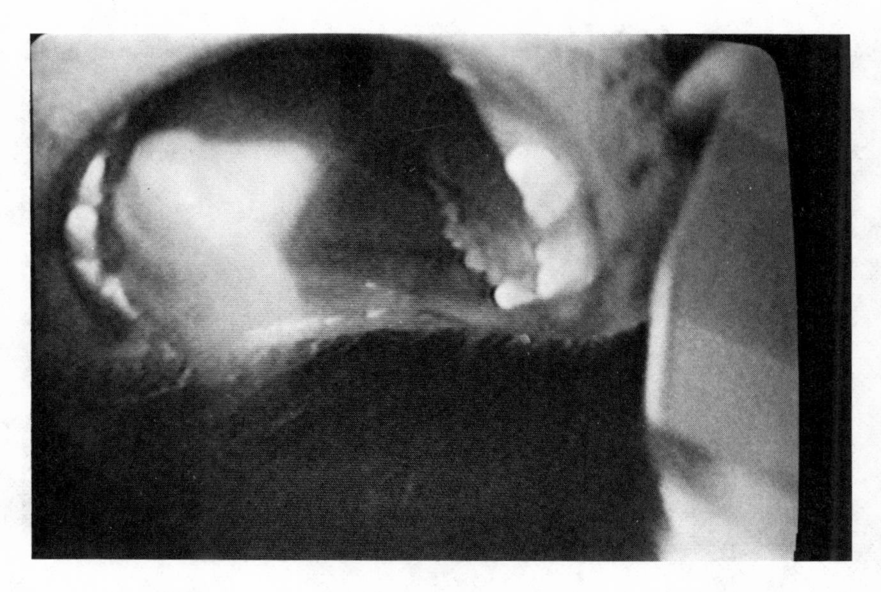

Figure 13.7. The cameraman helps Austin out by focusing right down inside his throat, giving Austin a perspective upon himself that he as never had before. Austin is clearly intrigued and waggles his tongue in and out and around in circles while watching the TV set. Photo from video tape.

along the grooves just above his lip. As he did this, he noticed that his teeth were stained pinkish red from the makeup. He meticulously attempted to remove all makeup from his teeth while vocalizing quietly to himself.

Food-Sharing by Television

Sherman and Austin's obvious interest in, and ability to identify, video images provided us with a unique opportunity to test their ability to communicate about things removed in time and space. Up to now, the communications between them had been limited to items that were physically present, such as foods or tools.

One of the most critical dimensions of symbol use in our own species is the ability to employ symbols to evoke a representation in the mind of the listener. It is then possible to ask the listener to search for and find an object for you even if you are not there. If all objects are always present in the immediate visual fields of both parties, then symbols may not be needed. Pointing can suffice. Symbols may be used instead of pointing but they are not essential, as the food sharing sessions with Sherman and Austin demonstrated. When the table was between them they could communicate which food to give simply by pointing to it, and at times they did, though pointing tended to be used to clarify symbolic requests.

An interesting alteration of the food sharing setting was devised using the video camera to eliminate the communicative effectiveness of pointing. A television set was located in front of the chimpanzees instead of the food table. On the screen was a live video image of the real food tray which was now located in the adjacent room. No longer could the giver simply reach out for the requested food. Instead, he had to go into the next room and bring it back. The speaker could not point to a food, since the screen was so small (8″) as to render pointing (chimpanzee style) unclear.

Sherman and Austin could and did use the keyboard spontaneously to request food in this new setting. Now in order to give the requested food, the listener had to keep the request in mind while traveling to another location. At times, both chimps found themselves forgetting which food they had set out to retrieve once they arrived in the adjacent room. Formerly, they could look back up at the keyboard or they could help each other with pointing gestures. Now, alone in the next room, they could neither recheck the keyboard nor be aided by a pointing gesture. Austin, being the one who relied most heavily on gestures, showed the greatest decrement in accuracy in this new situation.

Another difficulty which now appeared was that Sherman and Austin did not monitor one another as closely as they had previously. Instead, they seemed to focus the majority of their attention on the TV screen. This was interesting since one would expect that a TV picture of foods would not be as attractive to chimpanzees as the real items located right in front of them. However, it appeared more difficult for Austin and Sherman to alternate their attention between one another and the television, than it was for them to alternate their attention between one another and the tray of real foods.

Austin became so intrigued with the TV that, at times, he focused his attention solely on the set and did not observe Sherman's keyboard request at all. Thus Sherman would ask Austin to retrieve a food and Austin would simply remain seated watching TV as though Sherman had said nothing to him. Sherman had less difficulty shifting his attention from the keyboard to the TV and back to Austin, but when he was attending to Austin, Austin often did not look at him.

Much of the interest in the television was presumably generated by the necessity to rearrange the foods between trials so as to make them visible on the monitor. Often, as the chimpanzees retrieved foods they knocked others a bit askew, thus blocking the camera's clear view of all choices. Consequently, as the chimpanzees watched the monitor, they would see a hand enter the picture and rearrange their food tray. This proved far more intriguing to them than we had suspected. We even observed times when Sherman lit a symbol, pointed to it, and then shoved Austin toward the adjacent room, all without ever taking his eyes off the TV screen. Such behavior reminded one of an avid football fan who is watching the game, eating pretzels, and gesturing with his beer can toward the kitchen to request a refill, all without ever taking his eyes off the screen.

Once Austin went into the other room, another interesting problem was encountered: the difficulty of carrying back two portions of food or drink. If drinks were to be retrieved, the chimpanzee had to be particularly careful not to spill any—and that's hard for an animal accustomed to walking on all fours. They also had to resist eating the food as they walked from one room to another—which was particularly difficult for them if they were attempting to carry food with their mouths.

On some occasions, Sherman solved the transport problem by carrying one portion of food in each hand while walking bipedally. Austin's sense of balance was not good enough to permit him to do this. He solved the problem by carrying both pieces of food in one hand and walking tripedally, or by stuffing both pieces in his mouth so that he could walk normally.

Sherman and Austin's communicative skills transferred surprisingly well to this new situation. In spite of the stated problems they were able to request correctly and retrieve foods on many of their first attempts in this new setting. It should be noted, however, that their accuracy levels initially decreased to 60 percent correct for Austin and 70 percent correct for Sherman. With practice, however, they became as proficient in the TV monitor task as they had been when the foods were present in front of them.

As Sherman and Austin's accuracy increased, we were able to film the video task *with no experimenter present.* These blind tests were conducted by setting up two video cameras, one in the room containing the food tray, and one in the room with the chimpanzees, the monitor, and the keyboard. The first camera displayed the image of the foods which were present in the adjacent room. The second camera taped the chimpanzees as they observed the monitor and requested food. Data were collected on a total of 172 exchanges in this blind format.

Sherman requested food correctly in 87 of 90 requests, and Austin retrieved the correct food following 80 of those requests. Austin requested the correct food on 73 of 82 opportunities, and Sherman retrieved the correct food on 72 of those occasions. Overall, Sherman and Austin were successful on 94 percent of their first trial video food exchanges. These data confirmed that Sherman and Austin could employ food symbols to request a food that was not physically present, perceive on a TV monitor the constantly changing array of foods in the adjacent room, and correspondingly alter their selection of symbols to request an appropriate (i.e., retrievable) food.

Although gestures were not effective indicators of specific foods in this task, the chimpanzees did not cease to gesture in this new situation. Instead, they utilized their gestural abilities in new and different ways—ways which reflected the demands of the new task. Sherman's gestures now often were to encourage Austin to proceed either to request a food or to go to the adjacent room and retrieve one. Both chimpanzees also used gestures to draw the other's attention to the keyboard and to the specific symbols they had selected. They also gestured toward the other room to direct the retriever toward the proper location or toward the TV monitor if he was not looking at it. The frequency and types of gestures used by both chimpanzees in this task are listed in table 13.2.

The ease with which Sherman and Austin transferred their former symbolic communicative abilities to this new situation provides support for the contention that they had not learned simple rituals or strict stimulus-response–chained routines. Instead they had learned general communicative skills, both gestural and symbolic, which

Table 13.2. Type and Frequency of Gesture

Gesture	*Frequency*
Sherman points to symbol on projectors	68
Sherman pushes Austin toward keyboard	52
Sherman pushes Austin toward room 3	27
Sherman points to symbol on keyboard	2
Sherman pushes Austin's hand toward a symbol	2
Sherman gestures toward room 3	4
Sherman gestures toward keyboard	1
Sherman points to TV	1
Sherman gestures for Austin to get off cube	2
Austin points to symbol on projectors	2
Austin points to symbols on keyboard	3
Austin points to TV	5

could be readily adapted to new tasks that required the novel behaviors of:

(a) retrieving food from the adjacent room;
(b) attending to a small odorless video image that was often difficult to discriminate; and
(c) returning from the adjacent room with two portions of food without consuming them on the way.

The nature of the gestures which began to appear in this task (such as Sherman's giving Austin a gentle push to encourage him to move off his cube and go to the adjacent room) illustrated that the gestural communications between them were readily modified in accordance with the situation. Gesturing for Austin to leave the room would have been totally inappropriate when the real food was present and, quite understandably, such gestures never occurred. When it did become appropriate for one chimp to leave the room, the appropriate gesture appeared quickly and spontaneously, thereby revealing that comprehension of new task requirements was rapid and accurate.

Knowing That the Camera Shows Them the State of the World Elsewhere

Not only did Austin and Sherman come to discriminate between taped and live images of themselves, but they also began to evidence clear discrimination of live versus taped images of events which were happening elsewhere in the lab. They were able to make such distinctions by pairing the live sounds which traveled through space with the electronic replications of these sounds on the TV. Thus, if they heard someone yelling in the other room, and simultaneously saw and heard this commotion on the monitor, they treated the monitor as a representation of what was happening now in the other area.

For example, we observed that when Sherman and Austin were watching live images they would go toward a locked door and wait for someone to enter if they saw a person approaching the door on the TV monitor. Likewise, if they saw a TV image of someone leaving the building they would rush to their outdoor caging area where they could see the parking lot and get a better "live" view.

Observations of such behaviors occurred regularly. Indeed, live video pictures of events and people located elsewhere in the lab (out of Sherman's and Austin's view) proved to be their most favored video material. On one occasion, they watched a Halloween party held for children on the other side of the building. Keenly intrigued by the costumes, they alternated between focusing intently on the TV screen and racing outdoors to get a "real glimpse" each time they saw a costumed figure head toward the front door to leave the building. When they later viewed reruns of the same event, they watched with interest, but did not go outside "for a better look."

We even used live video to help alleviate their fears when a large tree was struck by lightning and fell on their outdoor enclosure. The tree had to be sectioned and removed by tree surgeons. Sherman and Austin were extremely frightened both by the fallen tree and by the men who came with chainsaws and lifts to section and remove the tree. It was necessary to lock the chimpanzees inside while the tree surgeons were removing the tree. Accordingly, they could not see what was happening to their outdoor quarters. The inability to see what a large number of men with loud power tools were doing to their living area provoked even greater fears. Their behavior suggested that they feared these men were attempting to break into the building to "get" them. They alternately displayed with loud "Waa" calls and huddled in the corner of their caging shaking all over with pilo erect hair.

By putting a camera on the tree surgeons and a monitor in Sherman and Austin's area, we were able to greatly alleviate those fears. They could see on the TV monitor that, rather than trying to break into their quarters, the strange men were simply "attacking" the tree. They watched the entire process without removing their eyes from the TV set. Once the TV was made available, they ceased displaying and shaking uncontrollably.

Video Games

What does it mean that Sherman and Austin recognize themselves on television, that they discriminate between live and taped scenes, both of themselves and others, and that they ask that the camera and TV be turned on so that they can see themselves? At minimum, it surely implies that they understand that what happens on the television screen is caused, in some indirect way, by events happening elsewhere. They must comprehend that they cause the picture to alter by sticking out their tongues, by putting on makeup, by bobbing up and down with fur draped over their shoulders, etc. It is this comprehension of themselves as causal agents of change on the screen that lends meaning to video scenes and renders them more than just kaleidoscopic canvases.

The validity of this interpretation was strikingly confirmed by the introduction of video games to Austin and Sherman. In contrast to TV, video games do not present one with a direct image of oneself, but rather with a dot or blinking cursor under electronic control that moves about on the screen in response to the movement of a joystick. Thus one observes on the screen *only the abstract effect* of one's actions and not a pictorial representation of those actions. To play such games, it is essential that the actor knows that he or she is indirectly controlling the movement of the abstract objects on the video screen. It is also important to know that a one-to-one correspondence exists between the movement of the joystick and the movement of the image on the screen.

A simple video game was designed for Sherman and Austin in which the goal was to use a joystick to move a small arrow into a square. The size of this square could be made smaller and smaller until finally it occupied only two screen dots on the video monitor. When the square was that small, great precision and control were required to locate the arrow immediately on the square. As the arrow was located precisely over the square, a tone sounded and a vendor attached to the monitor dispensed a piece of fruit.

Nothing in their previous language training tasks and *nothing* in their previous exposure to television had taught Sherman and Austin how video games, joysticks, or cursors worked. They saw their first video game in the role of onlookers while a teacher played. Moving an arrow into a square by means of a joystick was entirely different from any previous experience that they had encountered with video screens and we did not expect them to understand that the joystick controlled the arrow, or that the position of the arrow relative to the square was the critical factor which caused the dispenser to operate. Moreover, we did not anticipate that they would be interested in a TV screen that did not display realistic images, but rather showed only a white square and a small white arrow on a black background. This was not the type of video image that typically piqued their interest.

To our astonishment, after watching only three brief games, both Sherman and Austin gesturally begged to be allowed to "do it" themselves. Sherman was given the joystick first and after wobbling the handle around a bit, he immediately guided the arrow into the square. As the square was made smaller he continued to guide the arrow accurately. He not only understood what he was to do, he evidenced excellent eye-hand coordination. When Austin was given a turn, he did exactly as Sherman had done.

The square was then replaced with a simple cross maze. After a bit of frustration in learning that the arrow could not travel through the sides of the maze, but rather had to travel in the open spaces, both chimpanzees were able to guide their arrows accurately to the target.

The rapidity with which both chimpanzees understood what to do based only on observational experience is strong support for the hypothesis that they view themselves and others as causal agents. By seeing how the teacher caused the images on the screen to move, they inferred that they could have a similar effect on the images if they had access to the joystick. They did not need to be taught either the goal of this task, or that the movements of the joystick guided and controlled movements of the arrow on the screen.

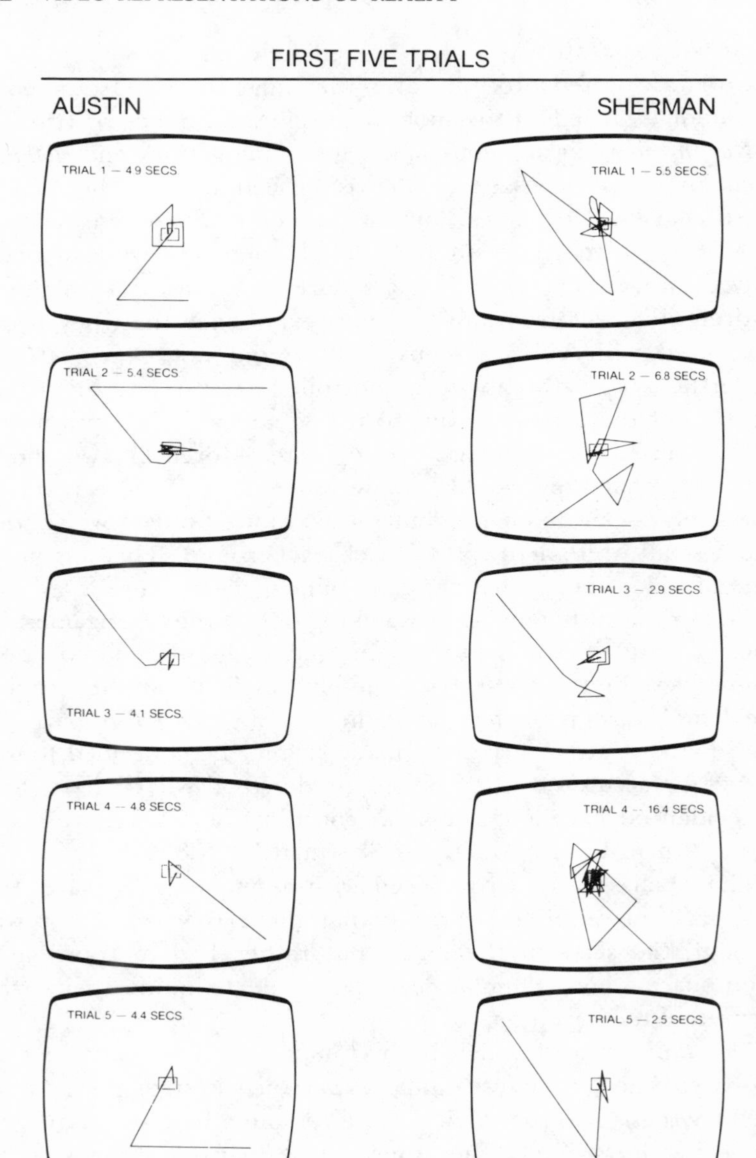

Figure 13.8. Tracings of the routes chosen by Sherman and Austin on the first video-game trials. These diagrams depict the movements of the cursor before it enters the square. As they illustrate, the chimpanzees learned to control the cursor very rapidly—indeed, within the first five trials for both chimpanzees.

This video game was presented to two other chimpanzees who had not received extensive language training. Both were unable to learn what to do by observing others. Further, attempts to teach them the task failed completely. They showed no evidence of understanding that the joystick controlled the movement of the cursor on the screen nor that the goal of the task was to get the arrow into the square.

The reader may wish that more details regarding how these behaviors were taught to the chimpanzees had been presented in this chapter. Yet such was not possible. The simple fact is that, aside from providing numerous opportunities to experiment with video equipment, we did nothing to teach these skills (in contrast to our extensive instruction regarding the use of lexigrams). It is possible that the symbolic tasks that these chimpanzees had encountered in other domains led to a sophistication in the general area of representational skills of all types. This possibility awaits future testing. Whatever the outcome, our initial results suggest that video images provide a powerful tool for studying the chimpanzee's mind.

Chapter 14

Making Statements

This chapter was written in collaboration with Elizabeth Rubert

Spontaneous Communication, or "A Way Out"

Do chimpanzees, when taught to associate a sound (sign, lexigram, etc.) with an object, discover—as do human children—that these symbols are communicative devices that stand in place of the real object or event? The data in the earlier chapters of this book indicate that Sherman, Austin, and Lana had difficulty in this regard.

By attempting to teach Sherman and Austin to use symbols as devices of communicative reference we have gained new insight into the kind of inferential processes the human child must bring to the problem of learning that names stand for things. Human children seem to grasp spontaneously what apes must learn in small steps. It is not necessary to teach children to speak or to comprehend, nor is it necessary to teach them the difference between naming and requesting. Quite unlike Sherman and Austin, children typically can understand words before they can use them productively. Does this mean that, once acquired, words are functioning differently for children than for chimpanzees?

This is a critical question. A negative conclusion would argue that unless a species, or individual, comes into the world genetically equipped to learn its skills spontaneously, it is doomed never to acquire them with sufficient competence to be truly functional. Thus learning and teaching would necessarily be regarded as relatively ineffective endeavors.

In our research effort, the question of competence to use lexigrams as true symbols applies both to apes and to mentally retarded human beings. Many severely and profoundly retarded individuals show little sign of understanding that symbols stand for things. Many of them, like Sherman and Austin, must learn about symbols in a piecemeal fashion if they are to communicate at all. Once they do use symbols to communicate their needs, desires, and thoughts to one another one must consider the same problems of interpretation of those symbols that we raised regarding Sherman and Austin.

What is needed is a simple set of rules that can tell us whether a behavior is truly language. Unfortunately, no such list exists. Yet there are some crucial aspects of symbol use that do seem to set it apart from behaviors that are typically thought of as merely conditioned responses. One of the most critical is that which is best characterized as "spontaneity" on the part of the user. That is, the teacher does not need to set the occasion for symbol usage by asking the user how a thing is called, what he or she wants to do, etc. Instead, it is the language user who initiates the dialogue and volunteers information spontaneously.

One of the recurrent problems in teaching language to mentally retarded individuals is their lack of ability to use the symbols they learn outside of the instructional setting (Stokes and Baer 1977). In other words, they do not become "spontaneous communicators." One would expect that if symbols had acquired true communicative value for the user, they would become generalized, and used in many contexts. On the other hand, if symbols are simply learned as appropriate responses to the teacher's proddings, or "a way out" as Terrace (1979a) has put it, then one would expect such communication to be confined to situations in which the teacher established the conditions needed to provoke the desired behavior.

The Appearance of Spontaneous Indication

Spontaneous indication was the first semantic skill that Sherman and Austin displayed without training. True indicative skill differs from their previous "labeling" capacity in that it does not require the teacher to set the occasion by showing an object and asking the chimpanzee to produce its name.

After the many training paradigms centered around giving, requesting, naming, retrieving, problem solving, and the complex tasks which drew upon these separate skills, Sherman and Austin began to produce truly indicative exchanges in which they announced what they were going to do. When a training task was begun, instead of waiting for the teacher to ask that certain items be given or labeled, the chimpanzees began naming items spontaneously and then showing the named item to the teacher. As the chimpanzees decided which objects were to be named and shown, they also incorporated many aspects of the teacher's role into their own behavior. They initiated trials, singled out objects, and actively engaged in behaviors designed to draw the teacher's attention to what they were saying. Moreover, these indicative behaviors, once they appeared, were not limited to training contexts. Recall, for example, Sherman's comment "Straw give scare," upon being frightened by a sparkler (chapter 12).

However, within the context of naming and giving tasks, the alterations in the chimpanzees' behaviors were quite striking. In these reviews, the teacher typically brought a group of objects or photographs into the room and, using the keyboard, asked that these objects be named or given one at a time. The chimpanzees' task was to attend to the teacher's requests and then select the requested object from among a group of 15 to 20 items.

The Initial Appearance of Indication in a Training Task

On one occasion, instead of waiting for the teacher to decide the object and issue the request, Sherman initiated the trial by choosing, naming, and then giving the teacher an object. Thus, while the teacher was still getting things ready, Sherman said "straw," and handed it to the teacher. Then he said: "blanket," picked it up, and showed it to the teacher. Next he briefly fingered the wrench, as though thinking of it, then said "wrench," and pointed to that tool while glancing back and forth between the teacher and the wrench, checking to see that she noticed the tool to which he was pointing.

The difference between this sort of approach to the task and the approach that Sherman had adopted previously was that he was now producing both productive and receptive behaviors on his own. That is, he was showing that he knew the names of these different items and that he did not need to wait to be asked. It was as if the pigeon in a Skinner box suddenly realized what the conditioned cues were for pecking various keys and said: "Oh you don't need to show me those colored lights one at a time anymore, you can just leave them all on and I will tell you which key belongs to which light."

When these behaviors first appeared, the teachers tended to discourage them, viewing them as attempts by the chimpanzees to interfere with the reviews. They would tell the chimpanzees to stop "naming items without being shown" and insist that the chimpanzees wait to respond until asked. Gradually it was realized that this reaction reflected a limited view of the significance of what was actually happening.

Spontaneous indications also began to occur in other contexts, such as the food-sharing bouts conducted between the chimpanzees. For example, on trials when Austin was having difficulty finding the correct lexigram, Sherman began to light the symbol in place of Austin, and then give Austin the food he indicated. As it became apparent that Sherman would often decide what they were going to have, announce it, and hand it to Austin, Austin started to encourage this. He did so by gesturing toward the keyboard as though to press a key and then drawing back and looking directly at Sherman. Sherman would take this as a cue to announce and give the next food.

Once the teachers recognized the value of this spontaneous indicative ability and began to encourage Sherman and Austin to take charge of their reviews, it became obvious that Sherman and Austin enjoyed doing things that way. When seated in front of a tray of 15 or 20 objects, they would proceed to name the objects one by one and give them to the teacher. Both chimpanzees seemed to *know* which items they tended to miss, as they would either save these for last, or simply not choose to deal with them at all. Having named all the items of interest to them, they would then ask for another set and discard the ones they did not wish to name. In this sense, they were running their own practice sessions.[1]

It is important to note that had we intentionally set out to teach the chimpanzees to do what they did, it would have surely been difficult, if not impossible. True indication requires the knowledge that an unvarying correspondence must exist between the item they select, the item they announce, and the item they actually give, point to, or show.

Errors in such a case are generated not by making a particular incorrect response, but rather by failing to display a consistency between the selected item, the announced item, and the given item. Moreover, it is not possible to inform the chimpanzee that he has made an error until all components of the selecting, announcing, and giving process are complete. A chimpanzee who does not have *a priori* knowledge of the correspondence requirement would, upon being informed that he was incorrect, likely presume that he had selected the wrong symbol or given the wrong object.

Once it became apparent that the chimpanzees were exhibiting indications comparable to those reported for human children, the difficulty of unequivocally and formally demonstrating that such indications occurred became apparent. For what we were dealing with was a chimpanzee making a statement about what it was going to do, had felt, or had seen. Consequently, we had no *a priori* control over what the chimpanzee would say at a given moment. This posed the following dilemma: How could we determine that a chimpanzee was behaving in a truly indicative manner when, in effect, we could not precisely determine when or what the chimpanzee would say, since the very lack of predictability was a defining criterion for the behavior.

We approached this dilemma by treating indicative behaviors as propositions or as statements of fact that could be judged *true* or *false.* None of Sherman and Austin's previous skills could be treated systematically in this way because they were only partial propositions. That is, if Sherman were asked to name a blanket and he called it a glass, this could be judged as "incorrect," but not as "false." However, if Sherman pulled a blanket from a group of objects, showed it to the teacher, and called it a glass, he would be making a statement about his own knowledge and beliefs regarding the proper names of the object. In that case, the teacher's judgment of true or false would represent her understanding of the same phenomenon.

Tests To Control for Random Symbol Lighting and Inadvertent Cueing

To be sure, it could be argued that Sherman and Austin were simply punching keys and then associating objects or behaviors with the keys, which they happened to light haphazardly. If so, the symbols they selected could not properly be interpreted as propositions or statements regarding their beliefs about the names of various objects.

In order to determine whether the chimpanzees were, in fact, selecting the object to be named, and not just lighting symbols haphazardly, we devised the following test. A group of five objects was placed on a tray outside one of the test rooms. These five objects were selected randomly and the entire group of five was changed each trial. Thus, the chimpanzee had to look at the five available objects for that trial, and decide which of that group he would give to the teacher. He could not just light any symbol and then look at the array to select that object, as the probability of a randomly chosen symbol corresponding to any given object in the array was low (probability of .17).

The chimpanzee came to the door, looked at the five objects in the tray, walked back into the room, used the keyboard to announce which object he was going to get, then returned and gave the teacher the named object. It was necessary that the chimpanzee remember the object he had selected, since once he moved to the keyboard he could no longer see the array. It was also necessary that he recall the object as he returned, since he could not see the symbol he had selected after he moved back out of the room to the array (see figures 14.1–14.3).

Inadvertent cueing was ruled out since the chimpanzee could not see the teacher either as he viewed the array or as he used the keyboard. Additionally, the teacher did not know which five objects were available in the array on any given trial. The teacher could tell by viewing the projectors located outside the room which item the chimpanzee had indicated on that trial; once that item was proferred, the teacher could verify its correspondence to the symbol illuminated on the projectors. If the object and the illuminated symbol were the same, the teacher praised and rewarded the

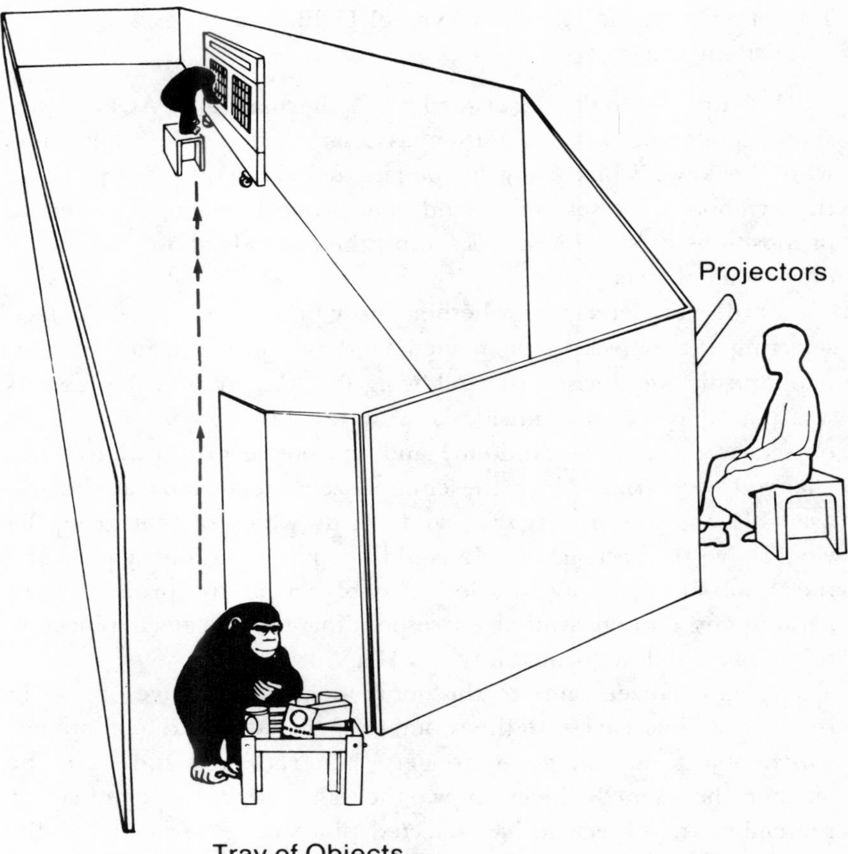

Figure 14.1. Sherman observes the table of items placed outside the door and goes to the keyboard where he lights a symbol to indicate which item he will select and give to the teacher. It is important that Sherman look at the table since the items are changed each trial. The teacher cannot see Sherman either when he is by the table or when he is at the keyboard. However, as soon as Sherman touches a lexigram symbol, that symbol will appear on the teacher's projectors.

chimpanzee. If they were not, the teacher expressed dismay and encouraged the chimpanzee to try again with five new objects.

Thirty different foods and tools were used in this test. Foods were represented by photos, while tools were represented by both

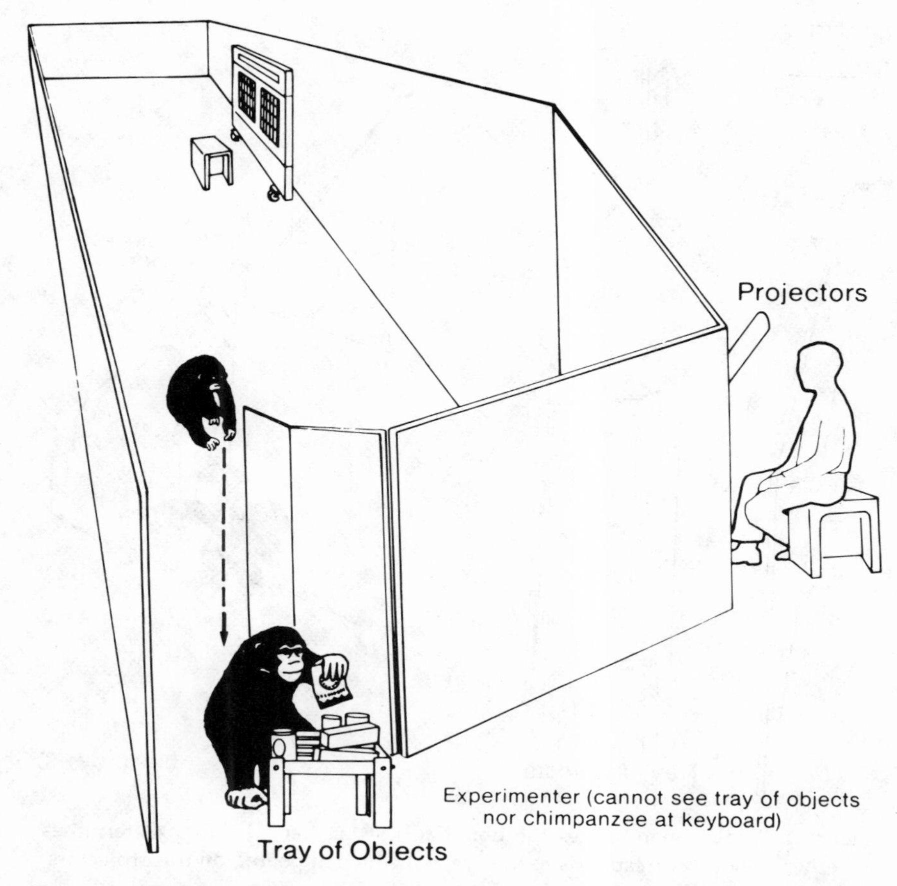

Figure 14.2. Sherman returns to the table and selects the object he has just symbolically indicated that he would give to the teacher.

photos and the real items.[2] Sherman was correct on 50 of 53 trials, while Austin was correct on 46 of 53 trials. (For a more detailed presentation of this data, see Savage-Rumbaugh et al. 1983). Out of the 30 items used in this test, Sherman indicated a total of 17 different items and Austin indicated a total of 23 different items.

Their errors were either of a perceptual or functional nature; for example, on one trial Austin viewed the following five alternatives: bread, key, straw, magnet, and orange drink. He said "strawberry—drink" and handed the experimenter the photo of

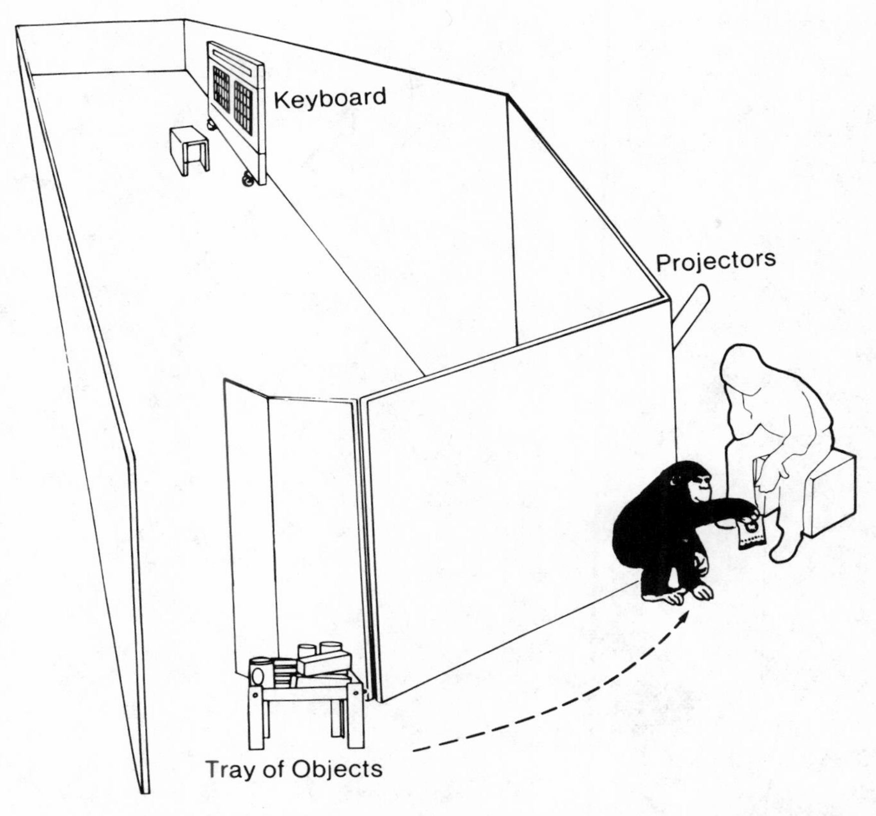

Figure 14.3. Sherman carries the object to the teacher who then determines whether or not it corresponds to the symbol which appeared on the projectors.

the orange drink. This was a perceptual error, since these two drinks are similar in taste and appearance, and neither of them are like the other alternatives available in that array. Thus Austin knew what it was he was naming, but he called it by the name of a similar item instead of by its correct name, much as you might say "Coke," when you really mean "Tab."

Sherman made the same sort of functional error on two trials; he called a syringe a "key." Both items are small and pointed and are inserted into other objects, thus their use or function is similar even though they do not look alike and the lexigrams which stand for them do not look alike. On both of these trials, the key, but not the syringe, was present as one of the five items on the tray.

Sherman said "syringe" and then handed the teacher the key. This error was not a true error of indication in the sense that the symbol and the ensuing behavior were random. Rather, Sherman selected the wrong name for the item he had chosen to give. The confusion of the names of these same two objects was often seen on other occasions as well.

The arrays used across trials, and the item selected from a given array by each chimpanzee, are listed in appendix 14.A. (Note that Sherman and Austin did not tend to respond to similar arrays by choosing the same object. Thus it can be concluded that there was nothing about the arrays per se which made one object more salient than others, and nothing about their common training history which would predispose them to respond similarly to arrays.

Ten of the arrays were shown twice to each chimpanzee to determine whether or not, when faced with the same group of items again, they would select the same item to indicate. If so, it would suggest that something about the perceptual characteristics of the array itself were functioning to single out a given item for that chimpanzee. Austin selected the same items on two trials (chance) and Sherman on five (slightly above chance, but not significant). Clearly, they were not responding to these displays in a conditioned manner, and indeed, we would not have expected that they would, as all arrays were novel except for these intentionally represented ones.

The blind procedure required that the chimpanzee indicate an item without being able to see the various alternatives as he used the keyboard. Thereby, he was forced to select the object he would have indicated before he lighted any symbol. Because the symbol had to be lighted without reference to the various alternatives, the chimpanzee was also forced to remember his choice and select the symbol based on this memory alone. Additionally, the position of the objects varied on each trial, as did the overall configuration of the display, and there was no relationship between an object's location within the display and the chimpanzees' tendency to choose that object.

These procedural constraints render inadequate a simple conditioned discrimination account of the behavior of the form—"See X, press key 1; See Y, press key 2." On any given trial, the chimpanzee saw five different alternatives sampled from a set of

30. Thus a conditioned-discrimination explanation could not adequately account for their behavior, since on some trials with X present, they chose to indicate X but on others, with X also present, they indicated one of the other four items.

It is also the case that they could not be making a conditioned discriminative response to entire arrays of the form "when presented with array A, choose item A_1; when presented with array B, choose item B_1, etc. If they were, they would not have selected different items when the same arrays were presented again.

Accordingly, it must be concluded that Sherman and Austin's choices of one object versus another on any given trial cannot be explained by principles of conditioned discrimination.

Indication Outside of Training Tasks

Although the constraints imposed by our test of indicative skill may seem artificial, the skills those tests revealed were not limited to such test situations. However, having demonstrated under tightly controlled conditions that they can use symbols in an indicative manner, we can proceed with greater certainty in attributing indication to them in other contexts.

For example, Austin had previously said "chase" or "tickle" only to request these activities. Subsequently, he began to actually interrupt a bout of social play in order to go to the keyboard and comment "tickle" or "chase" and then return to the game. He also began to announce at the keyboard that he was going to make a "funny face," before sticking out his tongue or pulling his lower lip down over his chin (a feat only apes can accomplish). Previously, his use of this symbol had been limited to occasions when the teacher would make a "funny face" and then ask Austin to tell her what she had done.

Not only were things announced to the teachers, but on occasion they announced things to one another. Sherman was observed to say "tickle" and then walk over and start tickling Austin who was displaying a play face in anticipation of being tickled. Austin did the same. Sherman has also told Austin what tool to use on several occasions. For example, when Austin asked Sherman for a wrench, and Sherman could see the baited tool site and knew that Austin needed a key, Sherman stated "key" and gave Austin the key instead of the wrench.

In general, apart from food-sharing bouts in which statements were common, instances of spontaneous indication remained relatively infrequent; they tended to occur only once or twice a week. For the most part, Sherman and Austin continued to limit their communications to requests, yet the instances of spontaneous indication or comment, when they did occur, were appropriate to the context and were not elicited by implicit or explicit demands of the teachers. The examples below, taken directly from the daily notes, illustrate the variety of this indicative usage.

1. During a food request session Austin is given coffee. He calls it "hot." The teacher tells Austin that it was "hot coffee" and encourages him to say "hot coffee." Austin continues to say only "hot." Finally the teacher gives Austin the entire pot. Austin then begins playing with the coffee on his own. He then comments to the teacher "coffee," and "hot coffee."

2. Austin and his teacher are working on a color sorting task when someone comes out of the play-yard and leaves the door open. Austin says "open play-yard" (presumably meaning that the door is open and he intends to go to the play yard) and then walks into the play-yard.

3. The teacher is attempting to teach Sherman the new action words "hug," "throw," and "bite." Sherman, having already learned the action word "talk" (which he could do by voluntarily producing a breathy bark), ignores the teacher's request to describe one of the new actions, and instead announces that he wants to "talk" and proceeds to make breathy barks.

4. Sherman is totally absorbed in watching a new video tape of wild chimpanzees. While watching, he repeatedly turns to the keyboard and says "TV."

5. Sherman and his teacher are working on a review task. He is bored and refuses to continue. He announces "Go sink," and walks to the kitchen where he begins to play with items on the counter.

6. Austin and his teacher have just finished a review session in which Austin has been asked to name some objects that he has had difficulty remembering. As a reward for work well done, the teacher gives him a large red apple. He munches it happily making food barks and commenting "Apple, apple," to the teacher.

7. Austin and his teacher are working on the tool task and the teacher has just baited the key site. While the teacher is baiting the site, Austin picks up the straw in the tool kit and begins playing with it. The teacher tries to orient Austin's attention to the food locked in the key box and encourages him to ask for the needed tool (a magnet). Austin comments "straw," gestures to the straw, then continues to play with his straw, ignoring the food in the locked box. Presumably he is informing the teacher that he would rather play with the straw than obtain food, as he makes no attempt to use the straw to open the box.

8. Austin and his teacher are grooming. Austin repeatedly stops grooming the teacher to lean over toward the keyboard and comment "groom" then continues to groom the teacher.

9. Austin sees Sherman and his teacher begin to play. He approaches them, announces "bite," and then joins in the play bout. (Austin has been perfectly free to join in the play without announcing his intention, or requesting. Biting is part of chimpanzee play.)

10. Austin sees a new flashlight in a bin of items that the teacher has brought into the room. He announces "flashlight" and then picks up a flashlight and a mirror and tries to look inside his throat.

11. A group of caretakers carry an anesthetized chimpanzee past Austin's outdoor play-yard. Austin is outside alone. When he sees this, he rushes inside, announces "scare" and pulls the teacher toward the window.

Why Did Requests Remain the Staple of Sherman and Austin's Language Use?

If we grant that Sherman and Austin had achieved the ability to produce spontaneous statements about their impending actions, or to comment on activities and objects, we must ask why did they not do so more frequently? They certainly requested activities, foods, changes of location, etc. spontaneously enough, and were even using combinations regularly to expand upon ambiguous single-word requests. Perhaps the simplest answer is that while they were interested in communicating their own needs, they rarely seemed interested in communicating other kinds of information.

Thus Austin might ask the teacher to play "chase" with him, but he never asked the teacher to play chase with Sherman, or for that matter with another teacher. Likewise, although Sherman would readily share food with Austin, he would never ask a teacher to give food to Austin, only to himself. It seemed that their difficulties in expressing things other than their own basic desires lay not in an absence of basic communicative ability but rather in a more general lack of the capacity to focus their attention on activities not connected with their own immediate desires. In Piagetian terms, they could be characterized as having a strictly egocentric view of the world.

Their behavior suggests that it is difficult for them to understand that others do not have access to the same information that they do. In the various paradigms used to encourage communication between them, it was always necessary for them to experience the roles of speaker and listener a number of times before their behaviors, as speaker, suggested that they knew that they had information which the listener did not have.

Conditioned Speaking Versus Speaking to a Listener Who Is Viewed as Needing Information

The very fact that Sherman and Austin needed experience in both the roles of listener and speaker before they began, as speakers, to engage in behaviors such as soliciting the attention of the listener, repeating requests, etc. implies that perhaps they never really understood the plight of the listener. Perhaps the behaviors which made it appear as though they did were instead conditioned on the basis of their effectiveness in the situation.

In order to investigate this possibility, a test situation was designed in which, by virtue of previous experience in the roles of listener and speaker, Sherman and Austin would have had ample opportunity to become familiar with what the listener and the speaker could be expected to know. However, aside from their respective roles and the general contextual setting, none of the previous means of communicating the needed information to the listener were provided to the speaker before the test.

One chimpanzee was allowed to see what kind of food was placed in a container. Both chimpanzees were then required to ask for the food (chapter 6). However, *neither chimpanzee had access to a keyboard.* Instead, we gave them plaques which displayed the brand names of various foods (as taken from the display on commercial packaging). Since Sherman and Austin had never been trained with such labels, they should not have been expected to be able to use them as communicative devices, unless they understood the need to transmit information to the uninformed listener. No previous history, training routines, etc. could have prepared them, when faced with a communicative need, and no keyboard, to adopt an alternative system of indication based on commercial packaging.

The brand names differed amongst themselves considerably ranging from the giant M&M, to the Doritos logo, to the familiar yellow Velveeta package (see figure 14.4). Sherman and Austin seemed quite familiar with all these labels even though they had not been used previously in training or testing.

On the very first occasion on which the keyboard was turned off, both Sherman and Austin spontaneously used the brand name labels to continue to enable them to pass information to the uninformed listener (see figures 14.5 to 14.7). Seventeen different food labels were used during the thirty trials of this test and only five errors were made.

Figure 14.4. These are some of the actual food-label plaques used in this study.

Figure 14.5. Austin shows Sherman the plaque with the Peter Pan food label on its surface. Photo from video tape.

Figure 14.6. Sherman hands Austin the lexigram for peanut butter. Photo from video tape.

Figure 14.7. The experimenter opens up the container to determine whether or not it contains peanut butter. Photo from video tape.

The level of success and the ease with which it was achieved shows clearly that, in this task, Sherman and Austin knew the importance of providing information to the uninformed listener. Moreover, their respective abilities to use a novel communication medium confirms our previous observation (chapter 11) that they appreciated the representational power of symbols. However, in this test, as opposed to the categorization test, they were not answering an implicit question posed by the experimenter. It is interesting to note that all of the five errors made involved selecting the incorrect lexigram rather than the incorrect food label.

Their errors and near errors illustrated the kinds of things about interindividual communication which they had not learned. For example, on one trial Sherman picked up the label for Welch's grape jelly, put it in his mouth, and carried it over to the window to show Austin. However, it happened that as he picked up the plaque, he placed it in his mouth with the label side down. The back of the plaque was blank. When Sherman got to the window he held the plaque in his mouth and waited for Austin to find the jelly lexigram. When Austin did nothing, Sherman did not recheck

Table 14.1. Test of Ability to Identify Food Labels

Trial	*Food*	*Informer*	*Correct Label Selected*	*Correct Lexigram Given*
1	Peter Pan Peanut Butter	Austin	Yes	Yes
2	Welch's Grape Jelly	Sherman	Yes	Yes
3	Velveeta Cheese	Austin	Yes	Yes
4	M & M Candy	Sherman	Yes	No
5	Minute Maid Lemonade	Austin	Yes	No
6	Sarah Lee Pound Cake	Austin	Yes	Yes
7	Dorito's Corn Chips	Sherman	Yes	Yes
8	Peter Pan Peanut Butter	Sherman	Yes	Yes
9	Jello Pudding	Austin	Yes	No
10	Velveeta Cheese	Sherman	Yes	Yes
11	SMA Milk	Austin	Yes	Yes
12	Roman Meal Bread	Austin	Yes	Yes
13	Jello Pudding	Sherman	Yes	Yes
14	Kool-aid Strawberry Drink	Austin	Yes	Yes
15	Avondale Peaches	Sherman	Yes	Yes
16	Vi-Daylin Vitamin	Austin	Yes	Yes
17	SMA Milk	Sherman	Yes	Yes
18	Libby's Peaches	Austin	Yes	Yes
19	Roman Meal Bread	Sherman	Yes	Yes
20	Pel Pak Raisins	Austin	Yes	No
21	Minute Maid Lemonade	Sherman	Yes	Yes
22	M & M Candy	Austin	Yes	Yes
23	Vi-Daylin Vitamins	Sherman	Yes	Yes
24	Dorito's Corn Chips	Austin	Yes	No
25	Coca Cola	Sherman	Yes	Yes
26	Thankyou Cherries	Sherman	Yes	Yes
27	Sara Lee Pound Cake	Austin	Yes	Yes
28	M & M Candy	Austin	Yes	Yes
29	Velveeta Cheese	Austin	Yes	Yes
30	Peter Pan Peanut Butter	Sherman	Yes	Yes

the orientation of the plaque. Rather, he seemed to presume that Austin was being uncooperative and he put the plaque down and walked away. In laying the plaque on the floor, Sherman inadvertently turned it face up; Austin then saw the Welch's symbol and selected the jelly lexigram.

On other trials on which errors occurred, the requester approached the window and placed the plaque on the floor face up. The window ledge was wide enough that a label placed just at the bottom of the window could not be seen by the chimpanzee on the other side. The speaker seemed not to realize that the line of visual regard of the listener was such as to prevent the listener from seeing the plaque. On some trials the speaker, having placed the plaque in this position and noted that the listener was not responding, pointed repeatedly at the plaque in a fruitless attempt to draw the listener's attention to it.

It should not be concluded from these examples that the chimpanzees understood nothing about the differences between their line of visual regard and that of the listener. On the majority of the trials they gave evidence of dealing with this problem readily by shoving the upright plaque right through the window or holding it up to the window. However, on trials where this did not work (either because they did not think to do it or where the listener was not attending at first), they were not able to negotiate through these communicative problems. In such cases, the tendency of Sherman and Austin was to presume that the listener, and not themselves as the speaker, was at fault. Accordingly, the speaker did not reassess what he might do to get his message across more directly.

None of their previous communicative experience with the keyboard had required that the speaker pay attention to the orientation of a symbol as it was shown to the listener. The keys, mounted in a panel, were always in the correct orientation. They could not be turned around, and if the listener were looking in their direction he could always see them. With continued practice in both roles using plaques, it is likely that Sherman and Austin would have come to be attentive to the orientation of symbols as an important dimension of communication with a visual symbol system.

Do All Chimpanzees Who Learn Symbols Use Them Indicatively?

Many readers who are familiar with the reports of ASL-tutored apes will recognize that Washoe, Koko, Nim, and Lucy are readily credited with the ability to use symbols indicatively. In fact, these apes reportedly began to sign indicatively as soon as they learned each new sign and they did not require the training paradigms used with Austin and Sherman to teach them (e.g., Fouts 1974b; Gardner and Gardner 1974a; Patterson 1978).

Having worked daily for two years with the signing chimpanzee Lucy (while her ASL vocabulary was approximately 100 items) I feel that some clarification is needed with regard to behaviors which might, in such a chimpanzee, be interpreted as evidence for indication. Consider, for example, Lucy's sign "flower" which she made when she saw a flower while walking about her home. Her human foster parents, who had raised her from birth, grew orchids and often had them in the house. Lucy was not supposed to touch these flowers, only smell them. She preferred to eat them. After signing "flower" Lucy would often take a bite of the orchid unless she were watched closely.

Lucy also signed "tea" as we made tea, "hot" as we turned on the stove, "open" as we went to the door, etc. In addition, she combined these signs with the touch signs "you" and "me," and ofter our proper-name signs. She liked to sit with a box of items which she had been taught to associate with signs and make a nest of them, picking them up one at a time and making the signs that went with each item, then returning to her nest and rearranging it over and over.

There are, however, some not so obvious but extremely important distinctions between this type of behavior and that reported for Sherman and Austin. First it must be pointed out that the specificity of Lucy's receptive capacity was not equal to that of her productive capacity. She could not, for example, search among a group of objects and hand the experimenter any one upon request. She could not retrieve objects out of sight or name objects which she could not see. In brief, her symbol usage was linked to the referent in a performative contextually dependent sense.

When she signed with objects in her nest or as she moved about the house, the signs were *not directed to* the experimenter, though they *were often done with* the experimenter. In fact, if the experimenter did not sign "flower" when Lucy did as she looked at a flower, she often pulled the experimenter toward the flower and waited to continue on about the room until the experimenter also signed "flower." However, when she signed with objects in her nest, she did not alternate her glance between the object and experimenter as though using the sign to draw the experimenter's attention to the object, nor did she give the object to the experimenter. Instead, she oriented the sign simply to the object itself as though signing "ball" was part of an internalized learned routine elicited by holding the ball.

It appears, therefore, that Lucy developed elaborate routines of looking at and interacting with objects while producing the signs for these objects. These behaviors were perhaps a way of acknowledging or recognizing the object [cf. Bates' (1979) discussion of similar behavior in human children prior to the onset of spoken words]. Yet there was no evidence to suggest that Lucy was using these signs to represent anything or to refer to anything. In fact, their tight linkage to the contextual situations suggested that they were as much a part of her routine as were the nonsymbolic components of the routine. Just as Lucy always got out the spoon, stirred the tea, and took out the bag—so she always signed "tea" as we did these things. There was no obvious functional aspect to repeatedly stirring the tea, just as there was no functional aspect to signing "tea."

This is not to suggest that Lucy's skills are unimportant or unrelated to the acquisition of language. Quite the contrary, human children frequently first acquire words in a performative manner, as did Sherman and Austin. Later they learn how these words function between individuals, both in terms of what they can expect when they direct these symbols to others and in terms of what others who direct these symbols to them expect. Still later, the symbols are freed from the interactional context so that they can be used to communicate about items which are not immediately present. At this point functional symbolic indication separates itself from routine bound symbolic usage.

It is not surprising that Lucy should have produced symbols more frequently in context-bound routines than did Sherman and Austin: it was in the execution of such routines that she was taught to produce the signs. During the preparation of tea, for example, her teachers were instructed to repeatedly stop and mold Lucy's hands into the sign for "tea." Thus the signing of "tea" became as much an aspect of getting the tea made as did the acts of heating water, stirring, etc.

Once such a sign has been linked thoroughly to the context and is readily performed, the question then becomes not how to get a chimpanzee to generalize that sign to similar contexts (this is relatively easy to do) but rather how to get the chimpanzee to use it to communicate information not inherent in the context. It is this form of communication that differentiated Lucy's symbol usage from that seen in Sherman and Austin. She was unable to communicate information that was not already obvious given the contextual setting and the embedding of routines in that setting. Moreover, she was not able to respond appropriately when others directed symbolically encoded messages to her. Thus Lucy could not go to an adjacent room and retrieve a food or object if asked to do so. Nor could she retrieve such an object if it was intermingled with a number of others directly in front of her. She could not make statements which conveyed what she was about to do and carry them out appropriately. She could not inform someone of a food or object that she had seen hidden elsewhere. Basically, her symbol use was limited to requesting or naming items that were physically present or activities that routinely occurred, such as walks, car rides, etc.

Perhaps an example will clarify this. One of the things Lucy enjoyed watching others do, and wanted to participate in, was window cleaning. As the windows were sprayed with Windex and wiped with paper towels, she repeatedly asked for these items so that she could do what she saw others doing. However, upon being given the Windex and the towels, Lucy unknowingly oriented the spray bottle opening toward her face. Consequently every time she pressed on the top and pointed the bottle at the window, she would squirt herself in the face. When this happened, she would screw up her face, wipe it off, then proceed to dab the window with the paper towel—just as if she had sprayed it. Lucy was essentially imitating the behaviors of pushing on the bottle and rubbing the window. She did not understand that Windex helped to clean the window, or even the difference between "rubbing" the window and truly cleaning. When the signs for "clean" were used in this situation, she imitated this also, but with no more comprehension than she had evidenced in the actual cleaning tools. That is, she knew *what* to do, but not why she was doing it.

As long as "clean" is what one says when one sprays, "tea" is what one says when one stirs tea, "open" is what one says just before one goes out the door, "apple" is what one says when the teacher holds up the fruit, etc., language function remains severely limited and tied to the here and now. To move beyond, receptive comprehension and cooperation must develop, as must the ability to refer to things that are absent in time and space. All of these things should appear—even if single words are all that are being produced.

Appendix for Chapter 14
Choice Sets and Selections

Trial	*Austin*	*Sherman*
1	Melon	Orange drink
	Pudding	*Stick*
	Coke	Pineapple
	Key	M&Ms
	Wrench	Blanket
2	String	Money
	M&Ms	Strawberry drink
	Sweet potato	Orange
	Sponge	*Coffee*
	Shot	Coke
3	Pudding	*Bean cake*
	Sponge	Cake
	Key	Carrot
	Corn	Key
	M&Ms	Corn
4	Money	Coke
	Sponge (said "string")	Blanket
	Lever	Banana
	Melon	*Straw*
	M&Ms	Melon
5	Strawberry drink	*Sponge*
	Sweet potato	Lever
	Straw	Orange drink
	Corn	Orange
	Coke	Stick
6	Peanut butter	Sponge
	Shot	Lever
	Orange drink	*String*
	Juice	Blanket
	Jelly	M&Ms
7	M&Ms	Cake

Appendix *(continued)*

Trial	*Austin*	*Sherman*
	Shot	*Stick*
	Jelly	Banana
	Coke	Coke
	Coffee	Corn
8	Chow	Corn
	Strawberry drink	*Sweet potato*
	Carrot	Bean cake
	Sponge	Money
	Money	Orange drink
9	Carrot	Melon
	Money	Banana
	Banana	Orange drink
	Lever	*Sponge*
	Chow	Carrot
10	*Coffee*	Orange drink
	String	Banana
	Orange drink	*Pudding*
	Banana	Orange
	Sweet potato	Coke
11	Straw	String
	Lever	Chow
	Juice	*Lever*
	Cake	Orange drink
	Pudding	Sweet potato
12	*Wrench*	*Key*
	Coke	Bean cake
	Banana	Coffee
	Cake	Blanket
	Jelly	Jelly
13	Blanket	*Lever*
	Strawberry drink	String
	Straw	Coke
	Orange	Money
	Lever	Cake
14	Key	Jelly
	Jelly	Cake
	Corn	Sweet potato
	Money	*Corn*

Appendix *(continued)*

Trial	*Austin*	*Sherman*
	Chow	Straw
15	Pineapple	Jelly
	Money	Juice
	Orange drink	*String*
	Coffee	Pineapple
	Strawberry drink	Sponge
16	Juice	Chow
	Bean cake	*String*
	Key	Lever
	Sweet potato	Orange drink
	Money	Orange
17	Bread	Juice
	Stick	*String*
	Coke	Strawberry drink
	Pudding	Bread
	Orange drink (said "strawberry drink")	Coffee
18	Orange drink	Chow
	Straw	*Carrot*
	Wrench	Juice
	Stick	Bread
	Corn	Coke
19	Pudding	Orange
	Coffee	Orange drink
	Peanut butter	*Straw*
	Blanket	Coke
	Straw	Blanket
20	Carrot	*Sponge*
	Pudding	Coffee
	Juice	Sweet potato
	Blanket	Bean cake
	String	Coke
21	*Stick*	Money
	Blanket	*Wrench*
	Lever	Cake
	Chow	Melon
	Jelly	Straw
22	Melon	Sponge

Appendix *(continued)*

Trial	*Austin*	*Sherman*
	Strawberry drink	M&Ms
	Straw	Coke
	Wrench	*Key*
	Orange drink	Shot
23	Key	Peanut butter
	Jelly	*Bean cake*
	Straw	M&Ms
	Orange	String
	String	Money
24	Pineapple	Cake
	String	Blanket
	M&Ms	Pineapple
	Chow	Carrot
	Coke	*Sweet potato*
25	*Blanket*	*Shot* (said "key")
	Stick	Corn
	Straw	Bean cake
	Coke	Banana
	Money	Money
26	Lever	Key
	Straw	Corn
	Coffee	Bread
	String	Stick
	Orange	*Wrench*
27	Bean cake	Corn
	Blanket	*Coke*
	Chow	Jelly
	M&Ms	Juice
	Sponge	Sponge
28	Bread	Sweet potato
	Bean cake	Stick
	Coke	*Wrench*
	Wrench	Coke
	Banana	Lever
29	*Cake*	Sweet potato
	Key	Lever
	Wrench	Stick
	Lever	*Cake* (said "string")

Appendix *(continued)*

Trial	*Austin*	*Sherman*
	Peanut butter	Wrench
30	Pineapple	Peanut butter
	Cake	Cake
	Pudding	String
	Bean cake	Pineapple
	Corn	*Stick*
31	Corn	Orange drink
	Orange drink	Straw
	Bean cake	Stick
	Lever	*Corn*
	Key (said key, gave something else)	Cake
32	Jelly	Strawberry drink
	Sweet potato	Corn
	String (said "sponge")	Key
	Strawberry drink	*String*
	Coke	Wrench
33	Stick	Carrot
	Peanut butter	*Pudding*
	Straw	Blanket
	Lever	Juice
	Blanket	String
34	Orange drink	Orange drink
	Strawberry drink	Orange
	Wrench	Banana
	Corn	Corn
	Stick	*Coke*
35	Bread	Chow
	Stick	Coke
	Orange drink	Juice
	Coke	Carrot
	Pudding	*Bread*
36	*Key* (said "shot")	Juice
	Juice	*Bread*
	Bean cake	Strawberry drink
	Money	String
	Shot	Coffee
37	Money	Chow
	Sweet potato	*Lever*

Appendix *(continued)*

Trial	*Austin*	*Sherman*
	Orange drink	Orange drink
	Pineapple	String
	Coffee	Orange
38	Jelly	Jelly
	Key	Juice
	Corn	*String*
	Money	Pineapple
	Chow	Sponge
39	*Blanket*	Jelly
	Orange	*Corn*
	Lever	Sweet potato
	Straw	Straw
	Strawberry drink	Cake
40	Chow	Cake
	Jelly	String
	Banana	Lever
	Cake	*Coke*
	Coke	Money
41	Straw	Jelly
	Pudding	*Key*
	Lever	Coffee
	Cake	Blanket
	Juice	Bean cake
42	Sweet potato	String
	String	Chow
	Coffee	Sweet potato
	Banana (said "pudding")	*Lever*
	Orange drink	Orange drink
43	Carrot	Orange drink
	Money	Pudding
	Banana	*Banana*
	Lever	Orange
	Chow	Coke
44	*Money*	*Shot* (said "key")
	Carrot	Coke
	Sponge	Jelly
	Strawberry drink	Coffee
	Chow	M&Ms

Appendix *(continued)*

Trial	*Austin*	*Sherman*
45	Corn	Shot
	Orange drink	Peanut butter
	Sweet potato	Jelly
	Money (said "pudding-lever")	Orange drink
	Bean cake	*Juice*
46	Cake	*Sweet potato*
	Corn	Orange drink
	Stick	Bean cake
	Coke	Money
	Banana	Corn
47	Stick	String
	Orange	Lever
	Orange drink	*Sponge*
	Lever	Blanket
	Sponge	M&Ms
48	Coke	Strawberry drink
	Blanket	*Sweet potato*
	Banana	Straw
	Melon	Corn
	Straw	Coke
49	Pudding	Corn
	Key	Cake
	M&Ms	Carrot
	Corn	Key
	Sponge	*Bean cake*
50	Strawberry drink	Juice
	Money	Orange drink
	Coke	*Bean cake*
	Coffee	String
	Orange	Pineapple
51	Orange drink	Money
	Blanket	Cake
	Pineapple	*Straw*
	Stick	Melon
	M&Ms	Wrench
52	Money	Jelly
	Cake	Blanket
	Wrench	Lever

Appendix *(continued)*

Trial	*Austin*	*Sherman*
	Melon	Chow
	Straw	*Stick*
53	*Peanut butter*	Melon
	Straw	Strawberry drink
	Cake	Orange drink
	Wrench	*Straw*
	Jelly	Wrench

Note. The italicized words indicate which item was selected in each set. A choice of a lexigram for an object that was not in the set is shown in parentheses.

Chapter 15

Implications for Language Intervention Research: A Nonhuman Primate Model

This chapter is written by Mary Ann Romski in collaboration with E. Sue Savage-Rumbaugh

Severely retarded persons are often deficient in comprehending, producing, and using language. (Bensberg and Seigelberg 1976; Keane 1972; Spradlin 1963). As previously indicated, an important implication of ape language research is its potential application to the development of language-intervention strategies for nonspeaking persons. Indeed, this has been one of the goals of the ape-language research described in this book, since its inception in 1971 (Rumbaugh 1977; Savage-Rumbaugh and Rumbaugh 1979). This chapter will describe the rationale for the use of a *nonhuman-primate model* to study severe language impairments. It will also detail the contributions of the model described in this book to research with nonspeaking severely retarded persons, and provide an overview of the subjects' symbolic communicative achievements.

Rationale for A Nonhuman-Primate Model

Traditionally, two treatment models have been espoused by language interventionists, a *remedial model* and a *developmental model.* Each is based on a different philosophy. The *remedial model* is behavioral in nature. It supposes that children "being taught lan-

The preparation of this chapter and the research reported within were supported by National Institutes of Health grants NICHD-06016 and RR-00165 to the Yerkes Primate Research Center, Emory University, Atlanta, Georgia.

guage relatively late in their lives, because they have failed to acquire it adequately in their earlier experiences, no longer possess the same collection of abilities and deficits that normal children have when they begin to acquire language" (Guess, Sailor, and Baer 1978: p. 105). This model focuses on goals that most quickly accomplish some improvement in the individual's communication skills. While such goals may be appropriate for nonspeaking language-impaired individuals, they do not consider the prelinguistic or cognitive skills the learner brings to the task. While the results of the training program may be similar to developmentally based programs, the route to the product differs.

The *developmental model,* based on developmental psycholinguistic theory, maintains that individuals who do not acquire language naturally *must still go through the same steps* as a normal child does. This model, then, takes information from normal language learning and devises teaching strategies from it (Weber-Olsen and Ruder 1984). Goals focus on skills that are consistent with the normal developmental process. Nonhandicapped children acquire language rapidly. The young child is transformed into a language user almost overnight (Lock 1980). Information about the sequence of acquisition provides a general framework within which a handicapped child should progress. However, it is difficult to devise specific teaching procedures based on phenomena that occur spontaneously in nonhandicapped children.

Neither of the traditional models provide language interventionists with information about the component steps necessary to teach symbolic communication skills to individuals who do not learn to talk. Regardless of the traditional model language interventionists espouse, additional information is needed if they are to provide efficient and effective services to their nonspeaking clients. One potential source for this type of information is language research with nonhuman primates (e.g., Carrier 1974; Deich and Hodges 1977; Parkel, White, and Warner 1977). While nonhuman primate models are a relatively new concept in behavioral research (Harlow, Gluck, and Suomi 1972), they offer a number of possibilities not offered by traditional models for language intervention. In their 1979 book, *Language Intervention from Ape to Child,* Schiefelbusch and Hollis describe some of these possibilities:

> The chimps have taught us that experimentally viable language models can include alternative symbol sets characterized by flexible receptive and expressive modes, variable task functions, individualized behavioral topographies, and highly specialized pragmatic outcomes. The practical result may be the creation of new strategies and perhaps new models for teaching language to human children. (p. 5)

As illustrated throughout this present volume, nonhuman primates do not acquire a symbolic communication system naturally. The research process entails training them to do so. As such, the information gained from this research should—and, in fact, does—provide language-intervention researchers with an alternative model, specifically one based on *training information* gained via the use of a nonspeech communication system. Using this model does not prohibit the concurrent use of more traditional models. Rather, this model supplements them by providing detailed, procedurally based descriptions of the components necessary to learn and use language in a community environment. Moreover, the model (as it is described in this book) focuses not upon vocabulary competence, but upon functional symbolic communication with other individuals.

The Nonhuman Primate Model

Three major aspects of the research with Sherman and Austin are of particular interest to language interventionists: (1) the computer-based visual-graphic communication system, (2) the operational definitions of linguistic/communicative behaviors, and (3) the experimental tasks designed to assess or foster functional symbolic communication. Together these three components have formed the basis for the development of instructional procedures that have been employed in language-intervention research with severely retarded persons. Each of these components is described below, with specific reference to the reasons why they have been selected for teaching and research with human subjects who do not speak.

A Computer-Based Communication System

The system we have used to teach Sherman and Austin [and Lana before them (Rumbaugh 1977)] has a number of features that make it viable for severely retarded learners. Because lexigrams are a visual graphic symbol set, they alter the cognitive demands placed on the retarded learner. The nature of lexigrams places the communication interaction in the visual mode and symbols can then be recognized rather than being recalled to mind (Romski, Sevcik, and Joyner 1984).

The lexigrams, used in conjunction with the computer-based keyboard, form a dynamic system. That is, even though the symbols themselves are static, the computer provides visual and auditory feedback. Moreover, it allows for reliable documentation of each utterance and provides a permanent record of all communications. The computer also allows the teacher to randomly relocate lexigrams on any given trial. This feature eliminates the possibility of positional responding, which is frequently a problem for severely retarded persons. In essence, then, the computer linkup makes this system far more powerful than other visual-graphic symbol sets used on communication boards.

Operational Definitions of Symbolic Communicative Behaviors

The nonhuman-primate model also provides definitions which succinctly outline the skills required to communicate symbolically. These definitions are procedurally based and hence can serve as the functional framework in which symbolic communication may be taught. Symbolic communication consists of four major components: (1) arbitrary symbols, (2) stored knowledge, (3) interindividual intentional use, and (4) appropriate decoding of and response to the symbol by the recipient.

Nonspeaking severely retarded persons who attain these skills have achieved an important goal—one that results from a coordinated effort between teacher and learner. Each component of teaching is linked to every other, and the interrelationships of the components produce complex patterns of communicative behaviors which are integrated appropriately with environmental events.

Tasks Which Foster Functional Symbolic Communication

The majority of the successful training paradigms used with Sherman and Austin have been adapted for teaching and research with mentally retarded individuals. These subjects needed to have language learning broken down into small units, just as Sherman and Austin had. As will be shown, these tasks were an essential ingredient in the overall success of the language intervention research.

Application of The Nonhuman-Primate Model

Over the course of the last ten years, the mental retardation project at the Language Research Center has evolved from a feasibility study to its present form: a large research project involving over 30 subjects. Initially, the goal was to determine if the nonhuman-primate model, as described in the preceding chapters, was feasible for use in language-intervention research with severely retarded individuals. Indeed, it has proved to be a successful model (Romski et al. 1984). A more recent goal has been to describe the process that these individuals proceed through as they learn to use symbols for communication (Romski and Sevcik 1985).

Subjects

Fourteen institutionalized severely retarded persons from two residential facilities have participated in various phases of the research efforts to date. (Research is beginning with an additional 30 subjects.) These 14 persons all had severe impairments in the comprehension and production of spoken language. One bond that they all shared was failure to make significant progress in other language-intervention programs which were designed to teach them to talk, sign, or use Blissymbols. Table 15.1 provides a description of each subject at the beginning of the study. While the heterogeneous nature of their communicative impairments has made each subject a unique case, table 15.1 shows that the subjects could be divided into two groups. Some subjects (e.g., subjects 4 and 12) came to the task with little or no comprehension of spoken language ("beginning subjects"), while others (e.g., subjects 5 and 14) began with relatively good spoken language comprehension skills ("advanced subjects").

Eight of these 14 subjects participated in the research program for three years or longer. The other six subjects did not remain in the project for a variety of reasons.[1]

Lexigram Achievements and Byproducts

Given their previous failures, each of the eight subjects who participated in the study for three years or more developed communicative competencies far beyond any initial expectations. The descriptions of their communicative achievements are meant to

Table 15.1. A Description of the Subjects at the Onset of the Study

Subject	*Sex*	*Chronological Age*	*Stanford Binet/Cattell*[a]	*Leiter*[b]	*Speech-Language Treatment History*	*Speech-Language Skills*	
						Comprehension	*Production*
Subject #1	F	11 yrs., 1 mo.	2 yrs., 6 mos.	2 yrs., 8 mos.	Excluded from speech-language services	5 single words and simple commands in context	2 reported words: "cookie," "eat"
Subject #2	M	12 yrs., 6 mos.	1 yr., 10 mos.	3 yrs., 3 mos.	Speech-language treatment; Blissymbols were unsuccessful; services discontinued	5 single words and simple commands in context	Severe dysarthria, drooling, some single word approximations that were unintelligible to a naive listener.
Subject #3	M	16 yrs., 10 mos.	2 yrs., 6 mos.	1 yr., 9 mos.	Excluded from speech-language services	5 single words and simple commands in context	Several unintelligible single word approximations; stereotypic vocalizations (e.g., buzzing, humming)

Subject #4	M	18 yrs., 2 mos.	1 yr., 7 mos.	3 yrs., 5 mos.	Excluded from speech-language services	Inhibits to "no", 2 simple commands	No word approximations
Subject #5	F	16 yrs., 5 mos.	not tested	2 yrs., 3 mos.	Excluded from speech-language services	PPVT[c]-2 yrs., 6 mos.; 10 2-step commands	Motor sequencing impairment; some single word approximations
Subject #6	F	18 years	2 yrs., 2 mos.	would not perform task	Excluded from speech-language services	Responds to "no"; follows 2-step commands; Identifies pictures of body parts, objects, and actions	No word approximations; laughs
Subject #7	M	12 yrs., 8 mos.	1 yr., 8 mos.	would not perform task	Excluded from speech-language services	Responds to "no" follows simple commands; identifies 2 body parts and 6 objects	Does not imitate; inappropriate stereotypic phrases (e.g., "tore it up"; "all wet")

Table 15.1. *Continued*

Subject	*Sex*	*Chronological Age*	*Stanford Binet/Cattell*[a]	*Leiter*[b]	*Speech-Language Treatment History*	*Speech-Language Skills*	
						Comprehension	*Production*
Subject #8	M	18 yrs., 3 mos.	2 yrs., 1 mo.	2 yrs., 9 mos.	Excluded from speech-language services	Follows simple commands; identifies 2 body parts, objects, and actions	Imitates words; Several spontaneous word approximations
Subject #9	M	13 yrs., 3 mos.	1 yr., 6 mos.	would not perform task	Excluded from speech-language services	Follows simple commands; responds to "no"; identifes body parts, objects and actions	Stereotypic vocalizations (e.g. humming)
Subject #10	M	6 yrs., 2 mos.	1 yr., 3 mos.	could not perform task	Speech-language treatment; articulation and language stim-	Simple commands in context; 5-10 single familiar words	Severe dysarthria, drooling, a few single word approximations that

					ulation unsuccessful; Services discontinued due to behavior management problems		were unintelligible to a naive listener
Subject #11	F	19 yrs., 5 mos.	1 yr., 3 mos.	could not perform task	Speech-language treatment; articulation, language stimulation and sign language training unsuccessful	Simple commands in context; 5-10 single familiar words	Motor sequencing impairment; 2 idiosyncratic signs and 2 word approximations
Subject #12	F	18 yrs., 11 mos.,	1 yr., 9 mos.	could not perform task	Excluded from speech-language services	No comprehension of speech; simple commands in context with gestures	No word approximations, stereotypic vocalizations, (e.g. buzzing, humming, laughing)
Subject #13	M	14 yrs., 7 mos.	1 yr., 4 mos.	could not perform task	Excluded from speech-language services	No comprehension of speech	No word approximations stereotypic vocalizations (e.g. humming)

Table 15.1. *Continued*

Subject	*Sex*	*Chronological Age*	*Stanford Binet/Cattell*[a]	*Leiter*[b]	*Speech-Language Treatment History*	*Speech-Language Skills*	
						Comprehension	*Production*
Subject #14	F	17 yrs., 1 mo.	5 yrs., 3 mos.	4 yrs., 1 mo.	Speech-language treatment; Blissymbols were unsuccessful; articulation training unsuccessful	PPVT[c]-3 yrs., 9 mos.	Severe dysarthria, drooling, some intelligible single word approximations in context, unintelligible to a naive listener

[a] *Cattell Infant Intelligence Scale* administered only if basal score not achieved on the Stanford-Binet
[b] Arthur Adaptation of the *Leiter International Performance Scale*
[c] *Peabody Picture Vocabulary Test*

provide an overview of the type of gains that resulted from the intervention program they received, which was based on the nonhuman-primate model. Table 15.2 describes the communication achievements of each subject.

The number of lexigrams these eight individuals learned to use productively ranged from 19 to 75 (Mean = 41). These vocabulary items included names for actions, body parts, consumables, locations, modifiers, objects, and people. In addition to learning vocabulary items, some acquired other related skills, including comprehension of spoken words, vocalizations that approximated spoken words, multi-word utterances, and other social-communicative skills.

One documented byproduct of using nonspeech symbol sets paired with speech is that spoken words are vocally approximated (Lloyd and Karlan 1984). After they had developed a substantial lexigram vocabulary, three of the eight individuals (Subjects 2, 5 and 14) were able to pair their previously unintelligible vocalizations with the lexigrams; when they illuminated a lexigram, they made a vocal approximation of the spoken word. One individual (Subject 11), who did not make such vocal approximations, learned to comprehend five spoken words for five of the 20 lexigrams she had been taught.

As pointed out in chapter 16, the transition from one-word to multiword utterances is the third milestone in the acquisition of language (Bates 1979). Five of the subjects made this transition, combining known lexigrams into novel utterances of up to six lexigrams in length. One individual in particular, Subject 2, used his lexigram vocabulary to carry on extended conversations with other adults and even his peers. While these utterances were not always syntactically correct, they seemed to be telegraphic approximations of English word order. Table 15.3 is an excerpt from a conversation between Subject 2 and one of his teachers. It serves to illustrate his conversational skills and the spontaneous vocalizations that accompanied his lexigrams.

Retention of Symbol Skills

Generalized long-term use of a communication system by the learner is essential to the success of a language teaching program (Warren and Rogers-Warren 1984). Five of the eight individuals who participated in an early phase of this research were tested 18 months

Table 15.2. Description of Lexigram Achievements and Byproducts

		Lexigram Vocabulary		*Combinatorial Lexigram Usage*
Subject	*Years in Study*	*Comprehension*	*Production*	
1	3 yrs., 2 mos.	48	46	2 lexigram combinations
2	3 yrs., 7 mos.	70	67	1–5 lexigram combinations
3	3 yrs., 7 mos.	52	48	2 lexigram combinations
4	3 yrs., 7 mos.	19	14	none
5	4 yrs., 7 mos.	67	75	2 lexigram combinations
6	discharged from institution	—	—	—
7	dismissed after 1 mo. due to unmanageable behavior	—	—	—
8	dismissed after 10 mos. due to unmanageable behavior	—	—	—
9	dismissed after 9 mos.; was unable to learn the association between a food and a single symbol	—	—	—
10	transferred to another institution	—	—	—
11	3 yrs.; still participating	20	20	none
12	3 yrs.; still participating	20	20	none
13	dismissed after 12 mos.; was unable to learn the association between a symbol and food	—	—	—
14	3 yrs.; still participating	37	37	2 and 3 lexigram combinations

Table 15.3. Excerpt of a Transcript of a Conversation between Tom and Royce*

Transcript	*Commentary*
R: "Look what I've got for you. Do you want to see?"	Royce is showing Tom her hands that are covering an unopened can of pudding. Chocolate pudding is a prized food for Tom.
T: /jɛ/	Tom is looking at Royce and utters his approximation for "yeah".
R: "All right. What do you want me to do?"	
T: *OPEN*/bʌ/	
R: "Open"**	
T: *OPEN*/bʌ/, *HAND*/hæ/	Tom is attempting approximations of both words.
R: "Open hands.** What did you find?"	
T: *ROYCE*/wʌs/ *TOM*/tɪ/ *EAT*/ijə/ *PUDDING*/ʌm/	Unintelligible approximation of the word pudding.
R: "Royce Tommy eat pudding.** Hey, look, *ROYCE*/"Royce" *NO*/"no" *EAT*/"eat". Royce doesn't want any, Tommy.	
T: /wʌs dʌ/**	Tom shakes his head in a horizontal direction. He then searches the conversation board and moves his chair closer to it.
T: *TOM*/tɪ/	
R: "Tommy"	
T: *TOM*/tɪ/*HELP*/he/ *ROYCE*/wʌ/ *PUDDING*/hʌm/	
R: "Tom help Royce with pudding."	

Table 15.3. *(continued)*

Transcript	*Commentary*
T: /tε/*SCISSORS*	Tom gives an unintelligible approximation of the word, and points quickly to the lexigram that represents it.
R: "What?"	
T: /tε/*SCISSORS*	Tom produced an unintelligible approximation of the word. He is not able to pry the top off the pudding can with his finger and needs a tool.
R: "Oh, you want scissors, huh. OK. Look. *ROYCE/* ("Royce") *TOM/* ("Tom")/*SCISSORS/*("scissors")	
T: /wʌs tɪ tε/	Tom nods his head in a vertical direction.
R: "OK, here are the scissors."	Tom takes the scissors and grunts as he tries to pry off the tab. He eventually succeeds and gets to eat the pudding.

* All caps denote lexigram symbols while quotes denote the spoken word.
** Refers to the listener's vocalizations of the lexigrams the speaker is producing on the conversation board. Note that both Royce and Tom are doing this.

after their instruction had ceased. They retained mean proportions of .83 and .70 of their comprehension and production vocabularies, respectively (Romski, Sevcik, and Rumbaugh 1985). This finding further supports the effectiveness of the nonhuman-primate model.

Social Communicative Byproducts

Positive changes in other areas of development were a byproduct of lexigram learning. The staff at the institution where the subjects lived gave unsolicited reports of major changes in the subjects' social behavior. They threw fewer tantrums, took more interest in completing tasks, and behaved more appropriately with their peers, their families, and the staff (Romski et al. 1984). In a more recent study, some of these changes were systematically examined. As a result of this language-intervention approach, the subjects showed improved attention spans, outside of the training setting and general improvement in their ability to communicate.

Overall, the intervention program resulted in byproducts that can positively affect the lives of retarded youth (Abrahamsen, Romski, and Sevcik 1985).

Contributions of the Nonhuman-Primate Model

Given the descriptions of the communicative skills these subjects attained the reader might ask, how exactly did the nonhuman-primate model facilitate these behavioral changes? The procedures described in the previous chapters proved particularly effective in a number of different areas.

As the pattern of the learning process for these severely retarded subjects was reviewed, a striking similarity to that of Sherman and Austin was found for many, but not all, of the human subjects. The difficulties encountered by Sherman and Austin were virtually identical to many of those encountered by some of the mentally retarded subjects. Likewise, paradigms which proved successful in enabling Sherman and Austin to master their difficulties also proved successful with the retarded subjects. These performance patterns serve to validate the information gained from the nonhuman-primate language research. The following discussion shows some of the ways in which the nonhuman-primate model has benefited this language-intervention research program.

Beginning Subjects

For "beginning subjects"—those who came to the task with little or no comprehension of spoken language—the nonhuman-primate model had a dramatic impact on their learning. While the initial learning process was slow, they stayed motivated and learned how to learn. Since the major difference between this attempt and their previous language treatment programs is the instructional model, it is credited as the difference. Difficulties at various stages of the training were overcome by implementing information gained from the model.

As described in chapters 4 and 5, request training proved to be the most successful approach for teaching the chimpanzees initial symbol usage skills. Since the request paradigm offered a number of positive features that could provide severely retarded subjects with immediate success, it served as a beginning point. Request teaching was carried out in a social environment. The acquisition of request skills gave these individuals a new degree of control over their environment. This control facilitated the development and use of causality skills.

The subjects first learned how to bring the operation of "food machines" and the "food giving behaviors of others" under their own control. They learned how to operate a food vending device by using a lexigram, and then request, using lexigrams, up to four foods they observed being loaded into a food-vending device.

Once the subjects realized they could ask the machine to dispense food, they learned specific associations between each food and its corresponding lexigram (Lexigram A = Food A, Lexigram B = Food B, etc.) After this skill was established (to a 90 percent correct criterion level over two consecutive sessions), the foods were removed from the machine and the individual was asked to request the same foods from a person.

During the initial teaching within the request paradigm, Subjects 4, 11, and 12 encountered some difficulty learning lexigram-item associations. While they were able to master the use of a single lexigram-item pair, the introduction of a second lexigram paired with a second food resulted in a level of performance below chance even after 600 teaching trials. Cueing procedures did not facilitate their accuracy.

Following the paradigm described in chapter 5, a second vending device was introduced. All three individuals' percent correct performance increased above chance in a few sessions, and with additional teaching trials they learned the conditional association between the lexigrams and the foods. They encountered little, if any, difficulty making the correct food-lexigram association when all foods were then sequentially placed into one vendor. Each of these individuals went on to learn two additional associations (in one vendor) and eventually to request these foods from a person using the lexigrams. Once these skills were established, the number of trials to criterion for subsequently presented lexigrams decreased significantly (Romski and Sevcik 1985).

The next step beyond the request task was to expand the use of these symbols to communicative functions other than requests.

For example, if an individual can request peanuts, can she also retrieve them when asked by another person and name them without receiving them? For Austin and Sherman, the answers to these questions were not simple ones. Additional tasks had to be devised to teach naming and comprehension skills. These tasks promoted new skills by capitalizing on well-established ones.

Given information gained from Sherman and Austin, the comprehension and naming skills in these subjects were monitored from the onset of lexigram teaching. Like Sherman and Austin, request did not, for some of the subjects, automatically result in accurate naming and comprehension. They, too, behaved as if they should have been given the item they had named or had given to the teacher. If they were not given the item, they reverted to positional responding or refused to perform the task. Even after accurate naming performance (80 percent correct or better) was evident, comprehension was still lacking relative to both request and labeling, up until the acquisiton of lexigram 15 for some of the subjects (Romski and Sevcik 1985). Such information parallels findings from the nonhuman-primate model project.

While communicating about foods is a beginning point for severely retarded individuals, additional vocabulary items must also be incorporated into their repertoires. The addition of other classes of words (e.g., objects) is sometimes difficult, since the motivation to acquire objects may not be as strong as it is for desirable foods. A similar motivational problem was encountered with the chimpanzees. Object usage and object names were made salient for the subject by storing food in locations that could be reached with the use of specific tools. Tools were obtained by using the lexigram that represented the appropriate tool to request it from another individual. This task was also adapted to include interindividual coordination and cooperation. A modification of the tool task allowed these severely retarded learners to expand their vocabulary to include items other than foods by maintaining a high level of motivation. In addition, this task served as another avenue by which to promote interindividual lexigram usage.

Advanced Subjects

For "advanced subjects" (individuals who began with relatively good spoken language comprehension skills), the initial learning task was relatively easy. Even though they were unable to produce the spoken words, they already had established associations between items and spoken words. The keyboard communication system provided them with a functional way to communicate using an arbitrary symbol set and a direct selection response. In addition, for the individuals who were able to combine lexigrams, it highlighted the sequencing process for them. The sequenced lexigrams were visible on the display monitor in the order in which they were illuminated.

The second important component of the nonhuman primate model for these "advanced subjects" was the use of paradigms designed to promote interindividual communication using the non-speech symbols. Mentally retarded persons often exhibit marked deficits in their ability to interact socially with their peers (Guralnick and Weinhouse 1984). The establishment of communication between peers, then, is a critical, yet often neglected, teaching goal.

The food sharing task (chapter 8) provides a model for the development of symbol-mediated social exchanges between peers. These exchanges were designed to foster positive social interaction and cooperation, as well as symbolic communication. The model employs a structured symbolic communicative task in which peers interact cooperatively to attain a mutually desired goal successfully. The interindividual lexigram communicative tasks make an important contribution to the study of methods for facilitating the establishment of peer-peer communication. They provide a systematic way to teach peers to communicate with each other.

Beginning and Advanced Subjects

Perhaps the most striking demonstration of all of these individuals' achievements is, like Sherman and Austin (chapter 14), the subject-initiated use of their respective lexigram repertoires. Typically, severely retarded individuals are responders rather than initiators (Calculator and Dollegan 1982; Harris and Vanderheiden 1980). All of the individuals who learned lexigrams, however, sometimes used their symbols spontaneously during the research. While the quantity and quality of the spontaneous usage varied with the individual subjects, the appearance of such spontaneity indicates that the lexigrams functioned as words do for children learning spoken language. Further, it attests to the appropriateness of the nonhuman primate model for language intervention. Individuals who previously had not developed symbolic communication skills had become symbolic communicators as a result of their participation in this experimental intervention program. Figure 15.1 illustrates a sequence in which Subject 12 requests soda (a nonvisible item), travels across the room to the refrigerator, retrieves the soda from among an array of items, and returns to the keyboard with it.

Figure 15.1. (a) Subject 12 requests a nonvisible item, soda, (b) travels across the room to the refrigerator, retrieves the soda from among an array of items, and (c) returns to the keyboard with it. Photos by Rose A. Sevcik.

Summary and Conclusions

This particular nonhuman-primate model for language learning makes a number of contributions to the research on nonspeech language communication with mentally retarded persons. These contributions include a dynamic, interactive communicative system, operational definitions of target behaviors, experimental teaching tasks, assessments, and observational procedures. This model has so far been successfully employed with eight nonspeaking severely retarded individuals. In so doing, it has demonstrated that some severely retarded persons, previously unsuccessful in other language intervention programs, can indeed learn to communicate symbolically. Their potential was tapped via the implementation of this nonhuman-primate model. Dissemination of this information to speech-language clinicians, special educators, and clinical psychologists should benefit a larger number of nonspeaking children and adults.

Chapter 16

Ape-Language Research Beyond Nim

This chapter is written in collaboration with Duane Rumbaugh, William D. Hopkins, Jeannine Murphy, Elizabeth Rubert and J. Philip Shaw

What Has Been Learned?

The main lesson obtained from the extensive research with Sherman and Austin is an answer to a very important question: In what sense can a species other than humans have language? Humans as well as machines can "speak" or otherwise produce words without comprehension. The early poetic recitations of children are frequently only the production of words without comprehension of what is being said. In the same sense, then, if an ape or other animal produces "words" through use of geometric symbols, hand signs, or vocalizations, it might or might not be using those words with competence. The crux of the question is, When animals use symbols in ways similar to humans' use of words, do they know and mean what they are saying? For animals to know and to mean what they are "saying," the process of representation must be extant and operating.

As documented here, Sherman and Austin gave strong behavioral evidence for the conclusion that, at a minimum, a significant portion of their symbols came to represent items not necessarily present. They evidenced representational processing particularly clearly with their foods and tools in the categorization study (see chapter 11). Recall that in the last phase of that study they were shown a number of symbols for specific foods and tools and that they were almost error-free in classifying each symbol. They were able to declare their answers through the use of categorical symbols and they did so on the basis of trial-1 performance, under conditions which prevented inadvertent cueing. There was but one error on 33 test trials. Sherman classified the symbol "sponge" as a food—possibly because he tended to chew off and swallow portions of sponges.

The results of this study demonstrate, as convincingly as things can be demonstrated in behavioral research, that these chimpanzees had developed a capacity of fundamental importance to language—the ability to use arbitrary symbols representationally. This evidence alone should justify continued research with apes into questions regarding the nature of language—an effort that was severely challenged as a consequence of the negative research results published by Herb Terrace (Terrace et al. 1979).

It should be noted that although Terrace himself did *not* conclude that apes were unable to acquire language, his work was widely interpreted as incontrovertible evidence for the absence of this skill by others. Terrace, however, left open an avenue, saying that "Quite possibly a new project, administered by a permanent group of teachers . . . [who] have the skills necessary for such experiments, would prove successful at getting apes to create sentences" (Terrace 1979a: 76).

However, notwithstanding this guarded suggestion that apes might still be capable of language if reared under different circumstances, Terrace's findings shook the confidence of the entire area. In addition, the impact of his negative results was inadvertently heightened by the fact that Terrace had previously reported (in widely circulated progress reports) that Nim had evidenced both vocabulary and syntax. Terrace reversed this position in a 1979 article entitled "How Nim Chimpsky changed my mind," where he stated "Initially, the regularities I observed in thousands of Nim's communications in sign language suggested that he was, in fact, using a grammar. However, after analyzing videotapes of his 'conversations' with his teachers. . . . I could find no evidence confirming an ape's grammatical competence, either in my own data or those of others" (p. 66). Terrace's reversal was taken by many to indicate that the other researchers had been fooled by their apes. It resulted in a significant loss of funding resources to the field.

Moreover, although Terrace's work had originally focused exclusively upon syntax as the key which would determine whether or not apes were capable of language, his final conclusions went far beyond the issue of syntax alone. He stated (Terrace et al., 1979) that (1) Nim's manual signs were, to a very worrisome degree, imitations of signs made recently by his teachers, (2) the chimpanzees' signs in other language projects also evidenced high levels of imitation, (3) Nim's stringing of signs failed to add meaning above and beyond that of single words, (4) Nim tended to interrupt instead of taking turns in conversation, and (5) Nim basically signed only when he had to in order to get what he wanted. In sum, Terrace asserted that there was little or no meaning behind most of the symbols used by apes. He noted "What is important to recognize . . . is that neither the symbols nor the relationships between the symbols have specific meaning. Although the words and word order may be meaningful to an English speaker, they may be meaningless to the animal producing them" (Terrace 1979a: 68).

In light of the material presented in previous chapters, it is our opinion that considerable evidence has now been presented to support the view that the above conclusions no longer stand as representative of current data.

In refutation of points one and two above (that apes' signs are likely to be imitations of signs made recently by the teachers), we turn to a recent paper by Greenfield and Savage-Rumbaugh (1984). On the average, Sherman and Austin imitated only 2.0 utterances per session, while Sherman averaged 60.5 and Austin averaged 33.5. The number of expanded imitations (those in which there were symbols used in addition to the symbols previously employed by the teacher) averaged 5.5 for Sherman and 1.5 for Austin. Finally, the average number of *spontaneous* utterances per session was 53 for Sherman and 30 for Austin. Clearly, these data for Sherman and Austin do not correspond with the data from Project Nim; instead they show that the vast majority of Sherman and Austin's utterances were not dependent upon the teacher's preceding utterances.

The third conclusion offered by Terrace was that Nim's strings of signs failed to provide information beyond that of single words—i.e., the strings were neither phrases nor sentences but rather multiple single symbol efforts to obtain the item at hand. In the paper by Greenfield and Savage-Rumbaugh (1984), we explicitly focused upon this issue, using a corpus of Sherman and Austin's data gathered during regular tool-use sessions. The question we addressed was why, on any given occasion, Sherman and Austin selectively encoded only certain aspects of the situations they encountered in training. We found that their utterances served to differentiate alternative possibilities and to represent change or novelty. They tended to leave "unsaid" what was unchanging or repetitive. Furthermore, the majority of their symbol usage was nonimitative, spontaneous, coordinated across time with the utterances of their teacher, and of high information value. Even more important, the length of utterances was associated with the complexity of the immediate situation. A single dimension of change led to single-lexigram utterances, while multiple dimensions of change were associated with multilexigram utterances. Thus, the absence of formal grammatical structure does not imply that chimpanzees' multiple-lexigram utterances were merely rote strings or concatenations of single lexigrams, any one of which might be "good" to try given the situation at hand. Rather, multilexigram utterances were used to encode variability, just as their single-lexigram utterances were. They accomplished this in a manner very similar to the way a normal human child does. (Greenfield 1978). Sherman and Austin used their lexigrams, singly and in combination, to maximize effective and efficient communication. They avoided mention of the constant and the obvious. They focused upon the changing dimensions of the context. Thus, they used their language to partition alternatives, a hallmark of language as used by humans.

In reference to the fourth of Terrace's conclusions, Sherman and Austin were not, of course, using manual signs. It is quite possible that our keyboard technology is particularly powerful in discouraging imitation and that it facilitates the coordination of turn-taking in a dialogue (the fourth of Terrace's conclusions). Previous chapters in the book, as well as all reports of the Lana Project (Rumbaugh 1977), give clear evidence that these chimpanzees excelled at turn taking, both with one another and with their teachers.

We would also take issue with Terrace's fifth and final point—that chimpanzees only sign because they have to. Although Sherman and Austin were initially taught to use symbols by withholding items contingent upon request, they moved beyond this, particularly as the ability to make statements appeared. They then began to announce what they were about, for example, by saying "tickle" before they started tickling us, or commenting "scare" when something frightening happened. Although the major portion of their symbol use remained request oriented, they clearly became able to use symbols to announce and/or comment upon their actions.

In view of the new evidence which has become available since his 1979 paper, Terrace's interest has become focused upon the question, "What is a name?" (Terrace 1985). We welcome his colleagueship and his attention to a question that has been of significant interest to us throughout the course of research with Sherman and Austin. Both Terrace and we hold that ape-language research should have been focused upon this question, rather than upon questions of syntax, during the initial efforts of the 1960s and the early 1970s.

Using Symbols with Intent

The collective efforts of researchers in the field have produced data which make it apparent that apes have considerable capabilities for language. They can learn words spontaneously and efficiently, and they can use them referentially for things not present (see chapter 13); they can learn words from one another (see chapter 9); they can learn to use words to coordinate their joint activities and to tell one another things otherwise not known (see chapter 10); they can learn rules for ordering their words (von Glasersfeld 1977; Pate and Rumbaugh 1983); they do make comments (Greenfield and Savage-Rumbaugh 1984); they can come to announce their intended actions (Savage-Rumbaugh et al. 1983 and chapter 14); and they are spontaneous and not necessarily subject to imitation in their signs (Miles 1983: Greenfield and Savage-Rumbaugh 1984). In sum, they are communicative and they know what they are about!

To illustrate this, let us take an example from a food sharing situation, where Sherman and Austin take turns asking for a specific food or drink, and then serving the requested item (Savage-Rumbaugh, Rumbaugh and Boysen 1978c). Recall that Sherman frequently intervened on Austin's turn, but more importantly, when Austin appeared not to know the name of the item that Sherman wanted, Sherman would take Austin's hand and guide it to the appropriate key on the keyboard, thereby helping Austin to "ask" for the food. Sherman would then promptly serve this food, giving a piece to Austin and taking one (usually the larger portion) for himself. Despite alternative explanations that can be conjured, it seems clear that Sherman in a very real sense knew what he wanted and knew how to get it through directing Austin's hand at the keyboard.

As additional evidence that Sherman and Austin know what they are about, they visually monitored one another to determine when requests were being received and complied with accurately. For example, if Sherman needed a wrench from Austin to get food out of a box, he would often, when he observed Austin about to give him a key instead, recheck the symbol he had used and correct it if he had erred. He would also tap on the projector that portrayed the correct symbol to bring Austin's attention to the new message. In addition, he would not accept the key or other item that Austin might have given in error.

With Sherman and Austin, we have attempted to determine the critical factors which separate a ritualistic form of symbol production, in which the subject learns what to do in order to receive food, from a more humanlike form, which is meaningful, representational, and intentionally communicates messages to a receiver—a receiver viewed as sentient and whose future behaviors are viewed as potentially alterable in precise ways by the indirect means of symbolic utterances.

Many behaviorally oriented psychologists (Rachlin 1976; Skinner 1957) maintain that this distinction between symbol production to obtain reinforcement and symbol production to communicate intentional messages is meaningless. They prefer to ask whether a given set of environmental events increases or decreases the probability of a given response.

Psychologists who hold such views do not deny that symbols have referents or that interindividual communication is dependent upon common referents being jointly alluded to and conceived of by both speaker and listener. They do deny that scientific pursuit to understand these events is worthwhile. They emphasize that we must deal only with external measurable contingencies. "Referents," "meaning," "awareness," "intentionality," etc. are viewed as phenomena that are outside the boundaries of scientific endeavor because they occur only in the phenomenological world of perceived experiences. That is, they are thought to have no "reality" apart from the "felt-reality" of individuals who believe themselves to be experiencing these events.

We suggest that the time has come to break away from views which hold that meaning, reference, and intentionality are not measurable phenomena and hence are closed to scientific investigation. Intentionality must be accepted if we are to understand language in a functional sense and tracing the evolutionary emergence of intentionality is critical to the appearance and understanding of language as it is used by our species.

If one approaches the study of behavior not as the study of individual, single-unit phenomena, but rather as interindividual, multiunit sequenced phenomena as we did with Sherman and Austin, it becomes possible to operationally define reference and intentionality as the occurrence of a sequence of interindividual interactions across time. The nature of the sequence as well as the structure of given behaviors within the sequence can be defined and quantified.

Recent work dealing with the issue of intentionality in developmental psychology (Bates 1979; Bruner 1974/75; Lock 1980) has described a series of developmental phases through which human infants pass as their communicative skills move from self-oriented, opportunistic behavior to stylized intentional communicative patterns. The term intentional, in the framework of modern developmental psychology, is used not to refer to the child's internal needs, wants, and desires, but rather to an observable and *measurable sequence of complex monitoring responses* in which the child (a) checks to see that a listener is present before emitting a communicative signal, (b) engages the attention of the listener before emitting this signal, (c) emits a signal that requires a specific behavioral or verbal response on the part of the listener, (d) monitors the listener's response visually and auditorially, and (e) if the response is not in accord with the stylized "meaning" the child has learned for that signal, he modifies and re-emits the signal. This modification may or may not be said to reflect the child's "original intent" (depending on who is using the term); however, there is general agreement that it is not the elusive "original intent" that is important, but rather the interindividual nature of the behavior. In other words, meaning and intent are not to be found by looking "inside" a speaker; rather, they are to be understood as a *process* that develops between speakers and listeners. This interactive process assumes a character we term "intentional" when one party appears to be repeatedly engaging in behaviors that function to monitor and alter the behavior of another in a specific manner. Thus, there is an observable interindividual sequence of events that must exist before the term intentionality is applied by the modern developmental psychologists who work within the framework first laid out by Bruner (1974/75).

Likewise, albeit primitive by human standards, Sherman and Austin attributed intent both to the utterances of their teachers and to one another. Thus, if the teacher were pressing symbols simply to determine that the keyboard was functioning properly, Sherman and Austin never mistook the lighting of symbols in those circumstances for intentional communications. Similarly, if the teacher was lighting symbols to address Austin, Sherman did not respond instead simply because he saw symbols light up. He knew that the message was being directed to Austin. Also, if the teacher were lighting symbols to describe her own actions, as opposed to soliciting an action from Sherman and Austin, the chimpanzees understood that the intent of the teacher's message was to inform them of her future actions. To be sure, had Sherman and Austin not been able to interpret these various "intents" and respond accordingly, they would never have been able to differentiate between symbol as a "general stimulus" and symbol as a "message with a specific purpose."

Although words and names might start as operants, they can become more than that. They can come to represent things and events not present. Sherman and Austin came to use symbols that way.

The Ape-Language Controversy in Retrospect

In the early years of the efforts to investigate questions pertaining to apes and language, the wrong question was at the fore—in the press and in the public's thinking, if not also in the perspectives of many scientists. The question was, in essence, "Do the apes have full language competence as we know it in humans—or don't they?" That could not be answered—then, now, or ever: there is no concurrence, nor is there one likely in the offing, as to what the minima are for it to be said that language exists. Where there is a "lot of it," there is no problem; however, when one is dealing with traces of emergent language, what are the criteria? Individuals may assert their own minima, such as vocabulary and syntax, but even these assertions serve only to define new questions: What are the critical attributes of vocabulary, of syntax? "How much" must there be of them for it to be concluded that there is language?

A far more productive question is, "What facets of language are to be found in nonhuman forms?" Other related questions are, "What are the requisites to them?" and "What are the variables that influence their emergence and competence in use?"

Ape-language research has taught us a great deal about dimensions of language that likely would never have been teased out in the study of normal children, who acquire language too rapidly for fine-grained analysis. Ape-language research has also taught us a variety of things unrelated to apes and to language. It has taught us that scientists need to work harder both to establish and to retain a sense of community—particularly when the question is such as to generate a challenge to the definition of humanity and its origins. Early efforts on the part of some researchers to preempt or to preclude the entry of others into the field of ape-language research was misguided and unfortunate in that they put at severe risk the probability that "community" and the exchange of data between laboratories might ever be attained. Fortunately, that risk, though severe, has at long last been negated and the field can now go on beyond Nim.

This work has also served to remind researchers, through hard object lessons, that the press can be at times more interested in controversy than in the constructive resolution of controversies. It is not always a patient observer of the scientific process through which there is the sorting out of data and the emergence of changing perspectives—all necessary for progress.

Ape-language research, in perspective, should encourage all to be more tentative in what is said, to be more careful about what is committed to paper, to speak with precision, and also to listen carefully and without prejudice. After all, the main function of research is to achieve the education of scientists through communications and interactions. To the degree that those exchanges are imprecise, that key function of research is frustrated. Scientists must be prepared to redefine their individual positions in the light of new evidence. That is one of their primary responsibilities. To the degree that scientists fail in this, they confound the progress of scientific understanding and insight, at the minimal gain of possibly defending their own reputations and positions of "authority."

Ape-language research should have reminded everyone by now that no one can prove the null hypothesis. At this point in time no one should say, "Apes will never be able to do this or that!" Time and new evidence are notorious for embarrassing those who make such flat statements. After all, such conclusions are basically assertions of the unprovable null hypothesis.

Language Acquisition in *Pan paniscus*

To illustrate the above point, we will briefly examine the linguistic skills of a species of ape not previously studied, *Pan paniscus.* Work with this species has already shown us that we were premature in presuming that our earlier conclusions applied to *all* apes, as though, in spite of their significant anatomical differences, they all functioned at the same cognitive level.

In all previous ape language studies (ours and others) (Gardner and Gardner 1969; Premack 1972; Fouts 1973; Rumbaugh, Gill, and Von Glasersfeld 1973b; Savage-Rumbaugh et al. 1978a; Terrace et al. 1979; Patterson 1978; Miles 1983) apes were given special training sessions during which they experienced repeated pairings of a symbol with an object. Correct responses were reinforced, and incorrect responses were modified using traditional error-correction techniques. Most studies reported spontaneous generalization of symbol use beyond these training sessions and the appearance of limited, yet spontaneous acquisition of symbols by sophisticated subjects (Gardner and Gardner 1969; Fouts 1973; Rumbaugh et al. 1973b; Savage-Rumbaugh et al. 1978c; Terrace et al. 1979; Patterson 1978; Miles 1983). The necessity of using explicit reinforcement contingencies to generate item-symbol associations, or "vocabulary," raised controversy regarding the nature of the symbolic skills acquired using such explicit training (Seidenberg and Pettito 1981; Epstein, Lanza and Skinner 1980; Terrace 1979; Savage-Rumbaugh, Rumbaugh and Boysen 1980; Savage-Rumbaugh et al. 1983; Savage-Rumbaugh 1984a). Since normal children acquire language through observation, the issue of whether explicitly trained associations can become equivalent to spontaneously acquired words emerged as a central focus of the ape-language controversy (Terrace 1979; Savage-Rumbaugh et al. 1980; Savage-Rumbaugh et al. 1983).

Language Acquisition Without Training

We have recently found that at least one species of ape, the pygmy chimpanzee *(Pan paniscus)*, can acquire language skills without explicit training if exposed to human communication systems early in life. This species has not been involved in previous long-term language acquisition studies because of the difficulty of obtaining such apes. (A study of problem-solving using Premack type tasks was carried out with a pygmy chimpanzee in the Frankfurt Zoo by C. Jorden and H. Jorden, 1977. Although this animal did well, the work was of extremely limited duration and focus.) The pygmy chimpanzee is on the rare and endangered list and has only recently become available for research because of births at the Yerkes Primate Center. The parents of this subject are at the Yerkes Center.

Although lack of accessibility has limited research on the pygmy chimpanzee, there are reasons to expect that it might differ from other apes, specifically with regard to a greater capacity for language: its natural gestural repertoire is more highly developed; its vocal repertoire is more varied; and eye contact between individuals plays a more important role in conspecific communication (Savage-Rumbaugh, Wilkerson, and Bakeman 1977; Yerkes and Learned 1929; Yerkes 1925; Thompson-Handler, Malenky, and Badrian 1984; Savage-Rumbaugh 1984b). Cognitive studies of *Pan paniscus* suggest they are brighter than other apes (Jorden and Jorden 1977; Yerkes and Learned 1929; Yerkes 1925; Savage-Rumbaugh 1984).

One pygmy chimpanzee, Kanzi, was first exposed to the use of graphic symbols, gestures, and human speech at 6 months of age. Since that time he has been reared much like Sherman and Austin, but with considerably greater freedom, as we have moved to a larger facility. Kanzi remained with his mother for the first 2½ years of his life. During this time his mother was involved in language training studies, which he observed. His mother was separated from him and placed in a social breeding group for 4 months when he was 2½ years of age. She then returned to the Language Research Center, where she gave birth to Kanzi's younger sister, Mulika, in 1983. Kanzi is allowed to spend as much time with his mother as he wishes. (His mother has shown a very limited capacity for symbol acquisition, suggesting that age of first exposure to language is important in the pygmy chimpanzee.) Kanzi, who is never caged, has access to five large living areas and 55 acres of natural forest. One or more human companions are with Kanzi 24 hours a day, 365 days per year. Eight different people are presently involved in the raising of Kanzi and Mulika, and this staff has remained stable. Kanzi's human companions use lexigrams, gestures, and speech to communicate with him. The symbol system is similar to that employed with Lana, Sherman and Austin. Kanzi's companions provide him with communicative models and encourage him to communicate just as they did Sherman and Austin, but they do not train symbols or require that Kanzi use symbols.

At 18 months, Kanzi spontaneously began to use nonritualized gestures to communicate preferred directions of travel and actions he wished to have performed (see figure 16.1). For example, he used his outstretched arm and hand to point toward areas to which he wished to be carried, he made twisting motions toward containers when he needed help in opening twist on lids, or made hitting motions toward nuts he wanted others to crack for him (Savage-Rumbaugh 1984). Although his gestures were tied to the immediate context, they were frequently accompanied by affective vocalizations, as are the gestures of children before the onset of speech proper. Kanzi also began to display elementary comprehension of vocal speech at this time. One year later, he spontaneously started using lexigram symbols in a communicative manner. With the onset of lexigram usage, his communications broadened beyond the immediate context as he began to use lexigrams to refer to absent foods, objects, and locations.

A complete record has been kept of all of Kanzi's utterances since he began using lexigrams. Indoors, all lexigrams used by Kanzi and his companions are automatically recorded by computer-monitored keyboards similar to those used with Sherman and Austin. In the woods, a portable keyboard is used. These utterances are recorded by hand, then entered into the computer record at the end of the day along with contextual notes.

Each utterance is classified, when it occurs, as correct or incorrect and as spontaneous, structured, or prompted/imitated. Spontaneous utterances are those initiated by Kanzi with no prior prompting, querying, or other behavior on the part of human companions designed to elicit an utterance. Structured utterances are those initiated by questions, requests, or object-showing behavior on the part of the companion. Prompted utterances are those including any part of the companion's prior utterance. Although Terrace

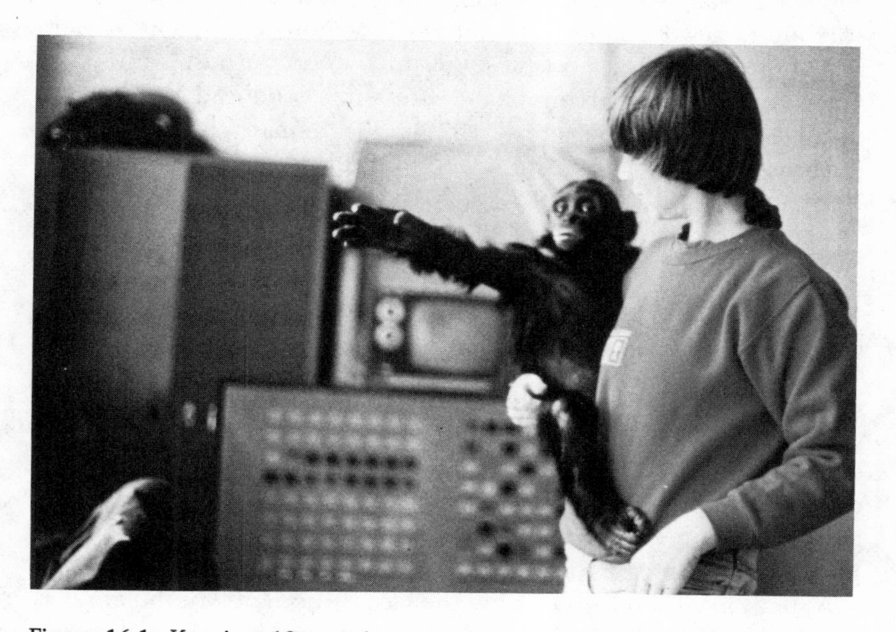

Figure 16.1. Kanzi at 18 months, using a spontaneous gesture to show the teacher where he wants to be carried.

(Terrace et al. 1979) found it necessary to perform an analysis of video material in order to decide when Nim was imitating his teachers, we have found it reasonable to determine this at the time of the utterance. Since the symbols are displayed visually, the teachers know which ones they have just used, and if Kanzi uses similar symbols, his utterance is scored as a prompted utterance. The reliability of such real-time scoring was checked by comparing 4.5 hours of videotaped sessions with real-time scoring. The scoring was done independently by two different observers, with one observer scoring the behavior real time and the other scoring the tape. At the time the scoring was done, the real-time observer did not know that the data would be used for a reliability check at a future date. Of the 37 utterances scored by both teachers, only one utterance was scored differently. In addition, nine utterances were noted on the video tapes that were not seen by the real-time observer. Of these nine unrecorded utterances, eight were spontaneous and one was structured.

Spontaneous utterances consistently account for more than 70 percent of Kanzi's single word and combinatorial utterances as fig. 16.2 illustrates. Prompted, imitated, or partially imitated utterances account for only 11 percent of Kanzi's total corpus, while they accounted for 39.1 percent of the utterances produced by the ASL-tutored Nim. (Terrace et al. 1979) Like human children, Kanzi tends to imitate mainly when he is learning new words (Bloom, Hood, and Lightbrown 1974).

The total number of combinations produced by Kanzi (2805) is far smaller than the number produced by Nim during a similar 17-month period (19,000). Kanzi's mean rate of seven utterances per hour is low. However, the majority of utterances which Kanzi does produce are not situation redundant responses, solicited by queries from teachers. Instead, they are primarily innovative requests which tend to occur whenever Kanzi wishes to change his current activity. They may also be comments about surrounding events, as when Kanzi notes things he observes while watching television—for example, noting "ball" while watching a tape of himself playing with a new ball.

A new means of measuring the acquisition of vocabulary has been used with Kanzi, since previous measures required only that symbol production appear to be appropriate to the context. Since

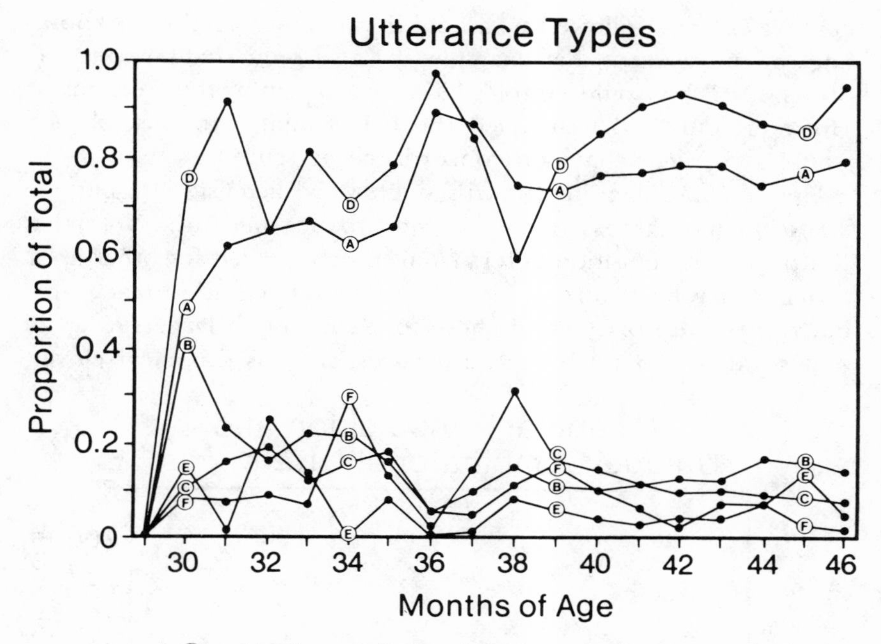

Ⓐ Single words used spontaneously

Ⓑ Single words elicited by companion's queries

Ⓒ Single words used in imitation or as a result of prompting

Ⓓ Combinations used spontaneously

Ⓔ Combinations elicited by companion's queries

Ⓕ Combinations used in imitation or as a result of prompting

Figure 16.2. This figure illustrates Kanzi's use of single words and combinations subdivided according to the categories of spontaneous usage, usage elicited by companions, and imitated usage.

such context-appropriate responses can occur without comprehension (Seidenberg and Pettito 1981; Terrace 1979; Savage-Rumbaugh et al. 1980; Savage-Rumbaugh et al. 1983), a behavioral concordance between symbol and action was required to validate comprehension of spontaneous occurrences. A symbol was classified as a member of Kanzi's vocabulary if and only if it occurred spontaneously on 9 of 10 consecutive occasions in the appropriate context *and* was followed by a behavioral demonstration of knowledge of the referent. For example, if Kanzi requested a trip to the "treehouse" he would be told, "Yes, we can go to the treehouse." However, only if he then led the experimenter to this location would a correct behavioral concordance be scored.

Figure 16.3 illustrates Kanzi's single-word acquisition from age 30 to 46 months based on the concordance measure. Note that Kanzi's word acquisition differs from those reported for other apes: it does not reflect words which experimenters chose to teach him, but rather the words Kanzi chose to learn from a larger group of symbols used around him. Many words, such as three-leaf clover,

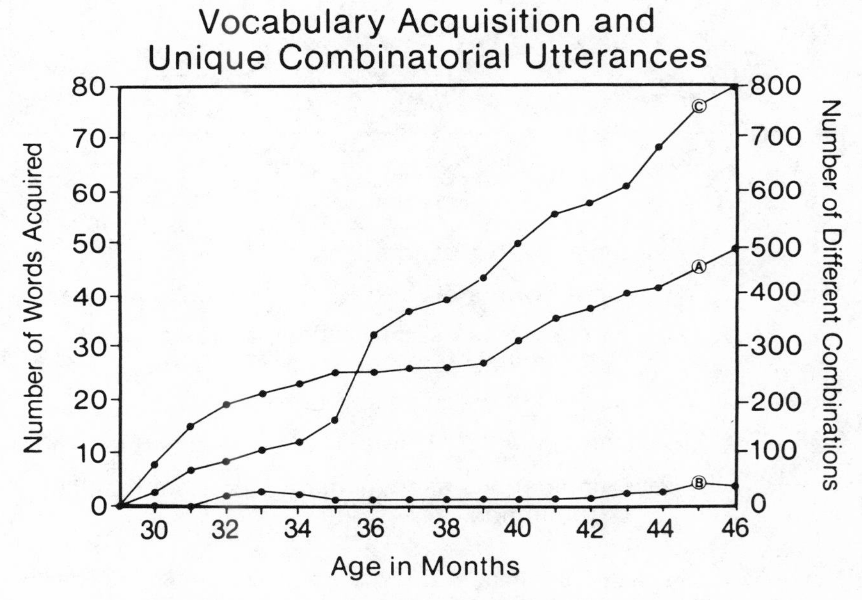

Figure 16.3. This figure illustrates Kanzi's rate of vocabulary acquisition and loss, and the number of unique nonimitated combinations he formed across the same 17 month period.

ice, Matata (his mother's name), Mulika (his half-sister's name), Mushroom trail (a favorite trail in the woods) are quite different from those other apes have acquired, and reflect Kanzi's access to his outdoor environment, his access to conspecifics, and the variety of his interests (figure 16.4).

Kanzi's nonritualized gestures and vocalizations continued to occur and in fact began to play a more elaborate communicative role after spontaneous use of lexigrams emerged. Gestures were used to reference point-at-able items, while vocalizations were used to convey affect, and lexigrams were used to convey information about items, events, animates, etc., that were not present. For example, if Kanzi desired to go to the "colony room" with Jeannine, he could convey this by saying "colony room" on the keyboard, then touching Jeannine to indicate that she was the person he wished to go with, all the while expressing his desire to go immediately by vocalizing "anngh."

Figure 16.4. Kanzi at 3 years of age using his portable keyboard to comment on the large round ball in the lower corner of the photograph. The teacher is observing, but has not withheld the ball or querried Kanzi about the ball to elicit the comment.

During the first 17 months Kanzi produced a total of 2540 nonprompted (nonimitated) combinations and 265 prompted (imitated) combinations. All but 10 of these nonprompted combinations were judged to be appropriate to the context and interpretable by his human companions. Of these 2805 combinations, 764 occurred only once.

Kanzi's combinations are less frequent than single words (they account for 6 percent of all utterances during this period) and, in their present form, should not be regarded as grammatical. Yet when they do occur, all of the symbols employed typically add new elements of information to the situation. Thus, instead of forming combinations such as "Play me Nim play" (Terrace et al. 1979), Kanzi produces combinations like "Austin hamburger go" (with "go" being conveyed by gesture) to ask a companion to take some of the hamburger Kanzi is having for supper to Austin.

Table 16.1 compares Kanzi's 25 most frequent 2- and 3-item combinations with Nim's. It is clear that Kanzi provided additional information when he used longer utterances. Moreover, 36 percent of Kanzi's most frequent symbol combinations specified individuals other than himself as the beneficiary of action while *none* of Nim's combinations functioned in this manner. It is also the case that 64 percent of Nim's most frequent 2- and 3-word combinations were food requests, while none of Kanzi's were.

Formal Tests

The possibility of inadvertent cueing of apes continues to be a frequent concern both within and outside the field of ape language (Terrace 1979; Savage-Rumbaugh et al. 1980; Savage-Rumbaugh et al. 1983; Savage-Rumbaugh 1984a). Since no symbol training tasks are employed at any time with Kanzi, this possibility is much reduced, if not eliminated entirely. However, tests to control for cueing were nonetheless conducted with Kanzi because of the importance of assuring the validity of these findings. These tests probed both productive and receptive capacities using spoken English and lexigrams, as the ease with which Kanzi acquired symbols and the accuracy of his usage appeared to be directly related to his ability to demonstrate receptive comprehension of the spoken English word. Patterson (1981) reports that Koko also spontaneously

Table 16.1. Kanzi and Nim's Most Frequent 2- and 3-Item Combinations

	Kanzi's 2 Item Combinations	*Nim's 2 Item Combinations*	*Kanzi's 3 Item Combinations*	*Nim's 3 Item Combinations*
1.	chase person	play me	chase person#1 person#2	play me Nim
2.	person(g) chase(g)	me Nim	person#1 pat(g) person#2	eat me Nim
3.	chase(g) person(g)	tickle me	person#1 person#2 pat(g)	eat Nim eat
4.	person(g) pat(g)	eat Nim	person#1 chase person#2	tickle me Nim
5.	chase bite	more eat	person#1 grab person#2	grape eat Nim
6.	chase Kanzi	me eat	person#1 chase(g) person#2	banana Nim eat
7.	person(g) come(g)	Nim eat	person#1 person#2 chase	Nim me eat
8.	tickle ball	finish hug	kanzi chase person(g)	banana eat Nim
9.	bite person(g)	drink Nim	chase bite person(g)	eat me eat
10.	come(g) chase(g)	more tickle	person chase kanzi	me Nim eat
11.	ball tickle	sorry hug	person#1 grab(g) person#2	hug me Nim
12.	chase sue	tickle Nim	chase grab person(g)	yogurt Nim eat
13.	kanzi chase	hug Nim	person(g) chase kanzi	me more eat
14.	suprise money	more drink	person#1 person#2 bite	more eat Nim
15.	bite chase	eat drink	chase kanzi person(g)	finish hug Nim
16.	pat(g) person	banana me	person#1 tickle person#2	banana me eat
17.	kanzi grab	Nim me	person(g) kanzi chase	Nim eat Nim
18.	grab person	sweet Nim	person#1 tickle person#2	tickle me tickle
19.	chase bite	me play	kanzi person(g) chase	apple me eat

Table 16.1. *(continued)*

	Kanzi's 2 Item Combinations	*Nim's 2 Item Combinations*	*Kanzi's 3 Item Combinations*	*Nim's 3 Item Combinations*
20.	pat(g) this(g)	gum eat	chase five kanzi	eat Nim me
21.	chase come(g)	tea drink	chase person(g) kanzi	give me eat
22.	person go(g)	grape eat	pat(g) person#1 person#2	nut Nim nut
23.	ball pat(g)	hug me	bite chase person(g)	drink me Nim
24.	person bite	banana Nim	person#1 person#2 chase(g)	hug me hug
25.	chase tickle	in pants	sue bite person(g)	sweet Nim sweet

Note that the combinations differ in topic and structure, with Kanzi rarely requesting food. Kanzi's 3-item utterances frequently have a beneficiary or agent other than himself, and do not have repetitive use of the same word in a given combination. The use of (g) following any word indicates the use of a gesture to convey this information as opposed to a lexigram. Kanzi typically indicates which person is to be the agent and which the recipient by a pointing or touching gesture. Thus. all such gestures have been translated as "person" rather than to use the person's name.

began to comprehend English. However, Patterson's test of this skill did not identify the specific words Koko could respond to under controlled test settings. Patterson notes that Koko signs as her teachers speak, but again, no specific tests of this skill have been reported. Also, Fouts, Chown, and Goodin (1976) trained a chimpanzee to produce signs in response to spoken English stimuli. However, this involved massed training presentations of the same stimulus. Only 10 items were used and no controlled tests of the chimpanzee's ability to produce the signs in response to randomly presented English words were conducted.

No training was required to get Kanzi to participate in symbolic tests or to comprehend the questions being asked of him. The tests reported here occurred between 39 and 41 months of age and all items that met the concordance criterion as well as other items which Kanzi seemed to know, but used infrequently, were tested. (A few items, starred in table 16.2, could be tested only in the English-to-lexigram conditions, since they could not be readily depicted in photographs in an unambiguous manner, for example, "chase").

During testing, three items were arranged face up on a flat booklike opaque surface by experimenter A. This "test book" was then closed and given to experimenter B. Experimenter B knew the identity, but not the location, of the item that Kanzi was to be asked to select. Experimenter B opened the "test book" so that only Kanzi had visual access to its contents and asked Kanzi to select the test item. After Kanzi had made his selection, experimenter B looked at the test book to determine if Kanzi was correct. All test items and alternatives were randomly determined in advance of each trial, with the requirement that neither alternatives nor test items were repeated on consecutive trials and that each item served as an alternative as well as a test item.

Kanzi's ability to select lexigrams in response to the spoken English word, to select photos in response to the spoken English word, and to select photographs when shown lexigrams, were all tested in this manner. The results, by type of test and vocabulary item are shown in table 16.2. Out of a total of 185 stimulus and sample unique trials, Kanzi made only seven errors. Similar tests of comprehension were also given to Sherman and Austin. While they were able to select the correct photo when shown the lexigram (without error), they were at chance when the sample was a spoken English word and the alternatives were photographs or lexigrams. Sherman and Austin have been exposed to spoken English daily for eight years. (In familiar contexts and during established routines, Sherman and Austin convey the strong subjective impression of comprehending what is said to them. Yet controlled tests reveal that this capacity is far more limited than superficial impressions convey.) Twenty-four additional trials in the English-to-lexigram condition were administered to Kanzi, with experimenters A and B switching roles. The purpose of these additional trials was to make certain that Kanzi could respond to the English commands of more than one party. He was correct on all 24 trials.

Conclusion

These results indicate that the propensity of the pygmy chimpanzee to acquire primitive language skills exceeds that reported for other

Table 16.2. Kanzi's Performance During Blind Tests

Symbol	Test Type ABCD	Symbol	Test Type ABCD		Test Type ABCD
Apple	++++	Key	++++	Trailer	++−+
Austin	++++	M&M	++++	Treehouse	++++
Ball	++++	Matata	++++	Turtle	+++
Banana	++++	Melon	++++	Umbrella	+++
Bite	+**+	Milk	++++	Water	++++
Blackberry	++++	Mulika	++++		
Blanket	++++	Mushroom	++++		
Bread	++++	Mushroom trail	+**+		
Campfire	+++	Orange	++++		
Carrot	++++	Orange drink	++++		
Chase	+**+	Orange juice	++++		
Cheese	++++	Peaches	++++		
Cherry	++++	Peanut	++++		
Clover	++++	Pine needle	+++		
Coffee	+++	Pine cone	+++		
Coke	++++	Play-yard	+++		
Dog	++++	Raisin	++++		
Egg	++++	Refrigerator	+**+		
Grab	+**+	Sherman	+++		
Green bean	+++	Sour cream	+++		
Groom	+**+	Sue's office	+++		
Hamburger	++++	Surprise	+**+		
Hotdog	++++	Sweet potatoe	++++		
Ice	+++	Television	++++		
Jelly	++++	Tickle	+**+		
Juice	+++	Tomatoe	++++		

This table illustrates Kanzi's performance on the various items tested uder conditions which precluded cueing. In condition A, Kanzi was shown 3 lexigrams and was to select the lexigram that the teacher named aloud. In condition B, Kanzi was shown photographs and was to select the one that his teacher named aloud (items marked with asterisks were exluded because they were not readily depicted in a photograph.) In condition C, the teacher showed Kanzi a lexigram and asked him to select the picture which the lexigram represented. (The teacher did not say the name of the picture aloud.) Condition D lists all of the items that were in Kanzi's vocabulary at the time of the tests. It is noteworthy that many of the items were successfully responded to in the present test, even though they were not yet a part of Kanzi's vocabulary according to the stringent vocabulary criterion which was employed.

apes. Language acquisition in the pygmy chimpanzee seems to be accompanied (and presumably facilitated) by the ability to understand spoken English. By contrast, this ability is absent, or at best minimally present, in *Pan troglodytes,* while more adequate data is needed for the gorilla and orangutan. The reasons for these differences are not intuitively obvious. It is interesting, though, that the vocal repertoire of the pygmy appears to be more flexible than that of other apes (Savage-Rumbaugh 1984), which offers the possibility that learned vocalizations may play an important role in natural settings. Kanzi's linguistic skills continue to improve without training, but at a much slower rate than that seen in the normal child. These findings have important implications for the prevailing views regarding the evolution of language capacities.

Chapter 17

Ape-Language Research: Past, Present and Future

This chapter is written in collaboration with Duane Rumbaugh

Few research efforts within behavioral science, or within the history of science for that matter, have generated so much controversy as that associated with ape-language research. Strong differences of opinion, conclusion, and judgment have arisen both within and outside of the research laboratories. Notwithstanding these differences—which at times have been so strong as to have generated more heat than light—the data produced by the various ape-language research projects are, in sum, sufficiently powerful to compel their incorporation into the perspective of science in general and the fields of psychology, anthropology, and biology in particular.

There seems no doubt that the "language of the apes" will leave an indelible, and in balance positive, marker on the history of behavioral science.

Retrospectively, we can see that one of the most serious complications that affected the viability and effectiveness of the earliest ape projects was the lack of a standard for the definitions of a number of key words and concepts—*language, sign, symbol, word, vocabulary, grammar, reference, representation,* and *syntax,* to name but a few. Research on language acquisition by children was proceeding satisfactorily, or so it seemed, without researchers being overly concerned about whether or not children understood in a meaningful sense the words they used (e.g., Brown and Bellugi 1964). Scholarly interest was riveted upon the incorporation of words into syntactical structures, as though it were form rather than function that made human language unique.

This focus on form led many to trivialize the distinction between sign and symbol. For example, Passingham (1982) in a recent discussion of the evidence for ape language observes that: "The chimpanzee signals that it is [about] to attack by glaring and opening its mouth; the facial expression is a sign but not a symbol. But it is not so obvious that the distinction between them [the sign and the symbol] is important" (p. 224). While Passingham is right to suggest that either a "glare" or a symbol (such as "bite") may come to signal an attack, he overlooks a most important point: That is, while an angry chimpanzee will, without forethought, "glare" at an opponent when the feeling of anger provokes him, the same chimpanzee will not, also without forethought, produce the symbol "bite" simply because he is angry.

Indeed, the question must be raised as to why a chimpanzee would ever come to say "bite" at all. What would such a symbol add to the aggressive messages already displayed by the glare, bulged lips, raised shoulders, and stiff gait? It could be argued that saying "bite" would remind the recipient of what the sender might do, yet such reminding hardly seems necessary. A chimpanzee could, of course, say "bite" *before* he feels angry as a description of how he *will* feel should the recipient continue with the present course of action. If that did occur, the chimpanzee would be making a statement about his future feelings, not his present ones. This would in itself be no small feat and should not be viewed as simply another form of expressing one's mood and equivalent to glaring. Rather it would serve as a statement expressing possible future action. In such a case, symbolic encoding would be used specifically to portray a possible behavioral alternative *in advance* of feeling the provocation. As such, the difference between the use of a sign and a symbol is not only obvious; it is significant at a fundamental level. To signal "how one feels" is very different from symbolically expressing how one might come to feel. The use of a symbol in such a way reflects a degree of forethought that goes beyond the expression of pure emotional reactions.

A second source of confusion surrounding ape-language work was generated by the misleading assumption that the ape's natural modes of communication were simplistic, inflexible, and incapable of permitting simple intentional communications. By contrast, human communication (or language) was viewed as flexible, stimulus free, and intentional (Hockett and Altman 1968; Wilson 1975). When the Gardners first began their work with Washoe, it was thought that no communication system short of language could be open, flexible, or acquired to any significant degree.

The reports of Van Lawick-Goodall (1968), van Hoof (1974) and de Waal (1982) regarding the complex nature of the gestural communicative behaviors of group-living chimpanzees had not yet achieved wide impact. There did not seem to be a general awareness that a nonlinguistic form of communication could be intentional. The earlier work of Yerkes and Yerkes (1929), Crawford (1937), and Hayes and Hayes (1951) had certainly provided evidence that chimpanzees could produce communicative innovations, but these studies had been ignored by most discussions of the nature of primate communication (Altman 1967; Sebeok 1967). Even now it is not generally recognized that the degree of intentionality which characterizes ape communication goes far beyond that found in the more familiar mammalian species such as dogs and monkeys. Thus when it was reported that Washoe used an ASL gesture to initiate even something as simple as a game of tickle, this was viewed as significant.

Following these reports by the Gardners, the orderly progress of the field was made problematic by the uncritical acceptance of reports by ape-buffs. Desmond (1979) stated it well: "The overriding urge to assist apes in giving man his comeuppance, as if this were a Darwinian imperative, has boomeranged to the detriment of the ape, who is now judged according to an impossible human standard which should never have been set" (pp. 49–50).

Now, only a few years after Desmond's statements, it is clear that the ape can use symbols competently and with meaning. The ape is in the language domain, a behavioral domain that is a continuum—not a dichotomy.

Language learning studies with marine mammals have also shed light upon animal-language research. Herman (1980; Herman, Richards, and Wolz 1984) and his associates have presented strong data that dolphins are adroit at acquiring receptive language skills. They respond accurately to novel commands (given either by hand signs or by computer-generated sounds) by carrying out a variety of specific behaviors (for example, fetch the frisbee to the hoop or the hoop to the frisbee). Such work provides evidence of syntactical comprehension. Working with sea lions, Schusterman and Krieger (1984) report strong corroborating data.

More than 20 years have elapsed since Project Washoe rekindled current interest in questions regarding animals' capacities for language. More than any research with children, the research with the great apes, dolphins, and sea lions has served to stimulate attention and thought regarding the components, antecedents, and dynamics of language competence. It has also served to emphasize the merits of comparative research. Although any of a number of scientists (biologists, anthropologists, experimental psychologists, primatologists, etc.) might have been led to the field of ape-language research, it should not go without notice that the overwhelming majority of researchers in this field are comparative psychologists.

The generation of knowledge through empirical behavioral research can be as dry and as monotonous as the dust bowls of the 1930s or as exciting and as challenging as the line of research which gave rise to studies of animals' capacities for language. New lines of inquiry will necessarily encounter opposition, for they inherently will, and should, serve to contradict much of what has been "thought." New lines of inquiry will result in errors as well as in the generation of new insights.

The progress of science is not without both risks and mistakes. Not all new lines of inquiry and thought will survive the test of time. With that said, it should be emphasized that the current revived interest and study of animals' potential for language has survived better than two decades of effort, challenge, and criticism for a very good reason. The effort has taught humankind a great deal about the basic requisites of language, about the continuua which link operant learning to the competent use of symbols, about the ways in which nonverbal and verbal behaviors support the coordination of interindividual behaviors, and about apes as an advanced form of life that psychologically is not so different from us as their appearance might otherwise suggest. In the future, the field will be limited only by the numbers of persons who will make a career-long commitment and who will have availed to them the basic logistics for such commitment to be fruitful. Even if research in this field of inquiry were to suddenly stop today, we are confident that the data are sufficiently rich to entice the attentive thought and wonder of persons for the indefinite future.

We are confident that research regarding animals' potential for language is here to stay. Its veins of data will sustain the flow of the intellect upon which the vitality of all lines of scientific research ultimately depend. The fruits of the field will continue to feed substantively toward an understanding of language in its basic dimensions, and consequently will increasingly benefit those humans who, for various reasons, have an impaired ability to acquire language.

The enormous strides made within the field of child language learning in the past decade are beginning to have important ramifications for the studies of language in apes. Recent investigations of the cognitive and social development of normal children before and during the one-word stage are helping to elucidate the controversies surrounding ape language and are providing precise, meaningful, and measurable points of comparison between ape and child for future research (Bates 1976, 1979; Braumwald 1978; Greenfield and Smith 1976; Lock 1978, 1980; McShane 1980; Peters 1983).

Particularly important in this regard is work on the emergence of symbols in human children by Bates (1979) (see chapter 2). Bates (1979) has separated the acquisition of language into four major milestones:

(a) The emergence of intentional communication before speech (characterized by conventional gestures and preverbal sounds).
(b) The onset of the first words.
(c) The transition from one-word to multi-word speech.
(d) The transition from multi-word speech to true syntactic constructions.

The first two periods are the most relevant to the field of ape language. Bates (1979) presents strong evidence to support a model of language acquisition which posits that the co-development of the capacities of imitation, tool use, and communicative intent forges the *necessary prerequisite* for the emergence of the first words. Bates also points out that there is reason to believe that phylogenetically, as well as ontogenetically, these three capacities predate the emergence of language. She suggests that when each of these three capacities reaches a critical level, symbolic representation spontaneously and inevitably emerges (see figure 17.1).

In their natural state, common chimpanzees possess minimal behavioral capacities in each of these domains. Furthermore, as earlier chapters have illustrated, it is precisely in the domains of imitation, tool use, and communicative intent that we have attempted to expand the capabilities of Sherman and Austin. We believe that the general effect of the various types of social and cognitive stimulation that Sherman and Austin have received has

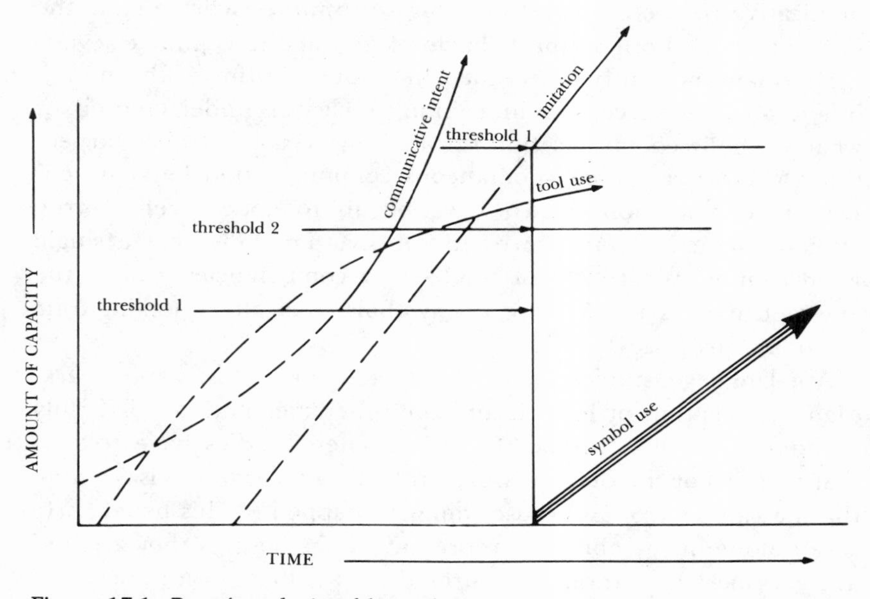

Figure 17.1. Bates' analysis of how the processes of imitation, tool use, and communicative intent emerge and develop independently until all three finally reach a crucial stage, at which time, they converge and symbolization proper emerges. The dotted lines represent subthreshold degrees of capacity and the solid lines represent degrees of capacity beyond threshold. The training which Sherman and Austin received focused formally on the development of tool use and communicative intentionality and informally upon imitation.

been to push these three capacities far beyond the level they would have reached under natural conditions. The result has been the spontaneous emergence of a symbolic capacity comparable to that which normally occurs between one and two years of age in our own species. By contrast, Kanzi, the pygmy chimpanzee, who is more advanced, at least in the domain of communication, needed only exposures, not training, to achieve a level of symbolic skill comparable to that attained by Sherman and Austin.

Clearly the previously held dichotomies between animal and man are blurred by the chimpanzee. As Sherman and Austin moved from the simplest discrimination tasks to complex spontaneous communications, it became increasingly apparent that they were continually learning to do far more than they were being taught. The issue of whether or not they had achieved "true" human language was never the goal. The goal was to improve their communicative competence and in doing so to more clearly define the skills involved, both at the behavioral and at the cognitive levels.

Sherman and Austin succeeded in communicating with one another, and we succeeded in gaining a clearer understanding of what symbolic communication is. Much more is left to be learned, though. How can more spontaneous communication be achieved? Do the combinations which have begun to appear reflect true attempts to use several words to convey something that a single word cannot? What sorts of single-word competencies lead to the emergence of syntax? How do symbol skills alter nonlinguistic reasoning processes?

Ape-language studies have already left an unmistakable mark. Man's conception of himself and the other members of his family of hominids will never be the same. These studies have told us that man is not the only creature on this planet who tends to think that he can control and cause things to happen for his benefit. He is not alone in the ability to represent things both as they are and as he wishes they were. He shares these attributes with apes, and even if apes are not able to express such complex things as man, we now know that the gap between ape and man is far narrower than we previously supposed. Our attempts to talk with apes have brought us much closer to them and we can for the first time, see them more clearly for what they are and, in turn, ourselves for what we are.

Notes

3. The Project and The System

1. Child Project is the companion project to the Animal Model Project. We and others have been studying similar language acquisition processes in profoundly and severely retarded children concurrently with the work using ape subjects since this project was established in 1975.

2. We have used both pointing boards and magnetic symbols with Sherman and Austin and both lack the simplicity, the attention orienting properties, the accuracy, and the interactive nature of the keyboard system. In general, the chimpanzees behave as though they find these other media cumbersome and unattractive, and when they use them they tend to become increasingly imprecise. Because the keyboard symbols light so effortlessly, response output itself is minimally demanding, an important factor when symbolic communication is not already a well developed capacity.

3. We are experimenting with different symbol types in the Child Project and have also used a printed English vocabulary with one human subject (Smith and White 1981). We are currently evaluating the relative ease with which different sorts of symbol systems can be acquired, but whatever the findings on this topic, it is still the communication system *per se,* not the symbols themselves, which remains the critical variable in our approach to language training.

4. Apes Who Don't Learn What You Try To Teach Them

1. The first lexigram was paired with four irrelevant lexigrams. These remained lit after the second and third lexigrams were introduced. They were rarely selected by the chimpanzees after the first lexigram was acquired. All later errors were between food lexigrams themselves rather than between food symbols and irrelevant lexigrams.

2. It is important to note here that the behaviors are not occasioned each time the chimpanzee sees or comes in contact with

the objects A, B, and C. Rather, they are occasioned by the teacher's acting upon the objects.

5. Talking To Teachers Instead of Machines

1. In Asano and colleagues' (1982) system, which was physically patterned after ours but procedurally functioned more like Premack's, the chimpanzees could respond only on experimenter structured occasions.

2. Later in training we did allow the chimpanzees to indicate which food they wanted to receive.

3. Note that, as previously, the banana was in the left dispenser and the beancake was in the right dispenser.

6. The Functions of Symbols

1. From this point in training onward we worked only with Sherman and Austin. Given the size of the teaching staff (4) we felt that more could be accomplished by concentrating on fewer animals. Kenton was sent to Stony Brook under the care of Dr. Randall Sussman who had been searching diligently for a cooperative chimpanzee subject who would wear a jacket that would permit the detailed collection of data on muscle physiology. Erika was returned to the main center, where she was housed with other chimpanzees her age and assigned to a long-term breeding program.

2. Software changes were made so that the machine no longer responded to requests. We made this change because we wanted the chimpanzee to view the dispensing of food as a reward for correctly naming the displayed food.

3. It might be objected that Sherman could still simply match his symbol selection to the teacher's on the basis of physical match-to-sample. However, the teacher knew at this point that neither Sherman nor Austin were able to solve a match-to-sample task using lexigrams as stimuli. While they readily handled match-to-sample tasks with three dimensional objects, and with colors, they performed at chance levels when the stimuli were lexigrams.

8. Giving, Taking, Sharing

1. Since Sherman and Austin were allowed to spend all the time they wanted together, being allowed to see one another through

a window did not seem particularly exciting to them, and they did not mind being apart at all as long as they had a teacher whom they liked with them.

9. Tools

1. Strawberry drink was called lemonade since lemonade was assigned, by Sherman, before strawberry drink. Pineapple was called melon. Both of these errors reflected a tendency to choose the symbol of the most recently assigned food or drink when presented with a second new food or drink.

2. Later, after learning tool names, they also spontaneously assigned the lexigrams for lever and string.

3. An accuracy level of 55% correct (overall) was taken by the Gardners as evidence for Washoe's naming capacity.

4. Of course, after the data in table 9.4 were gathered, Sherman and Austin continued to use their tools in many settings.

5. It will be noted that the control procedure of deactivating the keyboards has frequently been used with Austin and Sherman. The value and elegance of this control lies in the ability to completely eliminate, from a given communicative setting, the presumed mode of symbolic information transmission. Since this is the *only* thing that is eliminated from the setting, if there are any other extraneous variables operating to facilitate performance (inadvertent cueing by teachers, orders, gestures on the part of the chimpanzees, etc.) they should continue to operate. If, however, communication cannot take place, then the use of symbols by the chimpanzees can satisfactorily be presumed to be the responsible vector of success. Most communication systems used with apes do not have such a control available to them because of the nature of the modality.

6. When the "keyboard off" control is used, it is still possible for the chimpanzees to indicate the correct symbol by touching it—even though the symbol does not light. However, because of the small size of the symbols and the distance between the two rooms, the chimpanzees cannot tell which symbols are being touched unless they are displayed on the large projectors located above each keyboard.

7. Trials with the keyboard on alternated with the keyboard off trials during this test.

10. The Intermeshing of Gesture and Symbol

1. We discarded 27 exchanges because of keyboard failure, noise made by the teachers during an exchange, or other extra-task interference which could have potentially affected the nature of the nonverbal communication occurring between the chimpanzees. The accuracy of the discarded exchanges was equal to the overall accuracy rate.

11. When a Lexigram Becomes a Symbol

1. Lana was asked to review the names of all items in her vocabulary (shoe, box, etc.) several times a week.
2. Before training, blind tests, also shown in figure 11.2, were given on all vocabulary items to be used during categorization training.

12. Anecdotes

1. Sufficient data with regard to Sherman and Austin's capacity to use individual symbols have been presented in previous chapters.
2. This example is offered for those readers who might ask "What is the difference between what an ape does with signs and what my dog does when he brings me his lead and wants to go for a walk?" We have yet to see a dog who enjoys looking at *National Geographic,* let alone one who spontaneously uses pictures to communicate his wishes.

13. Video Representations of Reality

1. Perhaps Sarah did react clearly to these television images; in any case though, it would be helpful if some descriptions of her behavior were available, particularly of those behaviors which led Premack to believe that Sarah was responding to the television as a representation of real space and action.
2. Watching their lips manipulate food is a favored activity of chimpanzees. Their cranio-facial anatomy permits this in a way our own does not.

14. Making Statements

1. In order to preclude the possibility that the chimpanzees might simply fall into an order strategy, such as always naming the key first, the stick second, the blanket third, etc., the set of objects was shuffled and relocated each session. No order strategy—other than avoiding items of which they were unsure—appeared.

2. Foods were represented only by photographs because the presence of the actual foods resulted in occasional statements about which food was going to be eaten, rather than which item would be given to the teacher.

15. Implications for Language Intervention Research

1. Subjects 6 and 10 left the institution for reasons not associated with the research effort. Subjects 7 and 8 exhibited severe aggressive behavior towards themselves and/or others that the research staff was not equipped to handle. Subjects 9 and 13 failed to make any progress.

References

Abrahamsen, A., M.A. Romski, and R.A. Sevcik. 1985. Effects of symbol instruction on other domains of development. Manuscript in preparation.

Altmann, S.A. 1967. The structure of primate social communication. In S.A. Altmann, ed. *Social Communication Among Primates*, pp. 325–362. Chicago: University of Chicago Press.

Asano, T., T. Kojima, T. Matsuzawa, K. Kubota and K. Mutofushi. 1982. Object and color naming in chimpanzees *(Pan troglodytes). Proceedings of the Japanese Academy* 58:118–122.

Bates, E. 1976. *Language and Context: The Acquisition of Pragmatics.* New York: Academic Press.

Bates, E. 1979. *The Emergence of Symbols, Cognition, and Communication in Infancy.* New York: Academic Press.

Benedict, N. 1979. Early lexical development: Comprehension and production. *Journal of Child Language* 6:183–200.

Bennett, J. 1964. *Rationality.* London: Routledge and Kegan Paul.

Bennett, J. 1978. Some remarks about concepts. *The Brain and Behavioral Sciences* 1:557–590.

Bensberg, G.J. and C.K. Seigelberg. 1976. Definitions and prevalence. In L.L. Lloyd, ed. *Communication Assessment and Intervention Strategies,* pp. 33–72. Baltimore: University of Park Press.

Benson, F.D. 1979. Aphasia. In K.M. Heilman and E. Valenstein, eds. *Clinical Neuropsychology,* pp. 22–58. Oxford: Oxford University Press.

Bloom, L., L. Hood and P. Lightbrown. 1974. Imitation in language development: If, when, and why. *Cognitive Psychology* 6:380–420.

Braunwald, S.R. 1978. Context, word and meaning: Towards a communicational analysis of lexical acquisition. In A. Lock ed. *Action, Gesture, and Symbol: The Emergence of Language,* pp. 485–527. London: Academic Press.

Brodigan, D.L. and G.B. Peterson. 1976. Two choice conditional discrimination performance of pigeons as a function of reward expectancy, prechoice delay, and domesticity. *Animal Learning and Behavior* 4:121–124.

Brown, J.S. and U. Bellugi, eds. 1964. The Acquisition of Language. *Monographs of the Society for Research in Child Development,* vol. 29, series 92.

Brown, R. 1973. *A First Language: The Early Stages.* Cambridge: Harvard University Press.

Bruner, J.S. 1974/75. From communication to language: A psychological perspective. *Cognition* 3:255–287.

Bullowa, M. 1979. Introduction: Prelinguistic communication: A field for scientific research. In M. Bullowa ed. *Before Speech: The Beginning of Interpersonal Communication,* pp. 1–62. London: Cambridge University Press.

Calculator, S. and C. Dollegan. 1982. The use of communication boards in a residential setting: An evaluation. *Journal of Speech and Hearing Disorders* 47:281–291.

Carrier, J. 1974. Application of functional analysis and a nonspeech response mode to teaching language. In L. McReynolds, ed. *Developing Systematic Procedures for Training Children's Language* (Monograph #18), pp. 47–95. Rockville, Md.: American Speech and Hearing Association.

Chalkey, M.A. 1982. Language as a social skill. In S. Kuczaj, II. ed. *Language Development, Volume 2: Language, Thought, and Culture,* pp. 75–111. Hillsdale, N.J.: Lawrence Erlbaum.

Chevalier-Skolnikoff, S. 1973. Facial expression of emotion in nonhuman primates. In P. Ekman, ed. *Darwin and Facial Expression: A Century of Research In Review,* pp. 11–89. New York: Academic Press.

Chomsky, N. 1968. *Language and Mind.* New York: Harcourt Brace Jovanovich.

Crawford, M.P. 1937. The cooperative solving of problems by young chimpanzees. *Comparative Psychological Monographs* 14:1–88.

Deich, R.F. and P.M. Hodges. 1977. *Language Without Speech.* New York: Brunner/Mazel Publishers.

Desmond, A.J. 1979. *The Ape's Reflexion.* New York: Dial Press.

De Waal, F. 1982. *Chimpanzee politics. Power and Sex Among Apes.* New York: Harper & Row.

Epstein, R., R.P. Lanza and B.F. Skinner. 1980. Symbolic communication between two pigeons *(Columba livia domestica). Science* 207:543–545.

Ettlinger, G.A. 1982. A comparative evaluation of the cognitive skills of the chimpanzee and the monkey. Paper presented at the Harry Frank Guggenheim Conference, June 2–4, Columbia University, New York.

Fogel, A. 1977. Temporal organization in mother-infant face to face interaction. In H.R. Schaffer, ed. *Studies in Mother-Infant Interaction,* pp. 119–152. London: Academic Press.

Fouts, R.S. 1972. The use of guidance in teaching sign language to a chimpanzee. *Journal of Comparative and Physiological Psychology* 80:515–522.

Fouts, R.S. 1973. Acquisition and testing of gestural signs in four young chimpanzees. *Science* 180:973–980.

Fouts, R.S. 1974a. Capacities for language in the great apes. *Proceedings of the 18th International Congress of Anthropological and Ethnological Sciences,* pp. 371–390. The Hague: Mouton.

Fouts, R.S. 1974b. Language: Origins, definitions, and chimpanzees. *Journal of Human Evolution* 3:475–482.

Fouts, R.S. 1975. Communication with chimpanzees. In Kurth and Eibl-Eibesfeld, eds. *Hominisation und Verhalten.* Stuttgart: Gustav Fisher.

Fouts, R.S., W. Chown and L. Goodin. 1976. Transfer of signed responses in American Sign Language from vocal English stimuli to physical object stimuli by a chimpanzee *(Pan). Learning and Motivation* 7:458–475.

Fouts, R.S., W. Chown and G. Kimball. 1976. *Comprehension and Production of American Sign Language by a Chimpanzee (Pan).* Paper presented at International Congress of Psychology, 21st, Paris.

Fouts, R.S., D.H. Fouts and D. Schoenfeld. 1984. Sign language conversational interaction between chimpanzee. *Sign Language Studies* 42:1–12.

Fouts, R.S. and R.L. Mellgren. 1976. Language, signs, and cognition in the chimpanzee. *Sign Language Studies* 13:319–346.

Fouts, R.S., R. Mellgren and W. Lemmon. 1973. *American Sign Language in the Chimpanzee: Chimpanzee to Chimpanzee Communication.* Paper presented at Midwestern Psychological Association Meeting, Chicago.

Gardner, B.T. and R.A. Gardner. 1971. Two-way communication with an infant chimpanzee. In A. Schrier and F. Stollnitz, eds. *Behavior of non-human primates,* 4:117–183. New York: Academic Press.

Gardner, B.T. and R.A. Gardner. 1974a. Comparing the early utterances of a child and chimpanzee. In A. Pick, ed. *Minnesota Symposium on Child Psychology,* 8:3–23. Minneapolis: University of Minnesota Press.

Gardner, B.T. and R.A. Gardner. 1974b. Teaching sign language to a chimpanzee. VII: Use of order in sign combinations. *Bulletin of the Psychonomic Society* 4:264.

Gardner, B.T. and R.A. Gardner. 1975. Evidence for sentence constituents in the early utterances of child and chimpanzee. *Journal of Experimental Psychology: General* 104:244–267.

Gardner, B.T. and R.A. Gardner. 1978a. *Emergence of language.* Paper presented at the American Association for the Advancement of Science, Washington, D.C.

Gardner, R.A. and B.T. Gardner. 1969. Teaching sign language to chimpanzees. *Science* 165:664–672.

Gardner, R.A. and B.T. Gardner. 1972. Communication with a young chimpanzee: Washoe's vocabulary. In R. Chauvin, ed. *Modèles animaux du comportement humain,* pp. 241–264. Paris: National de la Recherche Scientifique.

Gardner, R.A. and B.T. Gardner. 1973. Teaching sign language to the chimpanzee, Washoe. (16mm sound film). State College, Pa.: Psychological Cinema Register.

Gardner, R.A. and B.T. Gardner. 1978b. Comparative psychology and language acquisition. In K. Salzinger and F. Dennaale, eds. *Psychology: The State of the Art. Annals of the New York Academy of Sciences* 309:37–76.

Goldin-Meadow, S. and H. Feldman. 1977. The development of language-like communication without a language model. *Science* 197:401–403.

Goodglass, H. 1980. Disorders of naming following brain injury. *American Scientist* 68:647–655.

Greenfield, P.M. 1973. Who is "Dada?" Some aspects of the semantic and phonological development of a child's first words. *Language and Speech* 16:14–43.

Greenfield, P.M. 1978. Informativeness, presupposition, and semantic choice in single-word utterances. In N. Waterson and C. Snow, eds. *Development of Communication: Social and Pragmatic Factors in Language Acquisition*, pp. 443–452. London: Wiley.

Greenfield, P.M. and E.S. Savage-Rumbaugh. 1984. Perceived variability and symbol use: A common language-cognition interface in children and chimpanzees *(Pan troglodytes). Journal of Comparative Psychology* 98:201–218.

Greenfield, P.M. and J.H. Smith. 1976. *The Structure of Communication in Early Language Development.* New York: Academic Press.

Grice, H.P. 1968. Utterer's meaning, sentence-meaning, and word meaning. *Foundations of Language* 4:226–242.

Griffin, D.R. 1976. *The Question of Animal Awareness. Evolutionary Continuity of Mental Experience.* New York: The Rockefeller Press.

Guess, D., W. Sailor and D. Baer. 1978. Children with limited language. In R.L. Schiefelbusch, ed. *Language Intervention Strategies,* pp. 101–143. Baltimore: University Park Press.

Guralnick, M.J. and E. Weinhouse. 1984. Peer related social interactions of developmentally delayed young children: Development and characteristics. *Developmental Psychology* 20:815–827.

Guthrie, E.R. 1952. *The Psychology of Learning.* New York: Harper & Row.

Harlow, H., J. Gluck and S. Suomi. 1972. Generalization of behavioral data between nonhuman and human animals. *American Psychologist* 27:709–716.

Harris, D. and G. Vanderheiden. 1980. Enhancing the development of communicative interaction. In R.L. Schiefelbusch, ed. *Nonspeech Language and Communication: Analysis and Intervention,* pp. 227–257. Baltimore: University Park Press.

Hayes, K. J. and C. Hayes. 1951. The intellectual development of a home-raised chimpanzee. *Proceedings of the American Philosophical Society 95*:105.

Hayes, K.J. and C. Nissen. 1971. Higher mental functions of a home raised chimpanzee. In A.M. Schrier and F. Stollnitz, eds. *Behavior of Nonhuman Primates,* 4:59–115. New York: Academic Press.

Heilman, K.M., R. Scholes and R.T. Watson. 1975. Auditory affective agnosia: Disturbed comprehension of affective speech. *Journal of Neurology, Neurosurgery and Psychiatry* 38:69–72.

Herman, L.M. 1980. Cognitive characteristics of dolphins. In L.M. Herman, ed. *Cetacean Behavior: Mechanisms and Functions,* pp. 363–429. N.Y.: Wiley.

Herman, L.M., D.G. Richards and J.P. Wolz. 1984. Comprehension of sentences by bottlenosed dolphins. *Cognition* 16:129–219.

Hill, J.A. 1978. Apes and language. *Annual Review of Anthropology* 7:89–112.

Hockett, C.F. and S.A. Altmann. 1968. A note on design features. In T. Sebeok, ed. *Animal Communication,* pp. 61–72. Bloomington: Indiana University Press.

Jorden, C. and H. Jorden. 1977. Versuche zur Symbol-Ereignis-Yerknupfung bei einer Zwergschimpansen (*Pan paniscus,* Scwarz, 1929). *Primates* 18:181–186.

Keane, V. 1972. The incidence of speech and language problems in the mentally retarded. *Mental Retardation* 72:3–5.

Kellogg, W.N. and L.A. Kellogg. 1933. *The Ape and the Child. A Study of Environmental Influence Upon Early Behavior.* New York: Whittlesey House.

Kendon, A. 1973. The role of visible behavior in the organization of social interaction. In M. van Cranach and I. Vine, eds. *Social Communication and Movement: Studies of Interaction and Expression in Man and Chimpanzee,* pp. 29–74. London: Academic Press.

King, M.C. and A.C. Wilson. 1975. Evolution at two levels in humans and chimpanzees. *Science 186*:107–116.

Kohler, W. 1925. *The Mentality of Apes.* New York: Harcourt, Brace Co.

Kohts, N. 1923. Untersuchungen uber die Erkenntnisfahigkeiten des Schimpansen. Aus dem Zoopsychologischen Laboratorium des Museum Darwinianum in Moskau.

Lieberman, P., E.S. Crelin, and D.H. Klatt. 1972. Phonetic ability and anatomy of the newborn and adult human, Neanderthal man and adult chimp. *American Anthropologist* 74:287–307.

Limber, J. 1977. Language in child and chimp? *American Psychologist* 32:280–295.

Lloyd, L. and G. Karlan. 1984. Nonspeech communication symbols and systems: Where have we been and where are we going? *Journal of Mental Deficiency Research* 28:3–20.

Lock, A. 1978. *Action, Gesture and Symbol: The Emergence of Language.* New York: Academic Press.

Lock, A. 1980. *Guided Reinvention of Language.* London: Academic Press.

McGrew, W.C. 1975. Patterns of plant food sharing by wild chimpanzees. In S. Kondo, M. Kawai, and A. Ehara, eds. *Contemporary Primatology.* Basel: Karger Press.

McShane, J. 1980. *Learning How To Talk.* Cambridge: Cambridge University Press.

Mason, W.A. and J.H. Hollis. 1962. Communication between young Rhesus monkeys. *Animal Behavior* 103:211–221.

Mathieu, M., N. Daudelin, Y.G. Dagenais and T.G. Decarie. 1984. Piagetian causality in two house-reared chimpanzees *(Pan troglodytes). Canadian Journal of Psychology* 34(2):179–186.

Meddin, J. 1979. Chimpanzees, symbols and the reflective self. *Social Psychology Quarterly* 42:99–109.

Menzel, E.W. 1974. A group of young chimpanzees in a one-acre field. In A.M. Schrier and F. Stollnitz, eds. *Behavior of Nonhuman Primates,* 5:83–153. New York: Academic Press.

Menzel, E.W. 1975. Natural language of young chimpanzees. *New Scientist* 65:127–130.

Menzel, E.W. and S.T. Halperin. 1975. Purposive behavior as a basis for objective communication between chimpanzees. *Science* 189:652–654.

Miles, H.L. 1983. Apes and language: The search for communicative competence. In J. De Luce and H.T. Wilder, eds. *Language in Primates.* New York: Springer-Verlag.

Nelson, K. 1977. First steps in language acquisition. *Journal of the American Academy of Child Psychiatry,* 16:563–583.

Newson, J. 1978. Dialogue and development. In A. Lock, ed. *Action, Gesture, and Symbol: The Emergence of Language,* pp. 31–42. New York: Academic Press.

Nissen, H.W., A.H. Riesen and V. Nowlis. 1938. Delayed response and discrimination learning by chimpanzees. *Journal of Comparative Psychology* 26:361–386.

Parkel, D., R. White and H. Warner. 1977. Implications of the Yerkes technology for mentally retarded human subjects. In D.M. Rumbaugh, ed. *Language Learning by a Chimpanzee: The LANA Project,* pp. 273–283. New York: Academic Press.

Passingham, R. 1982. *The Human Primate.* Oxford: W.H. Freeman.

Pate, J.L. and D.M. Rumbaugh. 1983. The language-like behavior of Lana chimpanzee: Is it merely discrimination and paired-associate learning? *Animal Learning and Behavior* 11:134–138.

Patterson, F. 1978. The gestures of a gorilla: Language acquisition in another pongid. *Brain and Language* 5:72–97.

Patterson, F. 1981. *Conversations with a Gorilla.* New York: Holt, Rinehart, and Winston.

Pea, R. 1982. Origins of verbal logic: Spontaneous denials by two- and three-year olds. *Journal of Child Language* 9:597–626.

Peters, A.M. 1983. *The Units of Language Acquisition.* Cambridge: Cambridge University Press.

Peterson, G.B. 1984. How expectancies guide behavior. In H.L. Roitblat, T.G. Bever, and H.S. Terrace, eds. *Animal Cognition: Proceedings of the Harry Frank Guggenheim Conference, June 2–4, 1982* (pp. 135–148). Hillsdale, New Jersey: Lawrence Erlbaum Associates.

Peterson, G.B. and M.A. Trapold. 1980. Effects of altering outcome expectancies on pigeons' delayed conditional discrimination performance. *Learning and Motivation* 11:267–288.

Peterson, G.B., R.L. Wheeler, and M.A. Trapold. 1980. Enhancement of pigeons' conditional discrimination performance by expectancies of reinforcement and nonreinforcement. *Animal Learning and Behavior* 8:22–30.

Petitto, L.A. and M.S. Seidenberg. 1979. On the evidence for linguistic abilities in signing apes. *Brain and Language* 8:162–183.

Piaget, J. 1954. *The Construction of Reality in the Child.* New York: Ballantine.

Plooij, F. 1978. Some basic traits of language in wild chimpanzees. In A. Lock, ed. *Action, Gesture, and Symbol: The Emergence of Language,* pp. 111–132. London: Academic Press.

Premack, D. 1971. On the assessment of language competence in the chimpanzee. In A.M. Schrier and F. Stollnitz, eds. *Behavior of Nonhuman Primates,* 4:185–228. New York: Academic Press.

Premack, D. 1972. Language in chimpanzee? *Science* 172:808–822.

Premack, D. 1976. *Intelligence in Ape and Man.* Hillsdale, N.J.: Lawrence Erlbaum.

Premack, D. and G. Woodruff. 1978. Does the chimpanzee have a theory of mind? *The Behavioral and Brain Sciences* 1:515–526.

Rachlin, H.C. 1976. *Introduction to Modern Behaviorism.* New York: W.H. Freeman.

Rice, M. 1980. *Cognition to Language: Categories, Word Meanings, and Training.* Baltimore: University Park Press.

Ristau, C.A. and D. Robbins. 1982. Language in the great apes: A critical review. *Advances in the Study of Behavior* 12:141–253.

Romski, M.A. and R.A. Sevcik. 1985. A framework for the establishment of symbolic communication in persons with severe retardation. Manuscript in preparation.

Romski, M.A., R.A. Sevcik and S.E. Joyner. 1984. Nonspeech communication systems: Implications for mentally retarded children. *Topics in Language Disorders* 5:66–81.

Romski, M.A., R.A. Sevcik and D.M. Rumbaugh. 1985. Retention of symbolic communication skills in five severely retarded persons. *American Journal of Mental Deficiency* 89:441–444.

Romski, M.A., R.A. White and E.S. Savage-Rumbaugh. 1982. Language training using communication boards: Some special considerations. Paper presented at the meeting of the American Association on Mental Deficiency, Boston.

Rosch, E. and B.B. Lloyd, eds. 1978. *Cognition and Categorization.* Hillsdale, N.J.: Erlbaum.

Rumbaugh, D.M. ed. 1977. *Language Learning by a Chimpanzee: The LANA Project.* New York: Academic Press.

Rumbaugh, D.M. and T.V. Gill. 1976. Lana's mastery of language skills. In H. Steklis, S. Harnad and J. Lancaster, eds. *Origin and Evolution of Language and Speeches. Annals of the New York Academy of Sciences* 280:562–578.

Rumbaugh, D.M. and T.V. Gill. 1976. Language and the acquisition of language-type skills by a chimpanzee *(Pan).* In K. Salzinger, ed. *Psychology in Progress. Annals of the New York Academy of Sciences* 270:90–123.

Rumbaugh, D.M., T.V. Gill and E.C. von Glasersfeld. 1973. Reading and sentence completion by a chimpanzee. *Science* 182:731–733.

Rumbaugh, D.M., E.C. von Glasersfeld, H. Warner, P. Pisani, T.V. Gill, J.V. Brown and C.L. Bell. 1973. A computer-controlled language training system for investigating the language skills of young apes. *Behavior Research Methods and Instrumentation* 5:385–392.

Savage, E.S. 1975. *Mother-infant behavior in group-living captive chimpanzees.* Doctoral dissertation. University of Oklahoma.

Savage, E.S. and B.J. Wilkerson. 1978. Socio-sexual behavior in *Pan paniscus* and *Pan troglodytes:* A comparative study. *Journal of Human Evolution* 1:327–344.

Savage-Rumbaugh, E.S. 1980. *Levels of communicative symbol use: Pre-representational and representational.* Paper presented at the T.C. Schneirla Conference, Wichita, Kansas.

Savage-Rumbaugh, E.S. 1981. Can apes use symbols to represent their world? In T.A. Sebeok and R. Rosenthal, eds. *The Clever Hans Phenomenon: Communication with Horses, Whales, Apes, and People. Annals of the New York Academy of Sciences* 364:35–59.

Savage-Rumbaugh, E.S. 1984. Verbal behavior at a procedural level in the chimpanzee. *Journal of the Experimental Analysis of Behavior* 41:223–250.

Savage-Rumbaugh, E.S. 1984. *Pan paniscus* and *Pan troglodytes.* Contrasts in preverbal communicative competence. In R.L. Susman, ed. *The Pygmy Chimpanzee: Evolutionary Biology and Behavior,* pp. 395–414. New York: Plenum Press.

Savage-Rumbaugh, E.S., J.L. Pate, J. Lawson, S.T. Smith and S. Rosenbaum. 1983. Can a chimpanzee make a statement? *Journal of Experimental Psychology: General* 112:457–492.

Savage-Rumbaugh, E.S. and D.M. Rumbaugh. 1979. Initial acquisition of symbolic skills via the Yerkes computerized Language Analog System. In R.L. Schiefelbusch and J. Hollis, eds. *Language Intervention from Ape to Child,* pp. 277–294. Baltimore: University Park Press.

Savage-Rumbaugh, E.S., D.M. Rumbaugh and S. Boysen. 1978a. Linguistically mediated tool use and exchange by chimpanzee *(Pan troglodytes). Behavioral and Brain Sciences.* 1:1–28, 539–554.

Savage-Rumbaugh, E.S., D.M. Rumbaugh, and S. Boysen. 1978b. Sarah's problems of comprehension. *Behavioral and Brain Sciences* 1(4):555–557.

Savage-Rumbaugh, E.S., D.M. Rumbaugh and S. Boysen. 1978c. Symbolic communication between two chimpanzees *(Pan troglodytes). Science* 201:641–644.

Savage-Rumbaugh, E.S., D.M. Rumbaugh and S. Boysen. 1980. Do apes use language? *American Scientist* 68(1):49–61.

Savage-Rumbaugh, E.S., D.M. Rumbaugh, S.T. Smith and J. Lawson. 1980. Reference: The linguistic essential. *Science* 210:922–925.

Savage-Rumbaugh, E.S. and R. Sevcik. 1984. Levels of communicative competency in the chimpanzee: Prerepresentational and representational. In G. Greenberg and E. Tobach, eds. *Behavioral evolution and integrative levels,* pp. 197–219. Hillsdale, N.J.: Lawrence Erlbaum Associates.

Savage-Rumbaugh, E.S., B.J. Wilkerson and R. Bakeman. 1977. Spontaneous gestural communication among conspecifics in the pygmy chimpanzee *(Pan paniscus).* In G.H. Bourne, ed. *Progress in Ape Research,* pp. 97–116. New York: Academic Press.

Schaffer, H.R., G.M. Collis, and G. Parsons. 1977. Vocal interchange and visual regard in verbal and preverbal children. In H.R. Schaffer, ed. *Studies in Mother-Infant Interaction,* pp. 291–324. London: Academic Press.

Schiefelbusch, R.L. and J. Hollis, eds. 1979. *Language Intervention from Ape to Child.* Baltimore, Md.: University Park Press.

Schusterman, R.J. and K. Krieger. 1984. California sea lions are capable of semantic comprehension. *The Psychological Record* 34:3–23.

Searle, J.R. 1969. *Speech Acts.* Cambridge: Cambridge University Press.

Sebeok, T.A. 1967. Discussion of communication processes. In S.A. Altmann, ed. *Social Communication Among Primates,* pp. 363–370. Chicago: University of Chicago Press.

Sebeok, T.A. and J. Umiker-Sebeok. 1980. *Speaking of Apes: A Critical Anthology of Two-Way Communication with Man.* New York: Plenum.

Seidenberg, M.S. and L.A. Pettito. 1979. Signing behavior in apes: A critical review. *Cognition* 1:177–215.

Seidenberg, M.S. and L.A. Pettito. 1981. Ape signing: Problems of method and interpretation. In T.A. Sebeok and R. Rosenthal, eds. *The Clever Hans Phenomenon: Communication with Horses, Whales, Apes and People. Annals of the New York Academy of Sciences* 364:115–129.

Shotter, J. 1978. The cultural context of communication studies: Theoretical and methodological issues. In A. Lock, ed. *Action, Gesture and Symbol: The Emergence of Language,* pp. 43–78. London: Academic Press.

Skinner, B.F. 1957. *Verbal Behavior.* New York: Appleton-Century-Crofts.

Smith, T.S. and R.A. White. 1981. *English word learning by mentally retarded children.* Paper presented at the annual convention of the Southeastern Psychological Association, Atlanta Ga.

Spradlin, J. 1963. Language and communication of mental defectives. In N.R. Ellis, ed. *Handbook of Mental Deficiency,* pp. 512–555. New York: McGraw-Hill.

Steklis, H.D. and M.J. Raleigh. 1979. Requisites for language: Interspecific and evolutionary aspects. In H.D. Steklis and M.J. Raleigh, eds. *Neurobiology of Social Communication in Primates: An Evolutionary Perspective,* pp. 283–314. New York: Academic Press.

Stokes, T.F. and D.M. Baer. 1977. An implicit technology of generalization. *Journal of Applied Behavior Analysis* 10:349–367.

Sugarman, S. 1983. Why talk? Comment on Savage-Rumbaugh et al. *Journal of Experimental Psychology: General* 112:493–497.

Tallal, P. 1978. An experimental investigation of the role of auditory temporal processing in normal and disordered language development. In A. Caramazza and E. Zurif, eds. *Language Acquisition and Language Breakdown,* pp. 25–61. Baltimore: Johns Hopkins University Press.

Taylor-Parker, S. and K.R. Gibson. 1979. A developmental model for the evolution of language and intelligence in early hominids. *The Behavioral and Brain Sciences* 2:367–408.

Teleki, G. 1973. *The Predatory Behavior of Wild Chimpanzees.* Lewisberg, Pa.: Bucknell University Press.

Temerlin, M.K. 1975. *Lucy: Growing Up Human: A Chimpanzee Daughter in a Psychotherapist's Family.* Palo Alto, Ca.: Science and Behavior Books.

Terrace, H.S. 1979a. How Nim Chimpsky changed my mind. *Psychology Today,* 65–76.

Terrace, H.S. 1979b. Is problem-solving language? *Journal of the Experimental Analysis of Behavior* 31:161–175.

Terrace, H.S. 1979c. *NIM.* New York: Alfred A. Knopf. Reprint. New York: Columbia University Press, 1985.

Terrace, H.S. 1985. In the beginning was the "name." *American Psychologist* 40:1011–1028.

Terrace, H.S., L.A. Petitto, R.J. Sanders and T.G. Bever. 1979. Can an ape create a sentence? *Science* 206:891–900.

Thompson-Handler, N., R.K. Malenky and N. Badrian. 1983. Sexual behavior of *Pan paniscus* under natural conditions in the Lomako Forest, Equateur, Zaire. In R.L. Susman, ed. *The Pygmy Chimpanzee: Evolutionary Biology and Behavior,* pp. 347–367. New York: Plenum Press.

Trapold, M.A. 1970. Are expectancies based upon different positive reinforcing events discriminably different? *Learning and Motivation* 1:129–140.

Trivers, R.L. 1971. The evolution of reciprocal altruism. *Quarterly Review of Biology* 46:35–57.

Tucker, D.M., R.T. Watson and K.M. Heilman. 1977. Discrimination and evocation of affectively intoned speech in patients with right parietal disease. *Neurology* 27:947–950.

Umiker-Sebeok, J. and T. Sebeok. 1981. Clever Hans and smart simians: The self-fulfilling prophecy and methodological pitfalls. *Anthropos* 76:89–165.

van Lawick-Goodall, J. 1968. The behavior of free-living chimpanzees in the Gombe Stream Reserve. *Animal Behavior Monograph* 1:161–311.

van Hooff, J. 1974. A structural analysis of the social behavior of a semi-captive group of chimpanzees. In M. von Cranach and I. Vine, eds. *Social Communication and Movement,* pp. 75–162. London: Academic Press.

von Glasersfeld, E.C. 1977. The Yerkish language and its automatic parser. In D.M. Rumbaugh, ed. *Language Learning by a Chimpanzee: The LANA Project,* pp. 91–130. New York: Academic Press.

Warden, C.J. and L.H. Warner. 1928. The sensory capacities and intelligence of dogs, with a report on the ability of the noted dog "Fellow" to respond to verbal stimuli. *Quarterly Review of Biology* 3:1–28.

Warren, S. and A. Rogers-Warren. 1984. *Teaching Functional Language.* Baltimore: University Park Press.

Weber-Olsen, M. and K. Ruder. 1984. Applications of developmental and remedial logic to language intervention. In K. Ruder and M. Smith, eds. *Developmental Language Intervention: Psycholinguistic Applications,* pp. 231–270. Baltimore: University Park Press.

Werner, H. and B. Kaplan. 1963. *Symbol Formation.* New York: Wiley.

Wilson, E.O. 1975. *Sociobiology. The New Synthesis.* Cambridge: Harvard University Press.

Woodruff, G. and D. Premack. 1979. Intentional communication in the chimpanzee: the development of deception. *Cognition* 7:333–362.

Yerkes, R.M. 1925. *Almost Human.* New York: The Century Co.

Yerkes, R.M. and B.W. Learned. 1929. *Chimpanzee Intelligence and Its Vocal Expression.* Baltimore: Williams and Wilkins.

Yerkes, R.M. and A.W. Yerkes. 1929. *The Great Apes.* New Haven: Yale University Press.

Zoloth, S.R., M.R. Peterson, M.D. Beecher, S. Green, P. Marler, D.B. Moody and W. Stebbins. 1979. Species-specific perceptual processing of vocal sounds by monkeys. *Science* 204:870–873.

Index of Names

Subject Index